Renewable Energy Resource Assessment and Forecasting

Renewable Energy Resource Assessment and Forecasting

Editor

George Galanis

MDPI • Basel • Beijing • Wuhan • Barcelona • Belgrade • Manchester • Tokyo • Cluj • Tianjin

Editor
George Galanis
Hellenic Naval Academy
Greece

Editorial Office
MDPI
St. Alban-Anlage 66
4052 Basel, Switzerland

This is a reprint of articles from the Special Issue published online in the open access journal *Energies* (ISSN 1996-1073) (available at: https://www.mdpi.com/journal/energies/special_issues/Renewable_Energy_Forecasting).

For citation purposes, cite each article independently as indicated on the article page online and as indicated below:

LastName, A.A.; LastName, B.B.; LastName, C.C. Article Title. *Journal Name* **Year**, *Article Number*, Page Range.

ISBN 978-3-03943-086-4 (Hbk)
ISBN 978-3-03943-087-1 (PDF)

© 2020 by the authors. Articles in this book are Open Access and distributed under the Creative Commons Attribution (CC BY) license, which allows users to download, copy and build upon published articles, as long as the author and publisher are properly credited, which ensures maximum dissemination and a wider impact of our publications.

The book as a whole is distributed by MDPI under the terms and conditions of the Creative Commons license CC BY-NC-ND.

Contents

About the Editor vii

Preface to "Renewable Energy Resource Assessment and Forecasting" ix

Francis M. Lopes, Ricardo Conceição, Hugo G. Silva, Thomas Fasquelle, Rui Salgado, Paulo Canhoto and Manuel Collares-Pereira
Short-Term Forecasts of DNI from an Integrated Forecasting System (ECMWF) for Optimized Operational Strategies of a Central Receiver System
Reprinted from: *Energies* **2019**, *12*, 1368, doi:10.3390/en12071368 1

Hideaki Ohtake, Fumichika Uno, Takashi Oozeki, Syugo Hayashi, Junshi Ito, Akihiro Hashimoto, Hiromasa Yoshimura and Yoshinori Yamada
Solar Irradiance Forecasts by Mesoscale Numerical Weather Prediction Models with Different Horizontal Resolutions
Reprinted from: *Energies* **2019**, *12*, 1374, doi:10.3390/en12071374 19

Mohammad Ehteram, Ali Najah Ahmed, Chow Ming Fai, Haitham Abdulmohsin Afan and Ahmed El-Shafie
Accuracy Enhancement for Zone Mapping of a Solar Radiation Forecasting Based Multi-Objective Model for Better Management of the Generation of Renewable Energy
Reprinted from: *Energies* **2019**, *12*, 2730, doi:10.3390/en12142730 37

Carlos Otero-Casal, Platon Patlakas, Miguel A. Prósper, George Galanis and Gonzalo Miguez-Macho
Development of a High-Resolution Wind Forecast System Based on the WRF Model and a Hybrid Kalman-Bayesian Filter
Reprinted from: *Energies* **2019**, *12*, 3050, doi:10.3390/en12163050 63

Chih-Chiang Wei
Evaluation of Photovoltaic Power Generation by Using Deep Learning in Solar Panels Installed in Buildings
Reprinted from: *Energies* **2019**, *12*, 3564, doi:10.3390/en12183564 83

Ekaterina S. Titova
Biofuel Application as a Factor of Sustainable Development Ensuring: The Case of Russia
Reprinted from: *Energies* **2019**, *12*, 3948, doi:10.3390/en12203948 101

Lilia Flores Mateos, Michael Hartnett
Incorporation of a Non-Constant Thrust Force Coefficient to Assess Tidal-Stream Energy
Reprinted from: *Energies* **2019**, *12*, 4151, doi:10.3390/en12214151 131

Julián Urrego-Ortiz, J. Alejandro Martínez, Paola A. Arias and Álvaro Jaramillo-Duque
Assessment and Day-Ahead Forecasting of Hourly Solar Radiation in Medellín, Colombia
Reprinted from: *Energies* **2019**, *12*, 4402, doi:10.3390/en12224402 149

Svetlana Ratner, Konstantin Gomonov, Svetlana Revinova and Inna Lazanyuk
Energy Saving Potential of Industrial Solar Collectors in Southern Regions of Russia: The Case of Krasnodar Region
Reprinted from: *Energies* **2020**, *13*, 885, doi:10.3390/en13040885 179

George P. Papaioannou, Christos Dikaiakos, Christos Kaskouras, George Evangelidis and Fotios Georgakis
Granger Causality Network Methods for Analyzing Cross-Border Electricity Trading between Greece, Italy, and Bulgaria
Reprinted from: *Energies* **2020**, *13*, 900, doi:10.3390/en13040900 . . . . **199**

Nikolaos Kampelis, Georgios I. Papayiannis, Dionysia Kolokotsa, Georgios N. Galanis, Daniela Isidori, Cristina Cristalli and Athanasios N. Yannacopoulos
An Integrated Energy Simulation Model for Buildings
Reprinted from: *Energies* **2020**, *13*, 1170, doi:10.3390/en13051170 . . . . **225**

Toby Green, Opio Innocent Miria, Rolf Crook and Andrew Ross
Energy Calculator for Solar Processing of Biomass with Application to Uganda
Reprinted from: *Energies* **2020**, *13*, 1485, doi:10.3390/en13061485 . . . . **249**

Jarosław Brodny, Magdalena Tutak and Saqib Ahmad Saki
Forecasting the Structure of Energy Production from Renewable Energy Sources and Biofuels in Poland
Reprinted from: *Energies* **2020**, *13*, 2539, doi:10.3390/en13102539 . . . . **263**

About the Editor

George Galanis holds a Professor position and leads the Mathematical Modeling and Applications Laboratory of the Naval Academy of Greece. The lab aims to develop and propose innovative solutions for real world problems, beyond the classical standards, by utilizing the rich scientific and technical framework that recent advances in pure and applied mathematics provide. Prof. Galanis' research interests include mathematical and statistical modeling, wave modeling, optimization, renewable energy site assessment and applications, as well as information geometry and applications for the optimization of simulation systems.

Preface to "Renewable Energy Resource Assessment and Forecasting"

Renewable Energy Resource Assessment and Forecasting; Guest Editor: Prof. George Galanis Mathematical Modeling and Applications Laboratory; Hellenic Naval Academy, Piraeus 18539, Greece; ggalanis@hna.gr.

George Galanis
Editor

Article

Short-Term Forecasts of DNI from an Integrated Forecasting System (ECMWF) for Optimized Operational Strategies of a Central Receiver System

Francis M. Lopes [1,2,*], Ricardo Conceição [1,2], Hugo G. Silva [1,2], Thomas Fasquelle [1], Rui Salgado [2,3], Paulo Canhoto [2,3] and Manuel Collares-Pereira [1,2,3,4]

1. Renewable Energy Chair, University of Évora, IIFA, Palácio do Vimioso, Largo Marquês de Marialva, Apart. 94, 7002-554 Évora, Portugal; rfc@uevora.pt (R.C.); hgsilva@uevora.pt (H.G.S.); thomasf@uevora.pt (T.F.); collarespereira@uevora.pt (M.C.-P.)
2. Institute of Earth Sciences, University of Évora, Rua Romão Ramalho 59, 7000-671 Évora, Portugal; rsal@uevora.pt (R.S.); canhoto@uevora.pt (P.C.)
3. Department of Physics, School of Sciences and Technology, University of Évora, Rua Romão Ramalho 59, 7000-671 Évora, Portugal
4. Portuguese Solar Energy Institute, IIFA, Palácio do Vimioso, Largo Marquês de Marialva, Apart. 94, 7002-554 Évora, Portugal
* Correspondence: fmlopes@uevora.pt

Received: 11 March 2019; Accepted: 4 April 2019; Published: 9 April 2019

Abstract: Short-term forecasts of direct normal irradiance (DNI) from the Integrated Forecasting System (IFS) and the global numerical weather prediction model of the European Centre for Medium-Range Weather Forecasts (ECMWF) were used in the simulation of a solar power tower, through the System Advisor Model (SAM). Recent results demonstrated that DNI forecasts have been enhanced, having the potential to be a suitable tool for plant operators that allows achieving higher energy efficiency in the management of concentrating solar power (CSP) plants, particularly during periods of direct solar radiation intermittency. The main objective of this work was to assert the predictive value of the IFS forecasts, regarding operation outputs from a simulated central receiver system. Considering a 365-day period, the present results showed an hourly correlation of ≈0.78 between the electric energy injected into the grid based on forecasted and measured data, while a higher correlation was found for the daily values (≈0.89). Operational strategies based on the forecasted results were proposed for plant operators regarding the three different weather scenarios. Although there were still deviations due to the cloud and aerosol representation, the IFS forecasts showed a high potential to be used for supporting informed energy dispatch decisions in the operation of central receiver units.

Keywords: short-term forecasts; direct normal irradiance; concentrating solar power; system advisor model; operational strategies; central solar receiver

1. Introduction

With the simultaneous increase of solar energy conversion units installed worldwide and computational technology, interest has been growing in using direct normal irradiance (DNI) forecasts in the field of solar power, at a regional or global scale, particularly for an efficient production of energy from concentrating solar power (CSP) plants. A strong reason for such an effort is the fact that CSP systems are able to provide high-quality dispatchable power at affordable prices, when compared to photovoltaic storage capacity, using molten salt as heat storage, a cheap, safe, and easily accessible material [1,2]. For a CSP plant operator, information concerning the day-ahead (up to 48 h) DNI values is required for an efficient energy planning and scheduling [3], allowing higher-penetration of commercial solar power, into the electricity market. In particular, it is during periods of direct solar radiation intermittency that CSP technologies demand accurate forecasts of DNI [4], since these periods are characterized by scattered clouds (which can differ in type and dynamic coverage [5]) and aerosols species [6], which are two primary factors that affect the direct solar resource at the ground level.

To accurately characterize DNI, a combination of the state-of-the-art monitoring and assessment techniques, with advanced numerical weather prediction (NWP) modeling is recommended. NWP models are based on the numerical computation of dynamic flow equations that allow solving the state of the atmosphere and its evolution, up to several days-ahead [7]. However, despite being able to provide satisfactory results [8], current models still demand developments towards DNI forecasting, particularly the parameterization of cloud cover [9] and the use of real-time aerosol information, considering that nowadays an aerosol climatology is still used, despite recent advances [10]. Moreover, an accurate conversion of predicted DNI to predicted energy output values from simulated power plant models is also necessary. In this context, user-friendly software such as the System Advisor Model (SAM) can be used to simulate a CSP plant. This method has been carried out by the authors in a previous work [11], where forecasted data from the IFS was used in the simulation of a linear-focus parabolic trough (PT) system, with a configuration similar to the Andasol 3, a 50 MW$_e$ power plant [12] located in Granada (Spain). Although the PT technology has dominated the solar thermal power industry in the last decades, central receiver (CR) units have been emerging, due to the potential that these have for higher efficiency and lower cost. This is possible because apart from having higher concentration ratios (300–800 suns versus only 25–30 in conventional linear concentration), modern CR technology uses molten salt as a heat transfer fluid (HTF) and, directly, as heat storage fluid. Most commercial PT solutions operate with thermal oils as HTF and even when heat storage is also performed with molten salts, the overall operating temperature is much lower (≈400 °C contrasting with 540 °C in the CR systems). In CR systems the higher temperatures place more stringent requirements on energy management and control of power block efficiency, than on lower temperature PT system [13].

Taking into account the aforementioned aspects, the present work uses day-ahead (24-h) forecasts of DNI from the Integrated Forecasting System (IFS), the global NWP model of the European Centre for Medium-Range Weather Forecasts (ECMWF) that possesses the highest scores regarding medium-range global weather forecast [14], together with a set of meteorological variables, in the simulation of a CR power plant. Moreover, an advantage in using the IFS, instead of higher resolution models, is that it allows the implementation of the present analysis and proposed method in any region of the world, with high prospects in the installment of CSP units. In this work, in order to convert DNI values to energy output forecasts of the modeled CSP system, the simulation of a CR power plant similarly configured as the 19.9 MW$_e$ Gemasolar thermosolar power plant [15] (located in the province of Sevilla (Spain)), was carried out. The obtained energy outputs based on DNI predictions and local measurements of the simulated CR power plant were assessed and then compared with the results obtained for a PT system [11]. This simulation used the same dataset, i.e., input variables (DNI and meteorological data), as for the PT simulation, being related to the same period and location in Southern Portugal, in which it showed substantial improvements towards the prediction of DNI, due to the new operational radiative scheme of the IFS.

The proposed work has been structured as follows. In Section 2, a description is provided regarding the measured and forecasted data, the CSP plant model, and the adopted methodology; results and respective discussions are given in Section 3; operational strategies for the three different weather scenarios are given in Section 4; and in Section 5, conclusions and future work perspectives are summarized.

2. Data and Methodology

2.1. Measurements

Measurements of DNI were used from a ground-based station located in Évora city (38.567686°N, 7.911722°W), from the Institute of Earth Sciences (ICT—Instituto de Ciências da Terra) in Southern

Portugal, a semi-arid region [16] with a high frequency of clear sky day occurrences and annual energy maximums around 2100 kWh/m^2 [17].

Pyrheliometers (model CHP1) from Kipp and Zonen instruments were used, being calibrated every 2 years. With an estimated daily uncertainty of <1%, these instruments follow the international organization for standardization, the 9059:1990 standard [18], as first-class instruments. To compare with NWP values, hourly mean values were obtained by averaging the sixty 1-min records. The Évora station (denominated EVO station) had a strict and regular code for the maintenance of the instruments, being subjected to quality control tests, prior to the analysis. The DNI at the EVO station showed only 0.003% of missing data for the considered year of continuous measurements. This showed how well-maintained the EVO station is, and why it was used in this work as a reference station. This station allowed us to provide high-quality data, showing only very small gaps that could have resulted from sudden power shut downs. To fill gaps, adopted filters for the location of study were used, including standard data quality filters, the Baseline Surface Radiation Network (BSRN) for Global Network quality check tests V2.0 [19] and gap-filling procedures. The latter, consisted in the use of hourly values from the nearest ground-based measuring station located at Mitra, MIT (38.530522°N, 8.011221°W), installed approximately 9.6 km from EVO, to fill gaps that have more than two hours of missing records. For the gaps with less than two hours of missing records, a linear interpolation between the previous and the next hours was then used to fill the missing periods.

Similarly, as performed in [11], continuous measurements of local atmospheric variables, such as air temperature, relative humidity, wind speed, and atmospheric pressure at ground level, were also acquired by nearby standard meteorological measuring equipment. Since atmospheric pressure was not measured at the EVO station and the local wind was disturbed by existing neighboring buildings, not being representative of the measuring location, hourly data from a nearby station (≈4 km apart)—maintained by the Portuguese Meteorology Service (IPMA—Instituto Português do Mar e da Atmosfera)—was used for the considered period of study.

2.2. Forecasts

The IFS is the atmospheric model and data assimilation system from the ECMWF (which is currently operational) that was used to perform global medium-range weather forecasts. The model is able to provide deterministic predictions of a large set of meteorology-related variables, including DNI. The radiative variables, in both short and longwave spectral bands, were computed using the Rapid Radiative Transfer Model [20]. Operational high-resolution (HRES) deterministic forecasts were set to have an issue time to start at 00:00 or 12:00 UTC (the latter option is used in this work). The current IFS cycle uses a triangular-cubic-octahedral global grid, with a horizontal resolution of 0.125° × 0.125° (≈9 km), 137 terrain-following vertical levels from the surface up to 1 Pa (≈80 km height), and a 7.5-min time step. The radiation scheme is updated every hour, on a grid with 10.24 times fewer columns than the rest of the model [21]. Contrary to the previous versions of the IFS, in which the DNI was not a direct output of the model, the current version was able to directly calculate hourly accumulated direct irradiation values (J/m^2), which were then converted to mean power values (W/m^2), in order to enable a straight comparison with measurements. The output of the IFS used here as representative of DNI is the *dsrp* parameter, i.e., the direct solar radiation, incident on a plane perpendicular to the Sun's beams.

To perform accurate forecasts of DNI, NWP models have to take into account several parameters that can affect such forecasts, for instance the local weather (e.g., air temperature, relative humidity, wind speed and direction, and surface pressure). Along with weather conditions, the forecast horizon can also affect the prediction of DNI, since it has an associated uncertainty that tends to be smaller with the use of shorter time horizons. However, these are closely linked to a high computational cost [22]. Forecast horizons can range from: (i) the intra-hour scale, where persistence models [23] and all-sky imagers [24] are used; to (ii) the intra-day scale, where artificial neural networks [25], and satellite-based and NWP models [26] are used; and (iii) up to several days (i.e., day or week-ahead forecasts) in which NWP models are able to perform [27]. Apart from the weather conditions and forecast horizons, initial conditions implemented in NWP models also play an important role [28]. These include the atmosphere, oceans, and ground surfaces physics, which are composed by a series of complex dynamical processes that comprise the spatial distribution of a large number of atmospheric parameters. Moreover, aside from these aspects that can hinder the prediction of DNI, particular attention has been given towards cloud microphysics and aerosol representation. The former is closely related to the complex parameterization of cloud cover and type [9], mainly during overcast periods, while the latter is usually based on monthly mean aerosol climatologies, which increases the errors of predicted DNI, especially during clear sky conditions. In particular, it is during very clean atmosphere periods that the implemented aerosol climatology affects the prediction of DNI more. This has been previously observed with day-ahead forecasts of DNI from the IFS [11,29], where the radiative effects of clouds and aerosols were, respectively, under- or over-estimated by the model, compared to local measurements. For instance, at the EVO station it was found that the predicted mean annual DNI had an overestimation of ≈7%, compared to local measurements [29], being essentially related to an underestimation of the cloud cover.

To improve DNI forecasts, the radiative schemes of NWP models have been constantly upgraded to new versions. One example is the current ecRad scheme that was recently implemented in the IFS [10], becoming operational in July 2017 (cycle 43R3). A detailed description of the ecRad and its use in the IFS can be found in [21]. Presently, the ecRad is composed of the following IFS atmospheric variables—pressure, temperature, cloud fraction, and the mixing ratios of water vapor, liquid water, ice, and snow. The cloud effective radius was computed diagnostically, using the parameterization stated in [30], for liquid clouds, and that stated in [31], for ice clouds. The optical properties for ice were computed using the scheme stated in [32] and that for liquid water were expressed in terms of a Padé approximation [33]. The mixing ratios for ozone, carbon dioxide, and an arbitrary number of aerosol species were computed from a climatology obtained from the Copernicus Atmospheric Monitoring Service (CAMS), being more realistic than the previous versions, in which the Tegen aerosol climatology [34] was implemented. The optical properties of aerosols were added to those of gases, with the assumption that aerosols were horizontally well-mixed, within each model grid box. Aerosol optical properties were computed off-line, using an assumed size distribution and the Mie theory, for 14 shortwave and 16 longwave bands. Moreover, in addition to an improved code that allowed us to reduce computational costs, ecRad was able to reduce numerical noise in cloudy periods, which enabled better DNI predictions than the previous radiative scheme [21]. A recent analysis [11] has shown that improvements of day-ahead forecasts of DNI from the ecRad were attained, in comparison to the previous version (McRad, cycle 41R2). Hourly and daily correlations of 0.87 and 0.91 between predicted and measured data in EVO were found for the same dataset used in the present work. Although the IFS still overestimated measurements, a relative difference of ≈1.2% was found regarding the annual mean values of DNI in EVO, which was much lower than the previous value obtained with the McRad (≈10.6%).

In this work, day-ahead forecasts produced by the ecRad were used to estimate the energy output from a CR power plant simulated through the SAM. Results were assessed by comparison with those obtained using the local measurements.

2.3. CSP Plant Model

The SAM software [35], version 2017.9.5, developed by the U.S. Department of Energy and National Renewable Energy Laboratory (NREL), was used here to assess the usefulness of DNI forecasts from the IFS, for the energy management of a CR power plant. Regarding the simulation of CSP systems, the SAM uses the transient system simulation (TRNSYS), comprising three components—(i) an interface where the setup of each simulation is performed in detail by the user; (ii) a calculation engine that implements discretization procedures in each simulation, and (iii) a programming interface. The power plant model calculates hourly performance values corresponding to a wide range of output parameters, providing an annual performance and financial metrics summary at the end of each run. DNI and other atmospheric variables (air temperature, relative humidity, wind speed, and surface pressure) were the necessary input parameters for the power plant model to generate local hourly performance data. The resulting hourly outputs represent a full year of annual electricity production of the considered CR power plant.

To simulate a CR power plant, it is important to know all the design and control parameters that are characteristic of such a system. A CR system, also known as a solar power tower, uses sun-tracking mirrors (heliostats) to focus the Sun's direct beam onto a receiver installed at the top of the tower. Within the receiver, a HTF was then heated, reaching temperatures up to 565 °C, allowing the generation of water steam, through a heat exchanger. The latter was then used by conventional turbine-generators, to produce electricity (Rankine cycle). Due to the higher temperatures of use and superior heat transfer and energy storage capabilities than other CSP systems, such as PT systems, current CR plants used molten salt, such as HTF. One example of this kind of power system is the 19.9 MW$_e$ Gemasolar thermosolar plant located in the Sevilla province (Spain), which has been operational since April 2011. This type of CR power plant possesses a 15-h storage capacity and is surrounded by 2650 heliostats (Figure 1), within an area less than 200 hectares. The Gemasolar was intended to produce 110,000 MW$_e$h/year [15], however, probably due to technical issues created by the new challenges that were addressed during the operation of the power plant, an annual generation of 80,000 MW$_e$h/year was reached [36]. In this work, in order to study the behavior of a CR solar power plant, a simulation with a similar configuration, such as the Gemasolar, was carried out. The criterion for selecting this power plant resulted from the fact that Gemasolar is considered to be a typical CR system, with the advantage of having considerable information available regarding the power plant operation input parameters, thus allowing to establish a case study for the CR power plants. Under Évora's conditions, this study used the same weather dataset as the SAM input parameters from the EVO station that were previously used for the simulation of a 50 MW$_e$ PT system [11], with configurations similar to the Andasol 3 located in Granada (Spain). Due to privacy reasons, full access to the complete configuration of the Gemasolar was not possible. Consequently, several design and control input parameters needed for the simulation were not provided by NREL, creating a limitation to the present analysis. However, in order to obtain the best performance results that corresponded close to the actual performance outputs of the Gemasolar power plant, some input parameters were needed for the simulation result from research-based assumptions made by the authors, regarding the operation of the CR systems. For more detailed information concerning the configuration input parameters used in the SAM simulation, see Appendix A.

Figure 1. Gemasolar thermosolar power plant located in the province of Sevilla, Spain (37.560613°N, 5.331508°W). All rights reserved (© Google Earth 2019).

3. Results and Discussion

In this analysis, electrical and thermal output parameters generated by the SAM simulations using forecasted and measured hourly values of DNI and meteorological variables. The outputs were selected according to their importance for the power generation and management of a CR power plant since the plant operator should analyze these parameters on a daily basis. In that sense, the total electric energy to the grid, E_P (MW$_e$h), and the stored thermal energy, TES (MW$_t$h), charge and discharge energies were analyzed for a 365 day-period (from 1st July 2017 to 30th June 2018) with the study location centered at the EVO station.

In Table 1, a statistical summary for the E_P and the respective TES charge and discharge energies, based on forecasts and measurements of DNI and meteorological variables, is shown. As expected, due to the IFS underestimation of cloud cover [29], the obtained results using the simulated hourly values showed a general overestimation of the IFS forecasts towards measurements. A total of ≈115,992 MW$_e$h/year and ≈121,668 MW$_e$h/year was obtained, respectively, for the E_P based in DNI measurements and forecasts, with a correlation coefficient (r) of ≈0.78, between both outputs. The representation of clouds performed by the IFS, significantly influenced the forecasted DNI values at the Earth's surface and, consequently, the respective E_P output from the CR power plant. Taking into account the parasitic power consumption during non-production hours and a constant derating (i.e., a decrease of the power plant output due to unusual environmental conditions, for instance, higher ambient temperature than design set point, or excess power within the electrical grid) value of 4%, for the simulated plant, the SAM results showed an annual energy generation of ≈111,353 MW$_e$h/year and ≈116,801 MW$_e$h/year, regarding measurements and predictions, respectively, i.e., a relative difference of ≈4.9%. Despite the fact that the objective of the present work was not a direct comparison with the Gemasolar's actual production values, the obtained annual values through the SAM simulations could differ from the values that would be obtained if an actual Gemasolar was operating in Évora, due to several reasons: (i) DNI and meteorological data from Évora was being used for a different period, comprising different inter-annual variations; (ii) lack of data regarding design and control parameters for the simulation of Gemasolar; (iii) start-up time (0.5 h) and stop operations of the simulated plant together with the internal temporal discretization, considered by the SAM; and (iv) daily operational strategies adopted for the plant power management.

Table 1. Statistical and descriptive analyses for the hourly values of electric energies into the grid, E_P (MW$_e$h), and stored thermal energy, TES (MW$_t$h), charge and discharge energies based on measurements (obs) and forecasts (ecmwf). The sum of the hourly values (Total) of E_P and TES corresponded to one year of data (from 1st July 2017 to 30th June 2018), produced by a central receiver power plant with configuration similar to the Gemasolar thermal power plant (Sevilla province, Spain), simulated through the System Advisor Model (SAM). Hourly statistical error metrics for the correlation coefficient (r), root mean square error (RMSE), and mean absolute error (MAE) are presented.

Energy	Total $_{obs}$ (MW$_{e,t}$h)	Total $_{ecmwf}$ (MW$_{e,t}$h)	r	RMSE (MW$_{e,t}$h)	MAE (MW$_{e,t}$h)
E_P	115,992	121,668	0.78	6.30	2.31
TES $_{charge}$	151,104	153,187	0.88	16.46	5.97
TES $_{discharge}$	148,399	150,465	0.83	12.32	4.09

The charge and discharge powers also showed an overestimation when using the forecasted inputs, in comparison with those obtained when using measurements, although with higher correlations. Simulation results showed annual charge and discharge energies of ≈151,104 MW$_t$h/year and ≈148,399 MW$_t$h/year, based on measurements, while ≈153,187 MW$_t$h/year and ≈150,465 MW$_t$h/year were obtained for the forecast-based outputs. Although the discharge energy had a lower r than the charge-hourly values (≈0.83), it was shown to possess less deviations between the measured and forecasted outputs.

A closer look at the hourly outputs generated by the SAM, based on the forecasted and measured DNI values, was presented in the scatter plots of Figure 2a, and Figure 3a,b, respectively, for the E_P and TES charge and discharge energies. In these plots, the red dashed line represents the identity line (y = x), in which the dots that are closer to the line depict higher correlations than the ones that deviate from it. Two green dashed–dotted lines (Figure 2a) bound an interval in which the predicted and measured E_P values had an absolute error (AE) less than the obtained mean absolute error (MAE) of ≈2.31 MW$_e$h. The total number of hourly values of E_P, within the established high and low thresholds corresponded to ≈85.94%.

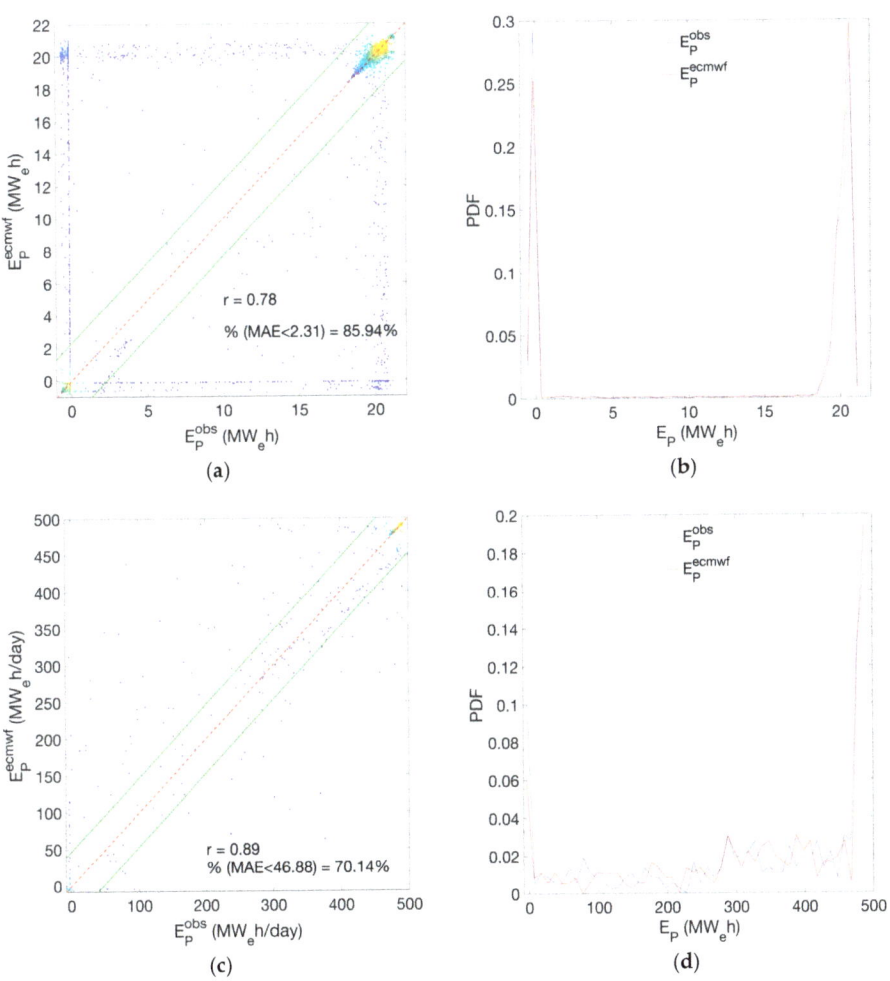

Figure 2. Estimated hourly (**a**, **b**) and daily (**c**, **d**) values of electric energies into the grid, E_P (MW$_e$h), and respective probability density functions (PDF), computed from forecasted (ecmwf) and measured (obs) data at Évora. Hourly values of direct normal irradiance (DNI) were used in the SAM to simulate the E_P from a central receiver (CR) power plant with configuration similar to the Gemasolar plant (Sevilla, Spain). In the scatter plots, identity lines (red dashed lines), corresponding correlation coefficients, r, and an interval defined by the calculated MAE (≈2.31 MW$_e$h), given by two green dashed–dotted lines, are shown. The period of study corresponds to one year, from 1 July, 2017 to 30 June, 2018.

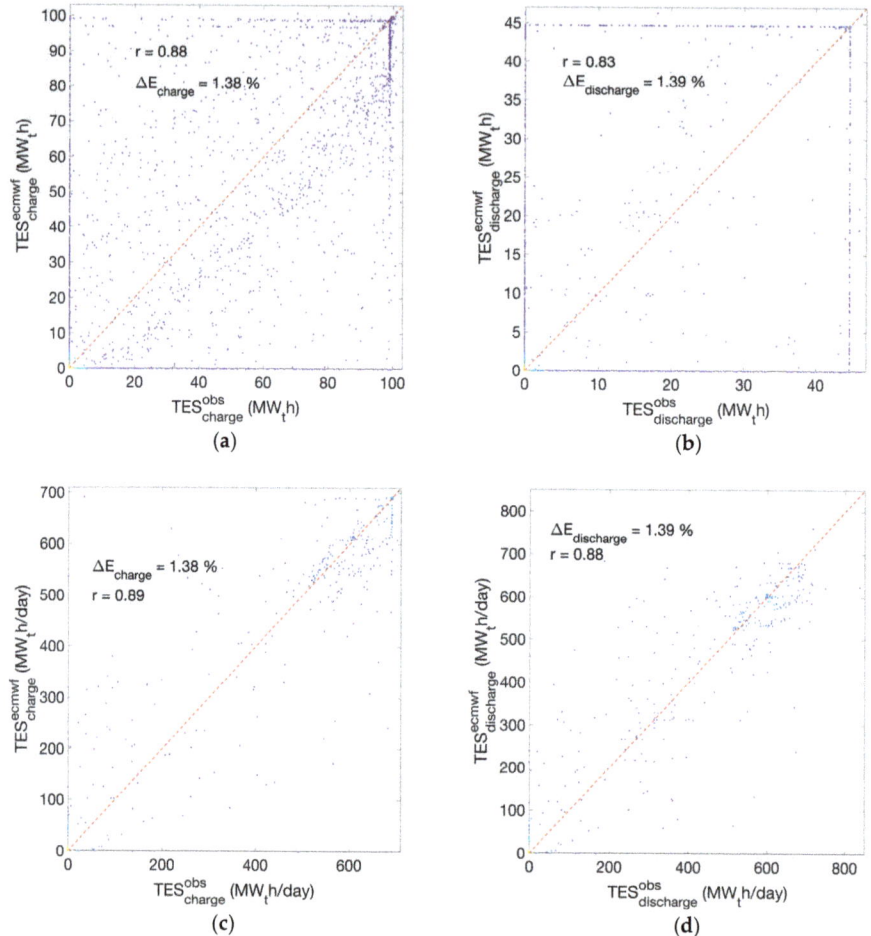

Figure 3. Estimated hourly values of stored thermal energy into the grid, TES (MW$_t$h)—(**a**) charge and (**b**) discharge energies, computed from forecasted (ecmwf) and measured (obs) data at Évora, while corresponding daily values are presented in (**c**) and (**d**). Hourly values of DNI were used in the SAM to simulate the TES from a CR power plant, with a configuration similar to the Gemasolar plant (Sevilla, Spain). Identity lines (red dashed lines), the corresponding correlation coefficients, r, and relative differences, ΔE, are shown. The period of study corresponded to one year, from 1 July, 2017 to 30 June, 2018.

A few features that were characteristic of CSP systems were observed. Most of the values were centered on the high values of E_P, between 18 and 21 MW$_e$h, which took place during periods of clear sky conditions. Outside these limits were the E_P values (including negative ones) that corresponded to the non-production hours in which electricity for parasitic power consumption needed to be purchased from the grid. During these periods, deviations between the forecasted and measured E_P values occurred, in particular for—E_P (obs) > 0 and E_P (ecmwf) = 0; E_P (obs) = 0 and E_P (ecmwf) > 0. During cloudy days with short periods of unobstructed solar beam radiation, predicted and measured E_P values also had deviations. If only non-negative hourly values of E_P were considered, the correlation between the forecasted and measured values would drop significantly to 0.37, showing the importance that non-production hours have in the correlations, since these periods correspond to shut-down and start-up operations carried out by the power plant. This meant that the predictions have a good correspondence with the measurements, during such periods. The respective probability density function (PDF) in Figure 2b clearly depicted the two observed features, as highlighted by the two peaks—the higher frequency of occurrence around the non-production hours (zero values), particularly by the E_P based in measurements; and the high frequency of occurrence for the higher values of E_P. Moreover, the hourly TES charge and discharge energies (Figure 3a,b) showed a slight improvement in correlation, for the charge periods ($\approx$0.88), in comparison to the discharge ones ($\approx$0.83), as these correlations were closely linked to the non-production (close to zero) and the high production periods ($\approx$100 MW$_t$h). Relative differences of $\approx$1.38% and $\approx$1.39% were found for the charge and discharge outputs, respectively. The hourly TES charge values depicted a tendency line (below the identity line), demonstrating that, less storage was gained with the forecasted based output, in comparison to the measured one. This was a consequence of the IFS underestimation towards measurements during days with very clean atmospheric conditions, in which the aerosol concentration was less than that in the prescribed climatology.

Daily values (i.e., calculated through the 24-h sum of each day) yielded higher correlations, as shown by the results in Table 2, despite overestimations from the forecasts, as depicted by the negative mean bias error (MBE) values. An r $\approx$0.89 was obtained for the daily E_P values (Figure 2c), with $\approx$70.14% of the total number of daily values having an AE below an MAE of $\approx$46.88 MW$_e$h. The respective daily PDF (Figure 2d) showed the same pattern as that for the hourly results, but with less frequency of occurrence, with two peaks, one for the non-production hours and another for the high values of E_P. Correlations of $\approx$0.89 and $\approx$0.88 were found between the daily TES charge and discharge energies, based on the measurements and forecasts (Figure 3c,d), respectively.

Table 2. Statistical analysis of the daily values (i.e., the sum of each 24-h values) of the estimated electric energy to the grid, E_P (MW$_e$h), and stored thermal energy, TES (MW$_t$h) charge and discharge energies computed from measurements (obs) and forecasts (ecmwf). Hourly values of E_P and TES correspond to one year of data (from 1 July, 2017 to 30 June, 2018) produced by a CR power plant with a configuration similar to the Gemasolar plant (Sevilla, Spain) simulated through the SAM. Daily statistical error metrics such as the correlation coefficient (r), root mean square error (RMSE), mean absolute error (MAE), and mean bias error (MBE) are presented.

Energy	r	RMSE (MW$_e$,$_t$h)	MAE (MW$_e$,$_t$h)	MBE (MW$_e$,$_t$h)
E_P	0.89	79.43	46.88	−15.55
TES charge	0.89	119.96	74.25	−5.70
TES discharge	0.88	111.66	71.37	−5.66

Since the same dataset (DNI and meteorological variables) from EVO station were used in both, the CR and the PT simulations, the performance of the 24-h predictions from the IFS in the operation of different CSP systems has been depicted in Table 3. The coefficient of variation regarding the RMSE and MAE, i.e., the normalized RMSE and MAE (nRMSE and nMAE, respectively), were obtained for the electric energy to grid outputs, from both Gemasolar and Andasol 3 simulations (E_P and E_G, respectively). The calculation of both nRMSE and nMAE are given in Equations (A1) and (A2) in Appendix A. The obtained hourly values of E_P and E_G show that forecasted data in the simulation of the Gemasolar power plant generates higher deviations than the ones obtained from the Andasol 3, with an increase of $\approx$7.3% for the nRMSE and $\approx$2.8% for the nMAE. Deviations were lower from the hourly to daily values, showing an increase of $\approx$2.9% for the nRMSE and $\approx$0.7% for the nMAE. These results indicated that the PT power plant considered (based on Andasol 3) was less sensitive to the DNI prediction than the CR one (based on Gemasolar). However, it must be taken into account that the considered PT system had less storage than the CR system, resulting in a larger number of non-production hours (i.e., zero values) for both forecasted and measured simulations, contributing to an apparent reduction of differences between them.

Table 3. Hourly and daily values of normalized root mean square error (nRMSE) and mean absolute error (nMAE) for the estimated total electric energy to the grid outputs, obtained from the 19.9 MW$_e$ Gemasolar and the 50 MW$_e$ Andasol 3 SAM simulations (E_P and E_G, respectively). The E_P and E_G simulated values are based on the same hourly dataset (DNI and meteorological data) of forecasted and measured input parameters acquired for Évora, for the same period of study (from 1 July 2017 to 30 June 2018).

Power Plant	nRMSE (%)		nMAE (%)	
	Hourly	Daily	Hourly	Daily
Gemasolar	28.48	15.88	10.43	9.37
Andasol 3	21.18	13.02	7.65	8.68

4. Operational Strategies for Typical Days

In order to maximize the energy efficiency of CSP plants, it is essential to adopt appropriated operational strategies, in accordance to the different weather scenarios (i.e., clear sky, partly cloudy, and overcast days), which could differently affect the CR power plant performance. For instance, for the CR systems, the advantage of knowing the energy availability for the day-ahead, allowed the operator to estimate the electricity generation in advance and sell it at the premium tariff [37], as an alternative to the fixed tariff option, thus, allowing the operator to have a direct role on the electricity market instead of being subjected to flat-rate prices.

As demonstrated in the previous study regarding solar assessment influence on a linear focus PT power plant operational strategies and production [11], the forecast model was able to generate satisfactory results for the days with clear sky conditions. However, results showed that such forecasts were hindered due to aerosol representation, particularly under very clean atmosphere conditions, in which the forecasts underestimated the DNI, or during overcast conditions, in which the IFS overestimated the DNI. The latter behavior can also be a result of extreme dust events, as shown by [38] and [39]. For cloudy days, the IFS was also reliable in predicting clouds, although temporal and spatial phase errors exist in the current cloud forecasting. For the case of the conventional 50 MW$_e$ PT power plant, results have led to three different operational strategies related to specific meteorological scenarios: a clear sky, a partly-cloudy, and an overcast period. An example of the implemented global strategy is to avoid power block start-up and shutdown, allowing to maintain the plant at a nominal power and a maximum efficiency. Another aspect is the possible full state of charge of the storage tank, during the day. In this scenario, the operator is advised to perform a partial charge in the early morning to handle a possible cloud passing over, except for predicted clear sky days, in which production is to be started as soon as possible. For the present case study, the high storage capacity of the CR power plant allows the easing of the operator decision algorithm.

4.1. Clear Sky Days

Full charge for a 15-h storage system can only be encountered during days that have very high solar irradiation levels. For such days, the best strategy is to maximize electricity production by starting the power block at the earliest moment. An example of this scenario is shown in Figure 4, where a constant power production is observed, due to the huge storage capacity and higher availability. Here, defocusing of the solar field is also shown, leading to a lower receiver output power after 14 h (production with predicted DNI) or 15 h (production with measured DNI).

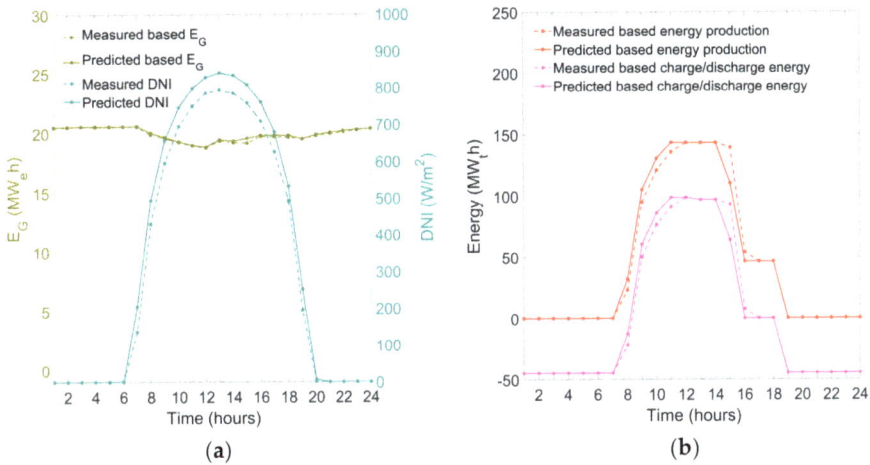

Figure 4. Comparison between the results of the SAM simulation for 21 August 2017 (prediction of high energy without clouds), with predicted and measured (**a**) hourly mean DNI (W/m^2) values (blue) and produced electrical energy E_G (MW$_e$h) values (green); and (**b**) thermal energy (MW$_t$h) of charge/discharge values (purple) and produced solar energy (red).

4.2. Cloudy Days

On cloudy days, different types of strategies can be applied—full shifting of the solar production to the evening, constant power generation during the entire day, among others. Since a solar power generation is easily higher than what the power block requires, this leads to a high amount of energy surplus in the system, even during the early morning periods. Under such conditions, in order to increase the safety of the power block continuous production, a small partial charge can be performed in the early morning. This partial charge does not lead to potential defocusing because the available energy is not sufficient to reach a 100% state of charge, during the day. Figure 5 shows an example of such an operation scenario, for a day, with low irradiation levels, because of passing clouds. Since the power block production can be started with a DNI higher than 300 W/m^2, a high amount of energy is charged during the morning hours. For the case of observed DNI, it drops to 300 W/m^2 after 12 h, due to the presence of clouds, leading to a stop of the charging process. After 16 h, DNI drops below 300 W/m^2 because of a second cloud and discharge is performed to maintain electricity production to a constant value.

In this example, the forecast model predicted only one long period of cloud obstruction, during the afternoon, with both forecasted and measured systems responding well, in terms of power production. It should be noted that this type of scenario tends to degrade the correlation between the forecasted and measured E_P, particularly under the hourly time scale, although for larger time scales, the differences between both outputs estimates are not so significant.

Other particular strategies can be given for the receiver protection of a CR system regarding thermal stress. For instance, avoiding periodic or sudden strong increases and decreases of the receiver temperature, due to a passing cloud. In such cases, the forecast model is by itself sufficient to warn the power plant operator that variations will take place but with a lower accuracy in the time of occurrence. Nonetheless, the available predicted information is already useful to apply thermal protection strategies on the receiver, or to strategize the energy management of the power plant for one day.

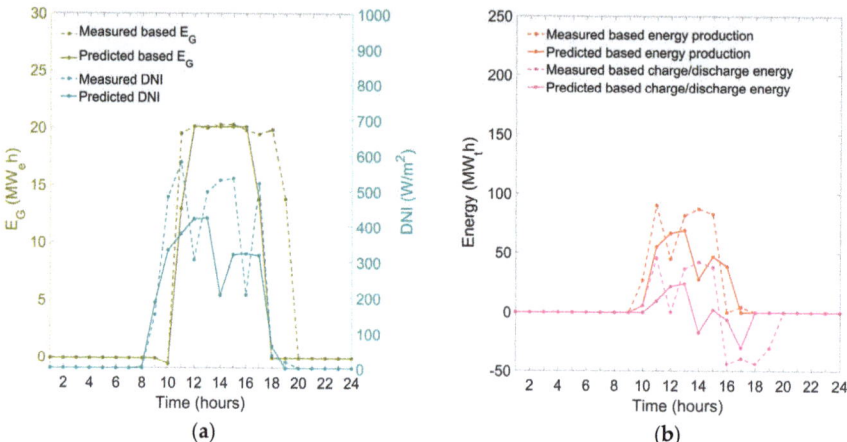

Figure 5. Comparison between results of the SAM simulation results for 11 January (prediction of low energy with clouds), with predicted and measured (**a**) hourly mean DNI (W/m^2) values (blue) and produced electrical energy E_G (MW$_e$h) values (green); and (**b**) thermal energy (MW$_t$h) of charge/discharge values (purple) and produced solar energy (red).

4.3. Overcast Days

For operational purposes, the analysis of a specific day without any, or near null, DNI income is also relevant, simply because when there is no DNI to be collected, then there is no production, consequently the power plant should be running at the lowest power generation possible (or in stand-by mode to keep the equipment warm and the salt in liquid state), depending on the available storage. In that sense, it is important to anticipate such long periods of no production, and the impact that these have on the energy management, and to accordingly implement the necessary strategies. During such periods, and in the case of no available storage, the power plant has negative production values, since the system needs to consume energy, in order to maintain the continuous function of basic equipment. For that reason, in this last section, the success that the IFS has in predicting periods of negative production, using the simulated values, is analyzed. In such a scenario, a dichotomous analysis is performed with the use of a contingency table (Table 4) created to evaluate the forecasts of E_P values. Moreover, as described in [40], an equitable threat score (ETS) skill score is calculated through Equations (A3) and (A4) (Appendix A) to measure the fraction of the forecasted and observed E_P events that were accurately predicted. The ETS is usually used in the NWP models to evaluate other meteorological variables, such as rainfall [41], since it allows us to equitably compare the obtained scores across different regimes.

Table 4. Dichotomous analysis for the total number of forecasted (ecmwf) and measured (obs) occurrences and non-occurrences of daily negative electrical production values (E_P) for the Gemasolar power plant simulation through the SAM. The obtained E_P simulated values were based on hourly DNI and meteorological data (forecasted and measured) input parameters acquired for Évora for the same period of study (from 1 July, 2017 to 30 June, 2018). Four different events of negative E_P values occurrences and non-occurrences have been depicted—'Hits' (E_P (ecmwf) < 0 and E_P (obs) < 0), 'False alarms' (E_P (ecmwf) < 0 and E_P (obs) > 0), 'Misses' (E_P (ecmwf) > 0 and E_P (obs) < 0) and the 'correct rejections' (E_P (ecmwf) > 0 and E_P (obs) > 0).

Electrical Production	E_P (obs) < 0	E_P (obs) > 0
E_P (ecmwf) < 0	16	6
E_P (ecmwf) > 0	19	324

Considering together, the daily values of the forecasted or measured production values, it was found that there were 41 days (in a total of 365 days) with partial or complete cloudy (overcast) conditions, i.e., depicting negative production values (consumption). Results showed that the forecast model predicts a total of 16 days of negative production, which coincided with the measurements, following the condition E_P (ecmwf) < 0 and E_P (obs) < 0, which denominated the 'Hits'. For cloudy days with short periods of no production, where the model predicted overcast, but that was not observed, i.e., when E_P (ecmwf) < 0 and E_P (obs) > 0 (which denominated 'False alarms'), a number of 6 days were found. The opposite occurred when the IFS did not predict overcast which was observed, i.e., when E_P (ecmwf) > 0 and E_P (obs) < 0 (which denominated 'Misses'), with a number of 19 days being found in such conditions. The latter was a clear result of the IFS general overestimation, due to cloud representation, as previously discussed in detail [29]. Moreover, the number of days in which the IFS and measurements did not show the occurrence of negative production values was 324, denominated here as the 'correct rejections'. Thus, the obtained ETS (Equation (A3)) for the occurrence of negative production forecasted by the IFS was ≈36%. Considering such a rate, the power plant operator was advised to not proceed with the electrical energy generation, when the forecast model predicted negative production (E_P (ecmwf) <0). In the case of a wrong prediction, if solar energy was available for collection during the day, then production was to be started but without spending any storage. If a success rate of ≈90% was to be found, then the operator would simply be advised not to produce during that day.

5. Conclusions

In this work, it was confirmed that the use of DNI forecasts and the implementation of control strategies could contribute to a more efficient energy management of a CSP plant, improving the local energy distribution from a solar tower system. Hourly and daily correlations of ≈0.78 and ≈0.89, respectively, were found for the SAM predictions of the total electric energy injected into the grid, based on forecasted and measured DNI and meteorological conditions, an important variable for the power plant operator to handle on a daily basis. In the case of the power plant stored thermal energy, charge correlations of ≈0.88 and ≈0.89 were found for the hourly and daily values, respectively, while ≈0.83 and ≈0.88 were found for the hourly and daily discharge values, respectively. Regarding the performance of the forecast model in the simulations of the two different types of CSP plants enforced with the same datasets, results showed higher deviations in the case of a CR system than in the previous simulated PT. Increases of ≈7.3% and ≈2.8% were found, respectively, for the hourly and daily normalized RMSE values of the generated electric energy. To improve the energy efficiency of CR plants, operational strategies have been proposed for the three different scenarios. Although there were still deviations due to the cloud and aerosol representation, the present analysis has shown that the IFS predictions are a valuable tool to be used in the daily energy dispatch operations of a CR power plant, potentially the main type of CSP systems to be used in the future, due to its advantages. With the continuous improvements that the NWP models have demonstrated in recent years, for the prediction of DNI, future versions of the IFS should also demonstrate an enhancement of the predicted production values from a power plant and, consequently, the energy management during solar intermittency periods.

Author Contributions: The concept of this work was made by F.M.L. and H.G.S. The applied methodology was implemented by F.M.L, H.G.S., R.C., R.S., and P.C. The use of the SAM software was carried out by F.M.L. and T.F. Analysis, validation, investigation, resources, data curation, writing, and editing procedures were done by F.M.L. All authors have reviewed the article prior to submission. H.G.S., R.S., and M.C.-P. have supervised this work.

Funding: F.M.L. and R.C. are thankful to the FCT scholarships SFRH/BD/129580/2017 and SRFH/BD/116344/2016, respectively. T.F. acknowledges the New StOrage Latent and sensible Concept for high efficient CSP Plants, the NewSOL (H2020, GA: 720985) project, and H.G.S. at the Integrating National Research Agendas on Solar Heat for Industrial Processes, INSHIP (H2020, GA: 731287) project. Co-funding from the European Union through the European Regional Development Fund was also provided, being included in the COMPETE 2020 (Operational Program

Competitiveness and Internationalization), through the ICT (UID/GEO/04683/2013, POCI-01-0145-FEDER-007690), DNI-A (ALT20-03-0145-FEDER-000011), ALOP (ALT20-03-0145-FEDER-000004) projects.

Acknowledgments: Available measured data provided by Afonso Cavaco (Instituto Português de Energia Solar, IPES) and Jorge Neto (Instituto Português do Mar e da Atmosfera, IPMA), as well as the forecasts from the ECMWF, is acknowledged.

Conflicts of Interest: The authors declare no conflict of interest. The funders had no role in the design of the study; in the collection, analyses, or interpretation of data; in the writing of the manuscript, or in the decision to publish the results.

Appendix A

To calculate the normalized error metrics (i.e., the nRMSE and nMAE) of E_P (%), the following equations were used:

$$\text{nRMSE} = \text{RMSE}/(E_{Pmax} - E_{Pmin}), \tag{A1}$$

$$\text{nMAE} = \text{MAE}/(E_{Pmax} - E_{Pmin}), \tag{A2}$$

where the RMSE and MAE between measured and forecasted E_P was divided by the difference between the maximum and minimum values of the measured E_P.

To evaluate the performance of the forecast model in predicting negative production values, the equitable threat score (ETS) could be calculate through:

$$\text{ETS} = (\text{Hits} - \text{Hits}_{random})/(\text{Hits} + \text{Misses} + \text{False alarms} - \text{Hits}_{random}), \tag{A3}$$

with

$$\text{Hits}_{random} = [(\text{Hits} + \text{False alarms}) \times (\text{Hits} + \text{Misses})]/\text{Total}, \tag{A4}$$

where 'Hits' represents the number of occurrences (i.e., number of days) with forecasted and observed negative E_P values, 'Misses' represents the number of days in which the forecast model did not predict the E_P values when these were actually observed, 'False alarms' corresponds to the predicted occurrences of E_P values that were not observed. 'Total' is the total number of occurrences, which also took into account the number of days of 'correct rejections' (i.e., when the forecast model did not predict the E_P values that were not actually observed). A perfect forecast (i.e., a perfect score of 1) would be characterized only by 'Hits' and 'correct rejections', without 'Misses' and 'False alarms'.

A detailed description regarding the input parameters for the SAM software for the simulation of a CSP power plant designed to run a CR system has been given in this section. For the Gemasolar thermosolar plant (Figure A1) case study, the available online information from NREL [36], was complemented with the default SAM inputs characteristic from this type of tower power plant, together with the research carried out by the authors, towards a few parameters that were taken into account, as presented in Table A1.

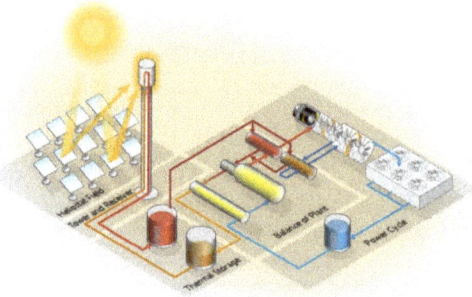

Figure A1. Schematic of the simulated Gemasolar thermosolar power plant in the SAM. The different components of a central receiver system are depicted. (© System Advisor Model Version 2017.9.5, SAM 2017.9.5).

Table A1. Input parameters for the SAM simulation of the Gemasolar thermosolar power plant during one year (from 1 July, 2017 to 30 June, 2018).

	General	
Name	Value	Reference
Single heliostat net area	115.7 m²	[42]
Ratio of reflective area	0.9642	[42]
Field gross collecting area	315,000 m²	2625 heliostats generated by SAM, 2650 according to [42]
Irradiation at design	700 W/m²	Chosen by authors
HTF	Solar Salt	[36]
Design loop inlet temperature	290 °C	[36]
Design loop outlet temperature	565 °C	[36]
Full load hours of TES	15 h	[36]
Storage HTF fluid	Solar Salt (direct storage)	[36]
	Receiver	
Name	Value	Reference
Tower height	140 m	[36]
Receiver height	10 m	[42]
Receiver diameter	9 m	[42]
Number of panels	14	Chosen by authors
Tube outer diameter	4×10^{-2} m	SAM standard value
Minimum receiver turndown fraction	0.25	SAM standard value
Maximum receiver operation fraction	1.2	SAM standard value
Receiver startup delay time	0.25 h	Chosen by authors
Estimated receiver heat loss	30 kW/m²	Calculated by authors (Equation (A5))
Piping length	360 m	Estimated by authors
Piping heat loss coefficient	1000 W$_t$/m	Calculated by authors (Equation (A7))
	Power block	
Name	Value	Reference
Design gross output	19.9 MW$_e$	[36]
Gross to net conversion factor	1	[36]
Rated cycle conversion efficiency	0.445	Calculated from storage and receiver capacities
Fraction of thermal power needed for standby	0.2	SAM Standard value
Power block start-up time	0.5 h	SAM Standard value
Fraction of thermal power for start-up	0.5	SAM Standard value
Maximum turbine over design operation (ratio)	1.05	SAM Standard value
Minimum turbine operation (ratio)	0.2	SAM Standard value
Boiler operating pressure	105 bars	[43]
Turbine inlet pressure control	Fixed-pressure	SAM Standard value

Heat losses from the receiver are due to radiation to the environment and convection. The equation of heat losses per square meter of a receiver is, therefore, given by the following equation:

$$P_{rec} = \epsilon_{rec} \cdot \sigma \cdot \left(T_{rec}^4 - T_{ext}^4\right) + h_{conv,ext} \cdot (T_{rec} - T_{ext}) \qquad (A5)$$

Using a receiver temperature (T_{rec}) of 565 °C (or 838.15 K), an external temperature (T_{ext}) of 20 °C (or 293.15 K), a receiver emittance (ϵ_{rec}) of 0.88 (input for the SAM), and a convection coefficient ($h_{conv,ext}$) of 10 W·m^{-2}·K^{-1}, Equation A5 can be solved as:

$$P_{rec} = 0.88 \times 5.67 \times 10^{-8} \cdot \left(838.15^4 - 293.15^4\right) + 10 \times (838.15 - 293.15), \tag{A6}$$

where the Stefan–Boltzman constant ($\sigma = 5.67 \times 10^{-8}$) is used. The obtained result can be approximated to 30 kW/m^2.

To calculate the heat loss from the pipes, the pipe loss coefficient is written as follows:

$$P_{loss,pipe}(W.m^{-1}) = \frac{T_f - T_{ext}}{\frac{\ln\left(\frac{r_{ext,p}+e_{ins}}{r_{ext,p}}\right)}{2\times\pi\cdot k_{ins}} + \frac{1}{h_{ext}\times 2\times\pi\cdot r_{ext,p}}} \tag{A7}$$

Assuming a pipe with an internal diameter of 800 mm and an external diameter of 812.8 mm, an external convection coefficient of 15 W/m^2·K and 15 cm of insulation (k_{ins} = 0.08 W·m^{-1}·K^{-1}), Equation (A7) can be solved as:

$$P_{loss,pipe}(W.m^{-1}) = \frac{565 - 20}{\frac{\ln\left(\frac{0.4064+0.15}{0.4064}\right)}{2\times\pi\times 0.08} + \frac{1}{15\times 2\times\pi\times 0.4064}} = 837 \text{ W.m}^{-1} \tag{A8}$$

Since pipe losses should take into account all heat bridges due to sensors, valves, etc., it has been decided to approximate the value to 1000 W·m^{-1}.

References

1. Mendelshon, M.; Lowder, T.; Canavan, B. *Utility-Scale Concentrating Solar Power and Photovoltaics Projects: A Technology Market Overview*; Technical report NREL/TP-6A20-51137; NREL U.S. Department of Energy: Washington, DC, USA, 2012.
2. Kearney, D.; Kelly, B.; Herrmann, U.; Cable, R.; Pacheco, J.; Mahoney, R.; Price, H.; Blake, D.; Nava, B.; Potrovitza, N. Engineering aspects of a molten salt heat transfer fluid in a trough solar field. *Energy* **2004**, *29*, 861–870. [CrossRef]
3. Coimbra, C.; Kleissl, J.; Marquez, R. Overview of solar forecasting methods and a metric for accuracy evaluation. In *Solar Resource Assessment and Forecasting*; Elsevier: Waltham, MA, USA, 2013; ISBN 9780123971777.
4. Schroedter-Homscheidt, M.; Benedetti, A.; Killius, N. Verification of ECMWF and ECMWF/MACC's global and direct irradiance forecasts with respect to solar electricity production forecasts. *Meteor. Zeitschrift* **2017**, *26*, 1–19. [CrossRef]
5. Stull, R.B. *An Introduction to Boundary Layer Meteorology, Chapter 13*; Atmospheric Science Programme: Dordrecht, The Netherlands, 1999; ISBN 9027727686.
6. Conceição, R.; Silva, H.G.; Collares-Pereira, M. CSP mirror soiling characterization and modelling. *Sol. Energy Mater. Sol. Cells* **2018**, *185*, 233–239. [CrossRef]
7. Tyagi, H.; Agarwal, A.K.; Chakraborty, P.R.; Powar, S. Advances in solar energy research. *Energy Environ. Sustain.* **2019**, *1*, 48–50.
8. Troccoli, A.; Morcrette, J.-J. Skill of direct solar radiation predicted by the ECMWF global atmospheric model over Australia. *Am. Meteor. Soc.* **2014**, *53*, 2571–2587. [CrossRef]
9. Tompkins, A.M. *The Parametrization of Cloud Cover*; European Centre for Medium-Range Weather Forecasts (ECMWF): Reading, UK, 2005.
10. Bozzo, A.; Remy, S.; Benedetti, A.; Flemming, J.; Bechtold, P.; Rodwell, M.J.; Morcrette, J.-J. *Implementation of a CAMS-Based Aerosol Climatology in the IFS*; European Centre for Medium-Range Weather Forecasts (ECMWF): Reading, UK, 2017; Volume 801.
11. Lopes, F.M.; Conceição, R.; Fasquelle, T.; Silva, H.G.; Salgado, R.; Canhoto, P.; Collares-Pereira, M. ECMWF forecasts of DNI for optimized operation strategies of linear focus parabolic-trough systems. *Appl. Energy* **2019**. In revision.

12. The Parabolic Trough Power Plants Andasol 1 to 3. The Largest Solar Power Plants in the World—Technology Premier in Europe. Solar Millenium AG. 2008. Available online: http://large.stanford.edu/publications/power/references/docs/Andasol1-3engl.pdf (accessed on 28 March 2019).
13. Rinaldi, F.; Binotti, M.; Giostri, A.; Manzolini, G. Comparison of linear and point focus collectors in solar power plants. *Energy Proc.* **2014**, *49*, 1491–1500. [CrossRef]
14. Haiden, T.; Janousek, M.; Bidlot, J.; Ferranti, L.; Prates, F.; Vitart, F.; Bauer, P.; Richardson, D.S. *Evaluation of ECMWF Forecasts, Including the 2016 Resolution Upgrade*; European Centre for Medium-Range Weather Forecasts (ECMWF): Reading, UK, 2016; Volume 792.
15. Burgaleta, J.; Arias, S.; Ramirez, D. Gemasolar, the first tower thermosolar commercial plant with molten salt storage. In Proceedings of the 17th SolarPACES Conference, Granada, Spain, 20–23 September 2011.
16. Salgado, R.; Miranda, P.M.A.; Lacarrère, P.; Noilhan, J. Boundary layer development and summer circulation in Southern Portugal. *Tethys* **2015**, *12*, 33–44. [CrossRef]
17. Cavaco, A.; Silva, H.G.; Canhoto, P.; Osório, T.; Collares-Pereira, M. Progresses in DNI measurements in Southern Portugal. *AIP Conf. Proc.* **2018**, *2033*, 1–7. [CrossRef]
18. International Organization for Standardization. *Solar Energy. Calibration of Field Pyrheliometers by Comparison to a Reference Pyrheliometer*; ISO 9059:1990(E.); International Organization for Standardization: Genève, Switzerland, 1990.
19. Long, C.N.; Dutton, E.G. BSRN Global Network Recommended QC Tests 2002, V2.0. Available online: 10013/epic.38770 (accessed on 20 February 2019).
20. Mlawer, E.; Clough, S. Shortwave and longwave radiation enhancements in the rapid radiative transfer model. In Proceedings of the 7th Atmospheric Radiation Measurement (ARM) Science team Meeting. U.S. Department of Energy ARM-CONF-970365, San Antonio, TX, USA, 3 March 1997; pp. 409–413.
21. Hogan, R.J.; Bozzo, A. A flexible and efficient radiation scheme for the ECMWF model. *J. Adv. Modell. Earth Syst.* **2018**, *10*, 1990–2008. [CrossRef]
22. Lara-Fanego, V.; Ruiz-Arias, J.A.; Pozo-Vázquez, A.D.; Gueymard, C.A.; Tovar-Pescador, J. Evaluation of DNI forecast based on the WRF mesoscale atmospheric model for CPV applications. *AIP Conf. Proc.* **2012**, *1477*, 317–322.
23. Voyant, C.; Notton, G. Solar irradiation nowcasting by stochastic persistence: A new parsimonious, simple and efficient forecasting tool. *Renew. Sustain. Energy Rev.* **2018**, *92*, 343–352. [CrossRef]
24. Richardson, W.; Krishnaswami, H.; Vega, R.; Cervantes, M. A low cost, edge computing, all-sky imager for cloud tracking and intra-hour irradiance forecasting. *Sustainability* **2017**, *9*, 482. [CrossRef]
25. Lauret, P.; Voyant, C.; Soubdhan, T.; David, M.; Poggi, P. A benchmarking of machine learning techniques for solar radiation forecasting in an insular context. *Sol. Energy* **2015**, *112*, 446–457. [CrossRef]
26. Aguiar, L.M.; Pereira, B.; Lauret, P.; Díaz, F.; David, M. Combining solar irradiance measurements, satellite-derived data and a numerical weather prediction model to improve intraday solar forecasting. *Renew. Energy* **2016**, *97*, 599–610. [CrossRef]
27. Larson, D.P.; Nonnenmacher, L.; Coimbra, C.F.M. Day-ahead forecasting of solar power output from photovoltaic plants in the American Southwest. *Renew. Energy* **2016**, *91*, 11–20. [CrossRef]
28. Smagorinsky, J.; Miyakoda, K.; Stickler, R.F. The relative importance of variables in initial conditions for dynamical weather prediction. *Tellus* **1970**, *22*, 141–157. [CrossRef]
29. Lopes, F.M.; Silva, H.G.; Salgado, R.; Cavaco, A.; Canhoto, P.; Collares-Pereira, M. Short-term forecasts of GHI and DNI for solar energy systems operation: Assessment of the ECMWF integrated forecasting system in Southern Portugal. *Sol. Energy* **2018**, *170*, 14–30. [CrossRef]
30. Martin, G.M.D.; Johnson, D.W.; Spice, A. The measurement and parameterization of effective radius of droplets in warm stratocumulus. *J. Atmos. Sci.* **1994**, *51*, 1823–1842. [CrossRef]
31. Sun, Z.; Rikus, L. Parameterization of effective sizes of cirrus-clouds particles and its verification against observations. *Q. J. R. Meteorol. Soc.* **1999**, *125*, 3037–3055. [CrossRef]
32. Fu, Q.; Yang, P.; Sun, W.B. An accurate parameterization of the infrared radiative properties of cirrus clouds of climate models. *J. Climate Atmos. Sci.* **1994**, *11*, 2223–2237. [CrossRef]
33. Edwards, J.M.; Slingo, A. Studies with a flexible new radiative code: 1. Choosing a configuration for a large-scale model. *Q. J. R. Meteorol. Soc.* **1996**, *122*, 689–719. [CrossRef]

34. Tegen, I.; Hollrig, P.; Chin, M.; Fung, I.; Jacob, D.; Penner, J. Contribution of different aerosol species to the global aerosol extinction optical thickness: Estimates from model results. *J. Geophys. Res.* **1997**, *102*, 895–915. [CrossRef]
35. Blair, N.; Dobos, A.; Freeman, J.; Neises, T.; Wagner, M.; Ferguson, T.; Gilman, P.; Janzou, S. *System Advisor Model, SAM 2014.1.14: General Description*; National Renewable Energy Laboratory: Golden, CO, USA, 2014. [CrossRef]
36. NREL, Concentrating Solar Power Projects-Gemasolar Thermosolar Plant. Available online: https://solarpaces.nrel.gov/gemasolar-thermosolar-plant (accessed on 27 March 2019).
37. Wittmann, M.; Breitkreuz, H.; Schroedter-Homscheidt, M.; Eck, M. Case studies on the use of solar irradiance forecast for optimized operation strategies of solar thermal power plants. *IEEE J. Sel. Top. Appl. Earth Obs. Remote Sens.* **2008**, *1*, 18–27. [CrossRef]
38. Kosmopoulos, P.G.; Kazadzis, S.; Taylor, M.; Athanasopoulou, E.; Speyer, O.; Raptis, P.I.; Amiridis, V. Dust impact on surface solar irradiance assessed with model simulations, satellite observations and ground-based measurements. *Atmos. Meas. Tech.* **2017**, *10*, 2435–2453. [CrossRef]
39. Mukkavilli, S.K.; Prasad, A.A.; Taylor, R.A.; Troccoli, A.; Kay, M. Mesoscale simulations of Australian direct normal irradiance, featuring an extreme dust event. *Am. Meteorol. Soc.* **2018**, *57*, 493–515. [CrossRef]
40. WWRP/WGNE Joint Group on Forecast Verification Research Website on Forecast Verification. Issues, Methods and FAQ. Available online: http://www.cawcr.gov.au/projects/verification (accessed on 28 March 2019).
41. Salgado, R. Interacção Solo—Atmosfera em Clima Semi-Árido. Ph.D. Thesis, University of Évora, Évora, Portugal, 2005.
42. Collado, F.; Guallar, J. A review of optimized design layouts for solar power tower plants with campo code. *Renew. Sustain. Energy Rev.* **2013**, *20*, 142–154. [CrossRef]
43. SIEMENS. Steam Turbines for CSP Plants. 2011. Available online: http://m.energy.siemens.com (accessed on 28 March 2019).

© 2019 by the authors. Licensee MDPI, Basel, Switzerland. This article is an open access article distributed under the terms and conditions of the Creative Commons Attribution (CC BY) license (http://creativecommons.org/licenses/by/4.0/).

Article

Solar Irradiance Forecasts by Mesoscale Numerical Weather Prediction Models with Different Horizontal Resolutions

Hideaki Ohtake [1,2,*], Fumichika Uno [1,2], Takashi Oozeki [1], Syugo Hayashi [2], Junshi Ito [2,3], Akihiro Hashimoto [2], Hiromasa Yoshimura [2] and Yoshinori Yamada [2]

1. Research Center for Photovoltaics, National Institute of Advanced Industrial Science and Technology, Ibaraki 3058568, Japan; uno.fumichika@aist.go.jp (F.U.); takashi.oozeki@aist.go.jp (T.O.)
2. Meteorological Research Institute, Japan Meteorological Agency, Ibaraki 3050052, Japan; shayashi@mri-jma.go.jp (S.H.); junshi@aori.u-tokyo.ac.jp (J.I.); ahashimo@mri-jma.go.jp (A.H.); hyoshimu@mri-jma.go.jp (H.Y.); yyamada@mri-jma.go.jp (Y.Y.)
3. Atmosphere and Ocean Research Institute, The University of Tokyo, Chiba 2778564, Japan
* Correspondence: hideaki-ootake@aist.go.jp; Tel.: +81-29-849-1526

Received: 7 March 2019; Accepted: 5 April 2019; Published: 9 April 2019

Abstract: This study examines the performance of radiation processes (shortwave and longwave radiations) using numerical weather prediction models (NWPs). NWP were calculated using four different horizontal resolutions (5, 2 and 1 km, and 500 m). Validation results on solar irradiance simulations with a horizontal resolution of 500 m indicated positive biases for direct normal irradiance dominate for the period from 09 JST (Japan Standard Time) to 15 JST. On the other hand, after 15 JST, negative biases were found. For diffused irradiance, weak negative biases were found. Validation results on upward longwave radiation found systematic negative biases of surface temperature (corresponding to approximately −2 K for summer and approximately −1 K for winter). Downward longwave radiation tended to be weak negative biases during both summer and winter. Frequency of solar irradiance suggested that the frequency of rapid variations of solar irradiance (ramp rates) from the NWP were less than those observed. Generally, GHI distributions between the four different horizontal resolutions resembled each other, although horizontal resolutions also became finer.

Keywords: solar irradiance forecasts; numerical weather prediction model; different horizontal resolution; forecast errors; validation; ramp rates

1. Introduction

Photovoltaic (PV) power forecasts are based on solar irradiance forecasts generated by a numerical weather prediction model (NWP) and/or machine learning predictions using data from a NWP. However, considerable errors are often included in the solar irradiance forecasts from NWPs (Schiermeier [1]). Previous studies (Ohtake et al. [2,3]) have investigated the relation between day-ahead global horizontal irradiance (GHI) forecasts and cloud types monitored in Japan Meteorological Agency's (JMA) operational model (a mesoscale model called "MSM"). These studies indicated that when certain types of clouds (i.e., stratus, altocumulus and cirrus) were monitored, GHI forecasts tended to have large one day-ahead forecast errors.

Inter-comparisons of solar irradiance and simulated cloud from NWPs with different horizontal resolutions have been performed (e.g., Mathiesen and Kleissl [4]; Perez et al. [5]; Lorenz et al. [6]), but no operational solar irradiance forecasts based on mesoscale NWPs with horizontal grid spacing finer than 2 km have been conducted. The potential usefulness of solar irradiance forecasts with a horizontal resolution finer than 1 km grid spacing is a subject of interest.

Gregory and Rikus [7] validated GHI, direct normal irradiance (DNI), and diffused irradiance (DIF) forecasts using the Australian Community Climate and Earth-System Simulator (ACCESS) operational model with a horizontal resolution of 0.11°. DNI forecasts were found to be overestimated and DIF forecasts were found to be under-estimated, depending on a two-stream approximation asymmetry factor in a radiation process in their NWP. Sosa-Tinoco et al. [8] compared solar irradiance forecast performance using Weather Research and Forecasting model (WRF) with different horizontal resolution models (27 and 9 km mesh models) for Mexico and suggested that a higher horizontal resolution model has little benefit; however, this is partly because their target region had not drastic topography change. Charabi et al. [9] performed a case study for Oman, which was characterized by high potential of solar energy (Sunbelt) and investigated the usefulness of a higher horizontal resolution NWP considering 40, 7 and 2.8 km horizontal resolutions for assessing PV energy. Charabi et al. [9] described that the downscaled NWP model gives much more accurate estimation of the monthly as well as the annual average solar radiation. Mathiesen et al. [10] validated hourly-averaged solar irradiance forecasts for marine clouds over the southern California coast using an NWP with a 1.3 km horizontal resolution. Similarly, Lin et al. [11] compared cloud forecasts generated by multiple nested WRF simulations with horizontal resolutions ranging from 20 to 0.8 km.

Solar irradiance forecasts using higher horizontal resolutions have a higher calculation cost; thus if we choose to use higher horizontal resolution models in the future, the costs and benefits will need to be carefully evaluated.

A previous study by Ito et al. [12] reported validation results on brightness, temperature, and surface precipitation from the JMA's non-hydrostatic model (JMA-NHM), with horizontal resolutions between 5 km and 0.5 km. They did not find a significant improvement in performance between horizontal resolutions of 2 km and 0.5 km. This study validates the same products as in Ito et al. [12]'s study, but the radiation processes (shortwave and longwave radiations) are verified using surface and satellite monitoring data.

For electric power operators, forecasting short-range variability, or ramp events, of solar irradiance is important. Ramp events frequently cause rapid increases and/or decreases of PV power generation. Wellby and Engerer [13] categorize city-scale (defined as a region size of approximately 30 km) ramp events according to meso-scale and synoptic-scale meteorological phenomena (e.g., cold fronts, thunderstorms, cloud bands). If the NWP can accurately predict these meteorological phenomena, then the ramp events can also be predicted.

The remainder of this paper is organized as follows: in Section 2, surface and satellite monitoring datasets for model validations are introduced. In Section 3, model setup and model performance metrics are described. Validation results for the outputs of radiation processes are described in Section 4. In Section 5, short-term variability (ramp events) of solar irradiance is analyzed, and Section 6 summarizes the findings.

2. Observation data

2.1. Global Solar Irradiance Data

JMA observes GHI at 48 stations on the Japanese islands. GHI, DNI, DIF, upward longwave radiation (ULW), and downward longwave radiation (DLW) were observed at the JMA Aerological Observatory (Tsukuba (Tateno) station; 36°3.4' N, 140°7.5' E) and included in the Baseline Surface Radiation Network (BSRN) integrated database. This data has a time resolution of one second and has been verified through the JMA quality check (Gilgen et al. [14]; Ohmura et al. [15], McArthur [16]; BSRN web site [17]). The location of Tsukuba station is shown in Figure 1. Photos of instruments that measure GHI and longwave radiation at Tsukuba station are shown in Figure 2.

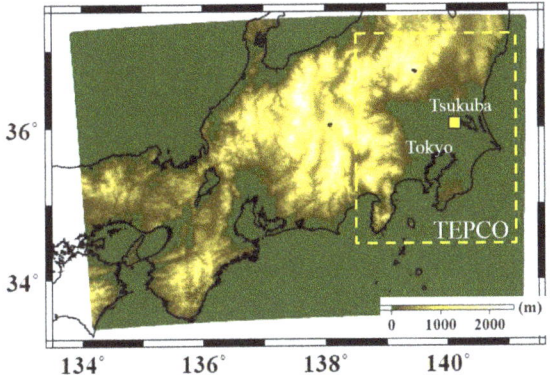

Figure 1. Topography of dx500m and the location of the Tsukuba station (yellow square). The yellow dashed rectangle indicates the Tokyo electric power company (TEPCO) area.

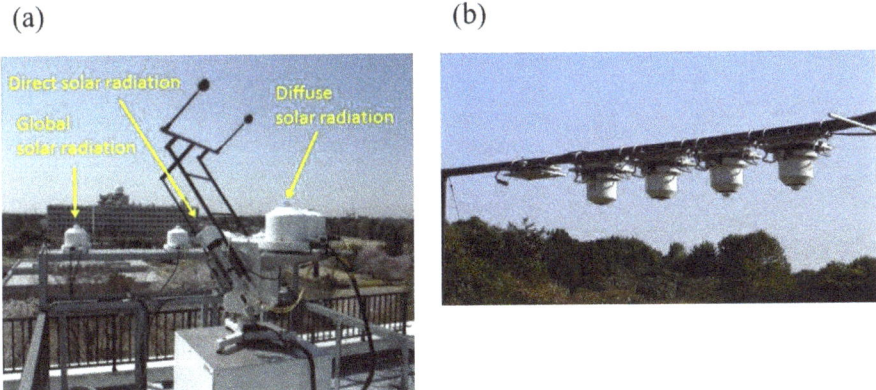

Figure 2. Photos of (**a**) BSRN instruments (GHI, DNI, and DIF, Source from the web site of JMA aerological observatory http://www.jma-net.go.jp/kousou/obs_third_div/rad/rad_sol.html) and (**b**) measurements of upward longwave radiation (ULW) instruments at Japan Meteorological Agency's (JMA) aerological observatory.

For GHI, DNI, and DIF observations, Kipp & Zonen CMP-21, Kipp & Zonen CHP-1 and Kipp & Zonen CMP-22 were used. Kipp & Zonen CGR-4 monitors DLW. Detailed descriptions of GHI monitoring instruments (pyranometers and pyrheliometers) in JMA stations can be found in Ohtake et al. [3]. For ULW and DLW, Kipp & Zonen CGR-4 and/or CG-4 were used.

2.2. Satellite-Derived Solar Irradiance

A geostationary satellite, Himawari-8, was launched on 7 July 2015 and is currently in operational use (Meteorological Satellite Center of JMA [18]). Detailed specifications of Himawari-8 have been reported in Bessho et al. [19]. Satellite-derived GHI datasets (including DNI and DIF), commonly referred to as "AMATERASS", is estimated based on the algorithm described in Takenaka et al. [20]. In this algorithm, cloud properties (cloud optical thickness, cloud particle size, and cloud top temperature) that influence the GHI were taken into consideration (e.g., Nakajima and Nakajima [21]; Kawamoto et al. [22]). However, aerosol optical depth (AOD) information was not sufficiently included in this algorithm. The data had a temporal resolution of 2.5 min and a horizontal resolution of 1 km for the Japanese islands and was provided by the Solar Radiation Consortium [23]. AMATERASS datasets from Himawari-8 were also used to validate spatial distributions in GHI model simulations. Validation of the satellite-derived GHI data with ground-based data resulted in a mean bias error (MBE) within the range of 20–30 W m^{-2} under all-sky conditions (10–15 W m^{-2} under clear-sky conditions) and a root mean square error (RMSE) of approximately 80 W m^{-2} (Damiani et al. [24]).

3. Numerical simulations

3.1. Model

The JMA developed the JMA-NHM (Saito et al. [25,26]; JMA web site [27]). In this study, the JMA-NHM was used to produce numerical simulations with horizontal resolutions of 5 km, 2 km, 1 km, and 500 m. Each simulation was referred to as "dx05km", "dx02km", "dx01km" and "dx500m", respectively. The experimental period in this study was the summer season from 1 July to 31 August 2015 (Exp-Summer) and the winter season from 12 January to 18 February 2016 (Exp-Winter).

Model settings for each numerical experiment are summarized in Table 1. The horizontal grid numbers (x, y) for each resolution were 220 × 180, 550 × 450, 1100 × 900, and 2200 × 1800, respectively. The number of vertical layers was 60. Model time steps were 15, 10, 5, and 3 s for each horizontal resolution, respectively.

Table 1. Model set-up for numerical experiments with four different horizontal resolutions (dx05km, dx02km, dx01km, and dx500m).

	dx05km	dx02km	dx01km	dx500m
Model grids (x, y, z)	220 × 180 × 60	550 × 450 × 60	1100 × 900 × 60	2200 × 1800 × 60
Forecast times (00UTC;09JST)	12 h	12 h	12 h	12 h
Time step	15 s	10 s	5 s	3 s
Cumulus parameterization	No	No	No	No
Cloud physics schemes	3 ice, 2 moment	3 ice, 2 moment	3 ice, 2 moment	3i ce, 2 moment
Turbulent schemes	MYNN2.5	MYNN2.5	MYNN2.5	MYNN2.5
Radiation schemes	partial saturated clouds	partial saturated clouds	partial saturated clouds	partial saturated clouds
Calculation time	25 min	84 min	282 min	450 min

The three-class bulk ice (3-ICE) microphysical scheme (Lin et al. [28]; Cotton et al. [29]) was used in these numerical experiments. The Mellor-Yamada-Nananishi-Niino 2.5 (MYNN2.5) level scheme (Nakanishi and Niino [30]) was used for the planetary boundary layer process. For radiation transfer processes of the JMA-NHM, downward shortwave processes for water clouds were based on Slingo [31], while processes for ice clouds were based on Ebert and Curry [32], respectively. Longwave radiation processes were based on Hu and Stamnes [33] for water clouds and Ou and Liou [34] for ice clouds. A partial condensation scheme (Sommeria and Deardorff [35]) was introduced to treat sub-grid scale condensation because small scale clouds could not be resolved. The radiative transfer processes were calculated at 60 s intervals for every two model grids in the x and y directions to reduce calculation costs. The finer horizontal resolutions (e.g., <2 km), which are often referred as convection-allowing models, are so fine that the parameterization of cumulus is thought to be no longer required. Cumulus parametrization schemes were not used in this study.

The target area was the Tokyo electric power company (TEPCO), which is located in the center of the Japanese islands. The topography of dx500m is shown in Figure 1. Model initialization and boundary conditions were given by the JMA meso-analysis data provided every 3 h. Initialization times were 00, 06, and 18 UTC (three times a day). Twelve hour time integrations were performed during Exp-Summer in 2015 and Exp-Winter in 2016. To evaluate GHI simulations with Himawari-8 satellite data, simulation data was output every 2.5 min (see Section 2.2).

3.2. Performance Metrics

To validate forecasted GHI values, we used several evaluation parameters; MBE, mean absolute error (MAE), and RMSE. They are defined by the following equations:

$$\text{MBE} = \frac{1}{N}\sum_{i=1}^{N}(FCST_i - OBS_i) \quad (1)$$

$$\text{MAE} = \frac{1}{N}\sum_{i=1}^{N}|FCST_i - OBS_i| \quad (2)$$

$$\text{RMSE} = \sqrt{\frac{1}{N}\sum_{i=1}^{N}(FCST_i - OBS_i)^2} \quad (3)$$

Forecasts (FCST) and observations (OBS) are forecasted GHI and GHI surface-observed values, respectively. N is the sample number. Standard deviation (SD) is defined by the following equation:

$$\text{SD} = \sqrt{\frac{1}{N}\sum_{i=1}^{N}(x_i - \bar{x})^2} \quad (4)$$

where x_i and $\bar{x}$ represent sample data and the mean value of the sample data.

4. Validation results

4.1. Validation of GHI, DNI and DIF

First, solar radiation simulations for dx500m were compared to the BSRN observations at the Tsukuba station (The location of Tsukuba is shown in Figure 1). Figure 3a,b show time cross sections of GHI, DNI, and DIF values for the Exp-Summer of 2015.

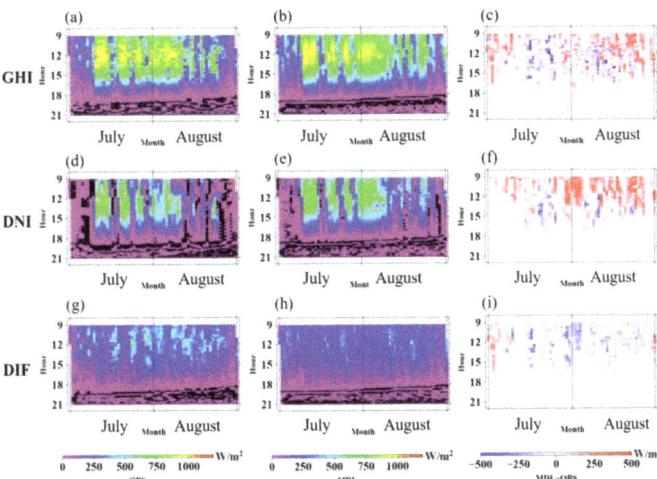

Figure 3. Time cross sections (each day from July to August (horizontal axis; Month)-hours in a day (vertical axis; Hour)) for GHI (top), DNI (middle), and DIF (bottom) for the Exp-Summer of 2015 at Tsukuba station. Left panels show GHI observations, center panels show simulations by dx500m with 00 UTC (09 JST) initialization time and right panels show the difference for dx500m (dx500m−observations). Color scales are shown at the bottom of each column.

The horizontal axis represents each day of the two month period and the vertical axis represents local hours, ranging from 08 JST (Japan Standard Time, JST = UTC + 9 h) to 22 JST. Temporal variations of GHI simulations from dx500m were like that of GHI observations (Figure 3a,b). Based on the difference between the GHI observations and the simulations, negative biases were generally found during the daytime (from 12 JST to 15 JST) for the latter half of July and the first half of August. In both the first half of July and the latter half of August, positive biases were also observed.

Figure 3d–f,g–i show time cross sections of DNI and DIF simulations (like Figure 3a–c). Positive biases for DNI dominate for the period from 09 JST to 15 JST. On the other hand, after 15 JST, negative biases were found. For DIF, weak negative biases were found. Similar results to this validation were identified for each of the four different horizontal resolutions (not shown).

Generally, for Exp-Winter, GHI, DNI and DIF biases were weaker than those in summer (Figure 4). Similar positive biases of GHI and DNI values were found during the daytime in the middle of February. GHI and DNI values in the daytime were lower than those in summer because of the lower altitude of the sun. Large simulation errors were found in GHI, DNI, and DIF values in both Exp-summer and Exp-winter, although the frequency was low.

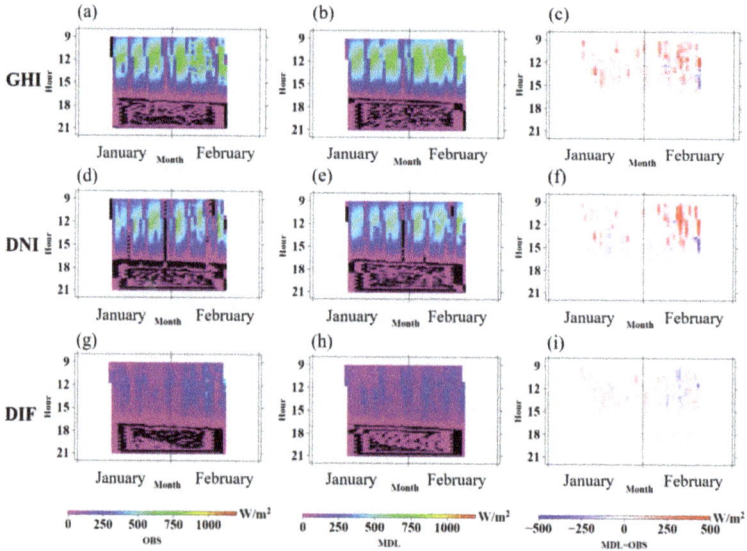

Figure 4. Same as Figure 3, but for the Exp-Winter of 2016.

Figure 5 shows hexbin plots for GHI, DNI, and DIF for the Exp-Summer and the Exp-Winter (e.g., Davy et al. [36]). The hexbin plots for GHI resemble a one to one line, suggesting that GHI simulations were close to GHI observations. Hexbin plots of DNI values for clear sky conditions (around 800 W m^{-2}) and optically thick conditions (under 200 W m^{-2}) were close to the observations. However, the middle level (from 300 to 500 W m^{-2}) of DNI (corresponding to optically thin conditions) exhibits larger simulation errors. DIF values did not exceed 500 W m^{-2}. DIF simulation errors tended to be large.

Results for the Exp-Winter in 2016 were also like those of the Exp-Summer in 2015 (Figure 5b). However, the frequency of GHI and DNI simulations under clear-sky conditions was higher around the one to one line when compared to the Exp-Summer. Under clear sky conditions, radiation processes in the JMA-NHM were validated. Figure 6 shows time series for GHI, DNI, and DIF simulations on 11 February 2016 at the Tsukuba station. On this date, the aerosol optical depth (AOD) was low enough (AOD < 0.2). In the case of clear sky conditions, GHI, DNI, and DIF simulations are almost consistent with surface observations, meaning that radiation processes under clear sky conditions are accurate.

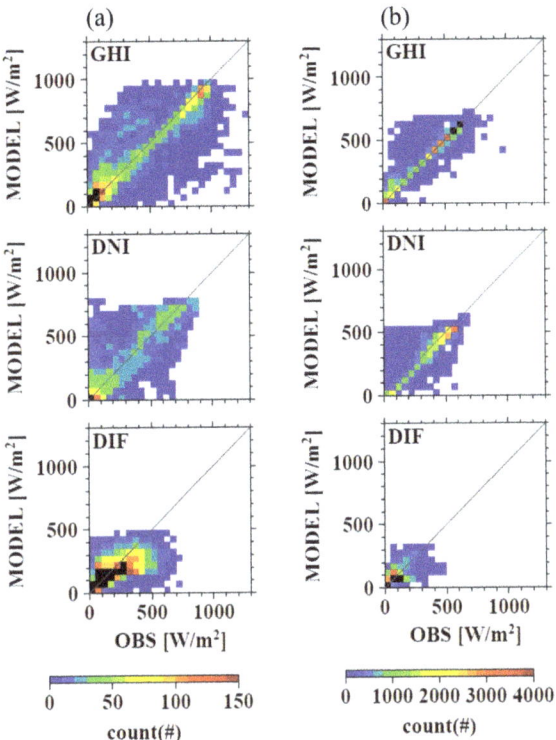

Figure 5. Hexbin plots of GHI (top), DNI (middle), and DIF (bottom) for dx500m in (**a**) the Exp-Summer of 2015 and (**b**) the Exp-Winter of 2016 at Tsukuba station.

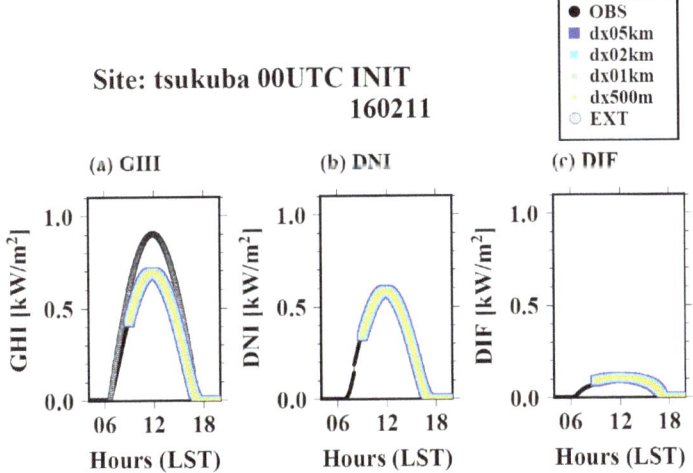

Figure 6. Comparison of (**a**) GHI, (**b**) DNI, and (**c**) DIF time series under clear sky conditions (AOD~0.2) at Tsukuba station on 11 February 2016 for dx05km (blue), dx02km (light blue), dx01km (green), and dx500m (yellow) with 00 UTC (09 JST) initialization time, respectively. Data was plotted every 2.5 min. Black circles indicate observations (OBS) and gray circles indicate extra-terrestrial solar irradiance (EXT).

Tables 2 and 3 show the three metrics (MBE, MAE, and RMSE) for simulation errors for GHI, DNI, and DIF, for Exp-Summer in 2015 and Exp-Winter in 2016. The simulations of 00 UTC initialization times are shown. From these metrics, no significant improvement was found as horizontal resolution increases. Validation results for dx05km, dx02km, and dx01km were similar to those of the dx500m.

Table 2. Forecast errors metrics (MBE, MAE, and RMSE) for GHI (left), DNI (center), and DIF (right) for Exp-Summer (July (upper tables) and August (lower tables)) in 2015. The forecasts with 00 UTC (09 JST) initialization time are shown. Model set-up for numerical experiments with four different horizontal resolutions (dx05km, dx02km, dx01km, and dx500m).

July 2015											[W m^{-2}]
GHI	MBE	MAE	RMSE	DNI	MBE	MAE	RMSE	DIF	MBE	MAE	RMSE
dx500m	8.9	88.9	152.5	dx500m	27.4	77.3	142.4	dx500m	−21.3	62.3	100.4
dx01km	7.2	88.1	151.7	dx01km	25.4	75.7	140.5	dx01km	−21.2	62.2	100.7
dx02km	4.6	85.9	149.2	dx02km	21.8	72.5	136.7	dx02km	−20.5	62.1	101.3
dx05km	−1.0	79.4	142.2	dx05km	11.6	63.0	123.9	dx05km	−17.1	62.4	104.4
August 2015											[W m^{-2}]
GHI	MBE	MAE	RMSE	DNI	MBE	MAE	RMSE	DIF	MBE	MAE	RMSE
dx500m	10.5	90.7	154.7	dx500m	30.1	79.4	145.1	dx500m	−22.2	62.2	99.8
dx01km	9.8	90.5	154.4	dx01km	29.7	79.0	144.7	dx01km	−22.5	62.2	100.1
dx02km	8.1	90.2	154.3	dx02km	28.4	77.9	143.5	dx02km	−22.9	62.1	100.3
dx05km	5.4	89.8	153.9	dx05km	26.0	76.1	141.6	dx05km	−23.2	62.1	100.6

Table 3. Same as Table 2, but for the Exp-Winter (January (upper tables) and February (lower tables)) in 2016.

January 2016											[W m^{-2}]
GHI	MBE	MAE	RMSE	DNI	MBE	MAE	RMSE	DIF	MBE	MAE	RMSE
dx500m	16.9	30.5	70.3	dx500m	6.8	40.5	82.6	dx500m	−1.0	23.1	45.0
dx01km	16.6	29.7	69.1	dx01km	6.1	39.6	80.8	dx01km	−0.8	22.7	44.2
dx02km	16.1	28.3	66.7	dx02km	4.6	38.2	77.9	dx02km	−0.2	22.0	42.8
dx05km	12.9	21.5	51.4	dx05km	−4.6	29.5	56.4	dx05km	3.9	18.7	35.7
February 2016											[W m^{-2}]
GHI	MBE	MAE	RMSE	DNI	MBE	MAE	RMSE	DIF	MBE	MAE	RMSE
dx500m	17.5	32.0	73.0	dx500m	8.5	42.4	86.7	dx500m	−1.8	24.0	46.7
dx01km	17.4	31.9	72.9	dx01km	8.6	42.3	86.5	dx01km	−1.9	23.9	46.5
dx02km	17.5	31.8	72.9	dx02km	8.6	42.3	86.7	dx02km	−1.9	23.9	46.4
dx05km	17.7	31.8	73.2	dx05km	9.2	42.7	88.2	dx05km	−2.2	23.8	46.6

4.2. A Case Study of Cumulus Cloud Simulation

Short range variability of solar irradiance is one of significant issues for energy management fields. Cumulus clouds (i.e., thunderstorms) often tend to develop rapidly and cause swift decrease of GHI values (i.e., ramp events, see Section 5). Wellby and Engerer [13]) identified weather types (thunderstorm, cold front, cloud band etc.) for city-scale negative ramp events in Australia. In this subsection, a case study of cumulus clouds simulation was performed. Cloud and GHI simulations with finer resolution simulation were investigated.

Figure 7a shows images from Himawari-8 Band-3 at 14 JST on 30 July 2015. The spatial resolution for the visible band is 1 km. Developed cumulus clouds over Tokyo can be seen in the satellite image (indicated by the yellow arrows in Figure 7a). Cloud regions (noted by the dashed ellipse region "A" in Figure 7a) prevail over the northern part of the TEPCO area. Figure 7b–e show comparisons of cloud field simulations (with close-up views for the TEPCO area) with 00 UTC initialization times for each of the four different horizontal resolutions (5 h ahead clouds simulations were calculated in the radiation process). Cloud distributions from dx05km simulations tend to be mosaic. The higher the horizontal resolution in the NWPs, the finer the cloud structures are. Locations of cumulus clouds on the northern part of the TEPCO region prevailed toward the northeastern region when compared to the satellite observations. The target cumulus clouds (shown by the yellow arrow) were not developed enough and the location of the target simulated clouds were prevailed to further west of Tokyo even in the finest resolution model, dx500m.

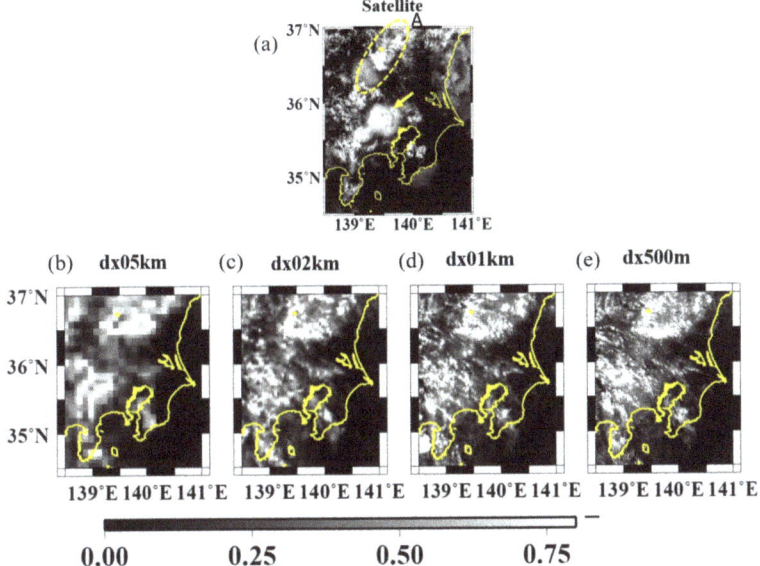

Figure 7. Comparison of the cloud horizontal distributions between the satellite observations and simulations. (**a**) Satellite visible image (Band-3) for the Tokyo electric power company (TEPCO) area at 14 JST on 30 July 2015. Cloud forecasts (5 h ahead forecasts of all cloud amounts used in radiation process) obtained from (**b**) dx05km, (**c**) 02km, (**d**) dx01km, and (**e**) dx500m with 00 UTC (09 JST) initialization time are shown, respectively. The yellow arrow in (**a**) indicates the location of the target cumulus clouds.

Figure 8 shows the validation results of GHI simulations performed at different horizontal resolutions using satellite-derived GHI observation provided by the Solar Radiation Consortium (see Section 2.2). Relatively low GHI regions (see location "A" in Figure 8a) found in the northern part of the TEPCO area were reproduced in the simulation results. As noted above, the false low GHI regions also prevailed in the northeastern region of the simulated GHI. In the location indicated by the red arrow in Figure 8a, a massive low GHI region (under approximately 200 W m^{-2}) comprised of the target cumulus clouds was observed over Tokyo and identified from the satellite data. Simulations using different horizontal resolutions have not reproduced this low GHI region (Figure 8b–e). GHI regions in each simulation were located further west of Tokyo than what was observed in the satellite data. Generally, GHI distributions between the four different horizontal resolutions resembled each other, although finer GHI distributions were simulated as horizontal resolutions also became finer.

To validate the simulated GHI values using the satellite-derived GHI ones, Figure 9 shows frequency distributions of both the satellite-derived GHI and the simulated GHI with the four different horizontal resolutions between July and August in 2015. The target area analyzed in this study was the area around TEPCO (shown in Figures 7 and 8). Here, satellite-derived GHI data collected every 30 min were analyzed over the period of 1 July to 3 August. After 4 August, satellite-derived GHI data collected every 10 min were used in this analysis. GHI values lower than 10.0 W m^{-2} were excluded. For the four different horizontal resolution models, 2.5 min time interval data were used.

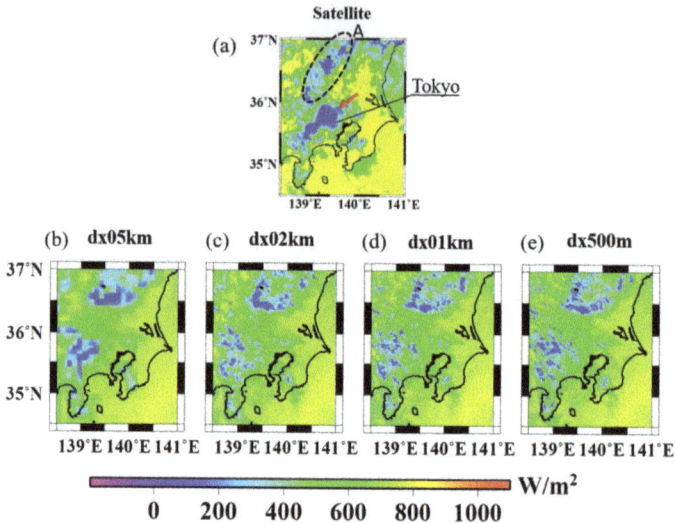

Figure 8. Comparison of the GHI horizontal distributions between the satellite observations and simulations. (**a**) Satellite-derived GHI distributions for the TEPCO area at 14 JST on 30 July 2015. GHI forecasts (5 h ahead forecast) obtained from (**b**) dx05km, (**c**) dx02km, (**d**) dx01km, and (**e**) dx500m with 00 UTC (09 JST) initialization time. The red arrow in (**a**) indicates the location of the target cumulus clouds.

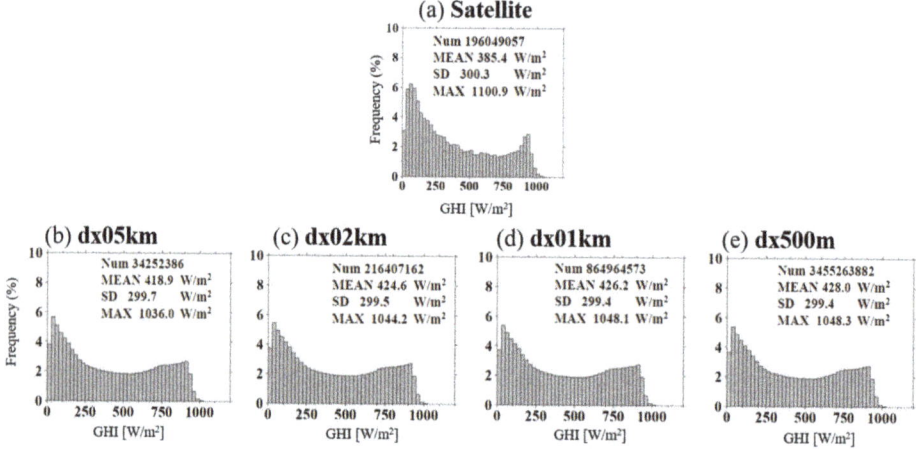

Figure 9. Frequency of (**a**) satellite-derived GHI values for TEPCO area during the two summer months of 2015. Frequency of GHI forecasts (3 h ahead forecast) obtained from (**b**) dx05km, (**c**) dx02km, (**d**) dx01km, and (**e**) dx500m with 00 UTC (09 JST) initialization time. "Num", "MEAN", "SD", and "MAX" in each panel represent data number, mean values, standard deviation and maximum values, respectively.

From this comparison, satellite-derived GHI shows a mean GHI ("MEAN") of 385.4 W m^{-2} and a standard deviation ("SD") of 300.3 W m^{-2} during the two summer months. From the dx05km through dx500m simulations, MEAN values were 418.9 W m^{-2}, 424.6 W m^{-2}, 426.2 W m^{-2}, and 428.0 W m^{-2}, respectively. Mean values in each simulation were generally larger than the satellite observations. SD values for the dx05km through dx500m simulations were 299.7 W m^{-2}, 299.5 W m^{-2},

299.4 W m^{-2}, and 299.4 W m^{-2}, respectively. SD values in each simulation were significantly closer to satellite observations. Maximum values of GHI in each simulation tended to be little lower than satellite observations. The maximum value of GHI ("MAX") identified via satellite is 1100.9 W m^{-2}. MAX values in each of the simulations were 1036.0 W m^{-2}, 1044.2 W m^{-2}, 1048.1 W m^{-2}, and 1048.3 W m^{-2}, respectively. Aerosol information (e.g., optical depth of aerosol) has not been considered in the current satellite algorithms for surface GHI value (see Section 2.2). For further verification of satellite-derived GHI values using surface GHI observations, see the previous publication by Damiani et al. [24].

4.3. Longwave Radiation

To discuss the error in terms of clouds and GHI simulations, validations regarding longwave radiation at the surface were also performed. Figure 10 shows time cross sections of ULW for both the Exp-Summer in 2015 and the Exp-Winter in 2016 at the Tsukuba station. Negative biases of ULW during the period from early morning to full daytime were significant in both summer and winter (Figure 10c,f). These results suggest that ULW (surface temperature) simulations could be under-estimated in the land surface processes of the JMA-NHM. During the mid-night (after sunset) in summer, the negative biases were significantly reduced. In case of clear sky conditions, negative biases of ULW for day time were relatively large in comparison to biases during cloudy conditions (not shown). For the Exp-Winter, positive biases were also found after 15 JST; however, daily variation of the biases was not observed. Systematic negative biases of ULW through both the Exp-Summer in 2015 and the Exp-Winter in 2016 were observed, although the validation was performed by a one-site validation in this analysis.

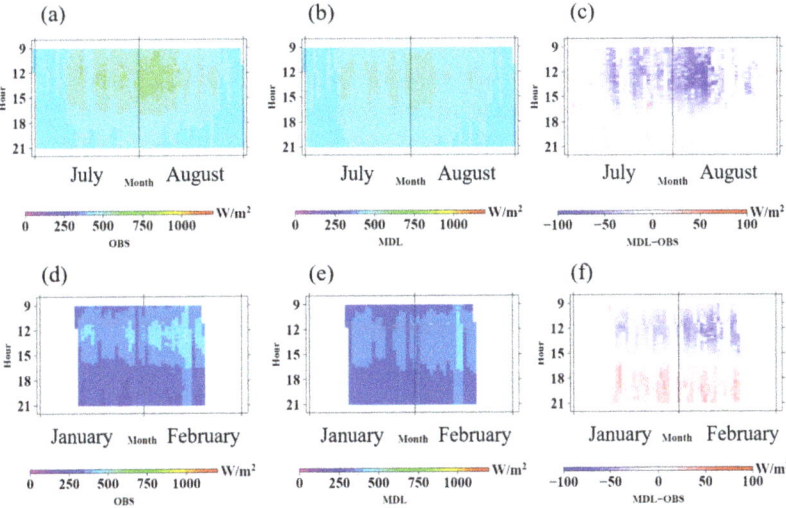

Figure 10. Time cross-sections of (**a**) ULW observations at the Tsukuba station (OBS), (**b**) dx500m forecasts (MDL) with 00 UTC (09 JST) initialization time, and (**c**) the difference (= MDL − OBS) for the Exp-Summer in 2015 (top panels) and the Exp-Winter in 2016 (bottom panels).

Figure 11 shows the validation results for DLW during the Exp-Summer in 2015 and the Exp-Winter in 2016. Generally, DLW biases tended to be weak negative biases during both summer and winter. It can be speculated that the clouds amounts simulated in the JMA-NHM were lower than they would be in a realistic atmosphere. Quantitatively, values for DLW negative biases were smaller than values for ULW biases (Figures 10 and 11).

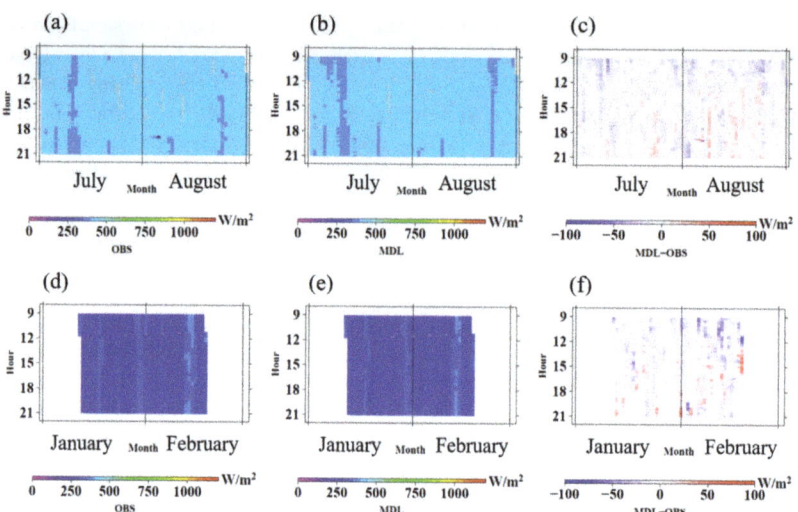

Figure 11. Same as Figure 10, but for DLW.

Tables 4 and 5 show the three metrics (MBE, MAE, and RMSE) for quantifying forecast errors of both ULW and DLW for the four different horizontal resolutions for the Exp-Summer in 2015 and the Exp-Winter in 2016. These results suggested that there was little difference in the simulation errors. ULW is given by the Stefan-Boltzmann law, $\sim \epsilon \sigma T^4$. Here, ϵ is emissivity (0.95 for ground-surface), T is temperature, and $\sigma = 5.67 \times 10^{-8}$ (W m^{-2} K^{-4}) is the Stefan-Boltzmann constant. With summer (T = 20 °C) and winter (T = 0 °C) air conditions, 1 K temperature difference would correspond to approximately 5.4 W m^{-2} and 4.5 W m^{-2} of ULW difference, respectively. The range of MBE values of ULW for the Exp-Summer in 2015 (−9.9 to −12.1 W m^{-2} for July and −12.2 to −12.4 W m^{-2} for August) corresponded to −1.8 to −2.2 K and approximately −2.3 K, respectively. For the Exp-Winter in 2016, MBE values of ULW for January (−1.1 to −3.6 W m^{-2}) and February (−3.9 to −4.1 W m^{-2}) corresponded to −0.2 to −0.8 K and approximately −0.9 K, respectively.

Table 4. Forecast error metrics (MBE, MAE, and RMSE) for ULW and DLW with the four different horizontal resolutions for the Exp-Summer (July (upper tables) and August (lower tables)) in 2015. The forecast uses a 00 UTC (09 JST) initialization time.

July 2015							[W m^{-2}]
ULW	MBE	MAE	RMSE	DLW	MBE	MAE	RMSE
dx500m	−12.1	17.3	24.4	dx500m	−3.8	8.9	18.0
dx01km	−12.0	17.2	24.3	dx01km	−3.7	8.9	17.6
dx02km	−11.7	17.1	24.0	dx02km	−3.8	8.7	16.6
dx05km	−9.9	16.2	22.1	dx05km	−4.6	8.4	10.9
August 2015							[W m^{-2}]
ULW	MBE	MAE	RMSE	DLW	MBE	MAE	RMSE
dx500m	−12.4	17.4	24.8	dx500m	−3.7	9.0	18.9
dx01km	−12.4	17.4	24.8	dx01km	−3.6	9.0	18.9
dx02km	−12.5	17.5	24.9	dx02km	−3.4	8.9	18.9
dx05km	−12.2	17.5	24.8	dx05km	−3.2	8.8	18.8

Table 5. Same as Table 4, but for the Exp-Winter season (January (upper tables) and February (lower tables)) in 2016.

January 2016							[W m^{-2}]
ULW	MBE	MAE	RMSE	DLW	MBE	MAE	RMSE
dx500m	−3.6	18.7	22.7	dx500m	−5.3	10.9	15.6
dx01km	−3.4	18.6	22.5	dx01km	−5.0	10.8	15.4
dx02km	−3.0	18.4	22.3	dx02km	−4.6	10.7	15.3
dx05km	−1.1	17.5	20.8	dx05km	−3.3	9.7	13.5
February 2016							[Wm^{-2}]
ULW	MBE	MAE	RMSE	DLW	MBE	MAE	RMSE
dx500m	−4.1	18.9	23.0	dx500m	−5.5	11.2	16.0
dx01km	−4.0	18.9	23.0	dx01km	−5.3	11.1	15.9
dx02km	−3.9	19.0	23.0	dx02km	−5.0	11.2	16.1
dx05km	−3.9	19.0	23.0	dx05km	−4.8	11.4	16.4

Previous studies (Hara [37]; Kusabiraki [38]) reported negative temperature biases when using JMA-NHM, an operational regional model with a horizontal grid spacing of 5 km, for summer, and suggested that there were research tasks on estimations of underground thermal transport and evaporation efficiency at the surface. Weverberg et al. [39] performed an intercomparison of NWPs on radiation processes and suggested that all models in the study had too warm near-surface temperature biases and that a warmer surface temperature ought to have a larger upwelling of longwave radiation. For DLW biases, Mocrette [40] suggested that the ECMWF formulation using ice cloud optical properties from Ebert and Curry [32] and this algorithm estimated an effective particle diameter from simply diagnosed temperature, which failed to capture the full effect of clouds on the DLW.

5. Ramp rate of solar irradiance

For the electrical power and renewable energy management fields, solar irradiance (or PV power generation) ramp events are an important research subject. In this subsection, we confirmed a possibility to reproduce rapid variation of GHI, DNI, and DIF based on a higher horizontal resolution model (dx500m) and BSRN data from the Tsukuba station.

Figure 12 shows a comparison of frequency of ramp rates for GHI, DNI, and DIF values, for both observations and simulations (dx500m) for July and August in Exp-Summer of 2015. Here, ramp rate, RR, is defined by the following equation:

$$RR = \frac{(x(t) - x(t-1))}{dt} \quad (5)$$

$x(t)$ and dt are solar irradiance and time interval data, respectively. dt was set to 5 and 10 min in this paper. MEAN and SD (Equation (4)) values in legends in each figure were calculated.

Although the maximum rates during 5 min intervals (positive ramp) of GHI, DNI, and DIF observations were 936.0 W m^{-2}, 878.0 W m^{-2}, and 592.0 W m^{-2}, respective positive ramp rates for simulations were 768.3 W m^{-2}, 606.6 W m^{-2}, 391.7 W m^{-2}. Similar tendencies for negative ramp rates during 5 min of GHI, DNI, and DIF observations were also found (−938.0 W m^{-2}, −844.0 W m^{-2}, −487.0 W m^{-2} for GHI, DNI, and DIF observation and −482.9 W m^{-2}, −353.5 W m^{-2}, −250.5 W m^{-2} for those simulations, respectively). SD values of ramp rates for GHI, DNI, and DIF simulations tended to be smaller than SD values for observations. Generally, maximum and minimum ramp rates for simulations tended to be smaller than ramp rates for observations.

Ramp rates during 10 min intervals were also investigated (not shown). From SD values for simulations and observations, distributions with 5 min intervals tended to be broader than that of 10 min intervals.

Ramp rates for the January and February (Exp-Winter in 2016) were also investigated (Figure 13). In winter season, weather condition tend to be stable because Tsukuba station locates on the leeward side of mountains in a winter monsoon. SD values for GHI, DNI, and DIF for observations (simulations) were 77.2 W m^{-2} (49.6 W m^{-2}), 150.5 W m^{-2} (43.8 W m^{-2}), 36.8 W m^{-2} (18.2 W m^{-2}), respectively. The SD values for GHI, DNI, and DIF for both of observations and simulations were smaller than the summer season.

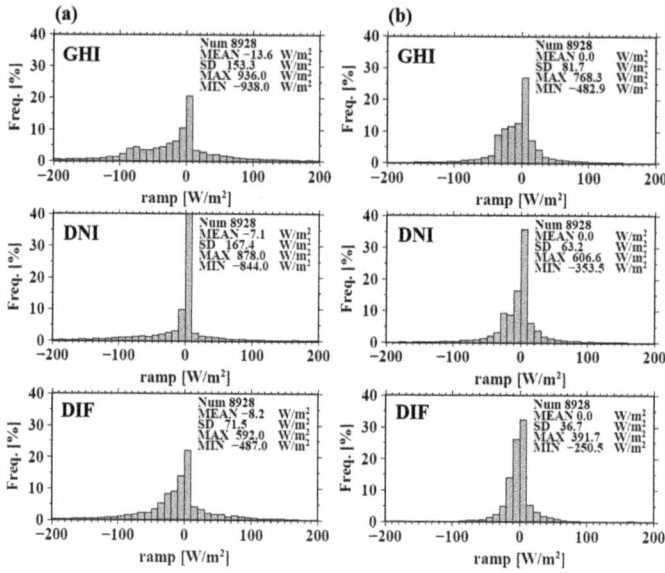

Figure 12. Frequency of solar radiation ramp rate (per 5 min) for (**a**) observations and (**b**) dx500m forecasts in July and August of 2015 at the Tsukuba station. Results for GHI (top), DNI (middle), and DIF (bottom). Num", "MEAN", "SD", "MAX", and "MIN" in each panel mean data number, mean values, standard deviation, maximum values and minimum values, respectively.

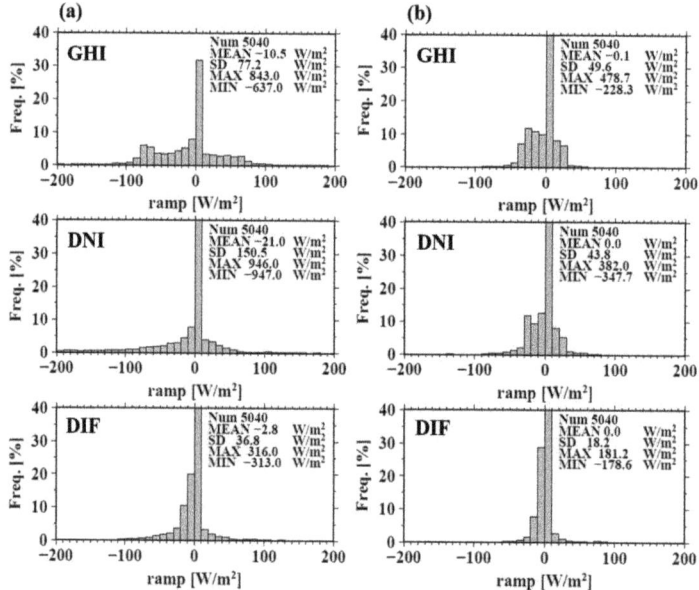

Figure 13. Same as Figure 12, frequency of solar radiation ramp rate in the winter season of 2016.

NWPs with different horizontal resolutions did not reproduce the developed cumulus clouds that cause ramp events of solar irradiance simulation (see Section 4.2). Above results suggested that time variations in GHI, DNI, and DIF simulations tended to be smoother than in observations.

6. Summary

In this study, solar irradiance simulations were created using NWPs (JMA-NHM) for four different horizontal resolutions (5 km, 2 km, 1 km, and 500 m). This study examines the performance of radiation processes (shortwave and longwave radiations) using NWPs. Results were validated using surface-observed solar irradiance datasets and satellite-derived observations. The validation period was during the summer of 2015 and the winter of 2016.

GHI, DNI, and DIF simulations were validated using BSRN datasets for the Tsukuba station in Japan. Validation results showed that GHI simulations and DNI values for both clear sky conditions (around 800 W m^{-2}) and optically thick conditions (under 200 W m^{-2}) tend to be close to observations. DIF simulation errors tend to be large.

In addition, a case study of developed cumulus clouds on 30 July 2015 was performed. The cumulus clouds were prevailed to further west of Tokyo compared with satellite observations even in the finest resolution model, dx500m. Differences in cloud distribution relating to horizontal resolutions were small. Statistical validations using satellite-derived solar radiation found that differences of frequency of GHI values between simulations and observations was small in the TEPCO area.

Validation results on ULW found systematic negative biases of surface temperature (corresponding to approximately - 2 K for summer and approximately - 1 K for winter). The negative temperature biases could not be improved even in the finer model, dx500m. DLW biases generally tended to be weak negative biases in the summer and winter seasons.

For electric energy utility, rapid variations in solar irradiance (called "ramp events") were also investigated, using higher horizontal resolution models. From the comparison of ramp rates for solar irradiance simulations between observations and simulations, frequency of ramp rates of GHI, DNI, and DIF tended to be narrower than those of observations.

From our numerical experiments on the four different horizontal resolutions of NWPs, it was found that differences between horizontal resolutions for clouds and solar irradiance distributions were not large. In addition to enhancing the resolution of the NWP, further improvements to cloud and/or radiation schemes will be required to realize more accurate cloud distributions and/or solar irradiance forecasts. These biases could be caused by our approximate radiative calculations. As a future work, the use of a three dimensional radiative transfer is worth attempting to improve accuracy of the radiation calculations.

Author Contributions: Data curation, S.H., J.I., A.H., and Y.Y.; Formal analysis, H.O.; Investigation, H.O.; Methodology, H.O.; Supervision, T.O.; Writing—original draft, H.O.; Writing—review & editing, H.O., F.U., T.O., S.H., J.I., A.H., H.Y., and Y.Y.

Funding: This study was supported by the Core Research for Evolutional Science and Technology (CREST) project, Grant Number JPMJCR15K1, provided by the Japan Science and Technology Agency.

Acknowledgments: We are grateful to Osamu Ijima of the Aerological Observatory of the Japan Meteorological Agency (JMA) for providing the BSRN datasets and giving us useful comments regarding radiation measurements. We are also grateful the Solar Radiation Consortium for providing the satellite-derived GHI values. This study was carried out in collaboration with the Meteorological Research Institute (MRI) of the JMA and simulations were performed using the FUJITSU PRIMEHPC FX100 super computer at their institute. Thanks to Generic Mapping Tools (GMT) for the software that was used to draw the figures in this paper.

Conflicts of Interest: The authors declare no conflict of interest.

References

1. Schiermeier, Q. And Now for the Energy Forecast. Germany Works to Predict Wind and Solar Power Generation. Nature.com, 2016. Available online: http://www.nature.com/polopoly_fs/1.20251!/menu/main/topColumns/topLeftColumn/pdf/535212a.pdf (accessed on 6 March 2019).

2. Ohtake, H.; Shimose, K.-I.; Fonseca, J.G.S., Jr.; Takashima, T.; Oozeki, T.; Yamada, Y. Accuracy of the solar irradiance forecasts of the Japan Meteorological Agency mesoscale model for the Kanto region, Japan. *Sol. Energy* **2013**, *98*, 138–152. [CrossRef]
3. Ohtake, H.; Shimose, K.-I.; Fonseca, J.G.S., Jr.; Takashima, T.; Oozeki, T.; Yamada, T. Regional and seasonal characteristics of global horizontal irradiance forecasts obtained from the Japan Meteorological Agency mesoscale model. *Sol. Energy* **2015**, *116*, 83–99. [CrossRef]
4. Mathiesen, P.; Kleissl, J. Evaluation of numerical weather prediction for intra-day solar forecasting in the continental United States. *Sol. Energy* **2011**, *85*, 967–977. [CrossRef]
5. Perez, R.; Lorenz, E.; Pelland, S.; Beauharnois, M.; Knowe, G.V.; Hemker, K., Jr.; Heinemann, D.; Remund, J.; Müller, S.C.; Traunmüller, W.; et al. Comparison of numerical weather prediction solar irradiance forecasts in the US, Canada and Europe. *Sol. Energy* **2013**, *94*, 305–326. [CrossRef]
6. Lorenz, E.; Kuhnert, J.; Heinemann, D.; Nielsen, K.P.; Remund, J.; Muller, S.C. Comparison of global horizontal irradiance forecasts based on numerical weather prediction models with different spatio-temporal resolutions. *Prog. Photovolt. Res. Appl.* **2016**, 1626–1640. [CrossRef]
7. Gregory, P.A.; Rikus, L.J. Validation of the bureau of meteorology's global, diffuse, and direct solar exposure forecasts using the ACCESS numerical weather prediction systems. *J. Appl. Meteorol. Climatol.* **2016**, *55*, 595–619. [CrossRef]
8. Sosa-Tinoco, I.; Peralta-Jaramillo, J.; Otero-Casal, C.; Lopez-Aguera, A.; Miguez-Macho, G.; Rodriguez-Cabo, I. Validation of a global horizontal irradiation assessment from a numerical weather prediction model in the south of Sonora-Mexico. *Renew. Energy* **2016**, *90*, 105–113. [CrossRef]
9. Charabi, Y.; Gastli, A.; Al-Yahyai, S. Production of solar radiation bankable datasets from high-resolution solar irradiance derived with dynamical downscaling Numerical Weather prediction model. *Energy Rep.* **2016**, *2*, 67–73. [CrossRef]
10. Mathiesen, P.; Collier, C.; Kleissl, J. A high-resolution, cloud-assimilating numerical weather prediction model for solar irradiance forecasting. *Sol. Energy* **2013**, *92*, 47–61. [CrossRef]
11. Lin, W.; Zhang, M.; Wu, J. Simulation of low clouds from the CAM and the regional WRF with multiple nested resolutions. *Geophys. Res. Lett.* **2009**, *36*, L08813. [CrossRef]
12. Ito, J.; Hayashi, S.; Hashimoto, A.; Ohtake, H.; Uno, F.; Yoshimura, H.; Kato, T.; Yamada, Y. Stalled improvement in a numerical weather prediction model as horizontal resolution increases to the sub-kilometer scale. *SOLA* **2017**, *13*, 151–156. [CrossRef]
13. Wellby, S.J.; Engerer, N.A. Categorizing the meteorological origins of critical ramp events in collective photovoltaic array output. *J. Appl. Meteorol. Clim.* **2016**, *55*, 1323–1344. [CrossRef]
14. Gilgen, H.; Whitlock, C.; Koch, F.; Muller, G.; Ohmura, A.; Steiger, D.; Wheeler, R. *Technical Plan for BSRN (Baseline Surface Radiation Network) Data Management (Version 2.1, Final)*; World Radiation Monitoring Center Tech. Rep. 1, WMO/TD No. 443, World Climate Research Programme; WMO, 1995; 57p. Available online: https://library.wmo.int/index.php?lvl=notice_display&id=11762#.XKv-kJj7Q1I (accessed on 9 April 2019).
15. Ohmura, A.; Dutton, E.G.; Forgan, B.; Fröhlich, C.; Gilgen, H.; Hegner, H.; Heimo, A.; König-Langlo, G.; McArthur, B.; Müller, G.; et al. Baseline Surface Radiation Network (BSRN/WCRP): New precision radiometry for climate research. *Bull. Am. Meteorol. Soc.* **1998**, *79*, 2115–2136. [CrossRef]
16. McArthur, L.B.J. *Baseline Surface Radiation Network (BSRN) Operation Manual (Version 1.0)*; WMO/TD-No. 879, World Climate Research Programme; WMO, 1998; 155p. Available online: https://library.wmo.int/index.php?lvl=notice_display&id=11769#.XKxPKSARXIV (accessed on 9 April 2019).
17. The Web Site of Baseline Surface Radiation Network (BSRN). Available online: http://www.bsrn.awi.de/ (accessed on 6 March 2019).
18. The Web Site of Meteorological Satellite Center. Available online: http://www.jma-net.go.jp/msc/en/ (accessed on 6 March 2019).
19. Bessho, K.; Date, K.; Hayashi, M.; Ikeda, A.; Imai, T.; Inoue, H.; Kumagai, Y.; Miyakawa, T.; Murata, H.; Ohno, T.; et al. An introduction to Himawari-8/9—Japan's New-Generation Geostationary Meteorological Satellites. *J. Meteorol. Soc. Jpn.* **2016**, *94*, 151–183. [CrossRef]
20. Takenaka, H.; Nakajima, T.Y.; Higurashi, A.; Higuchi, A.; Takamura, T.; Pinker, R.T.; Nakajima, T. Estimation of solar radiation using a neural network based on radiative transfer. *J. Geophys. Res.* **2011**, *116*, D8. [CrossRef]

21. Nakajima, T.Y.; Nakajima, T. Wide-area determination of cloud microphysical properties from NOAA AVHRR measurements for FIRE and ASTEX regions. *J. Atmos. Sci.* **1995**, *52*, 4043–4059. [CrossRef]
22. Kawamoto, K.; Nakajima, T.; Nakajima, T.Y. A global determination of cloud microphysics with AVHRR remote sensing. *J. Clim.* **2001**, *14*, 2054–2068. [CrossRef]
23. Solar Radiation Consortium 2015. Available online: http://amaterass.org/product.html (accessed on 6 March 2019). (In Japanese)
24. Damiani, A.; Irie, H.; Horio, T.; Takamura, T.; Khatri, P.; Takenaka, H.; Nagao, T.; Nakajima, T.Y.; Cordero, R.R. Evaluation of Himawari-8 surface downwelling solar radiation by SKYNET observations. *Atmos. Meas. Tech.* **2018**, 2501–2521. [CrossRef]
25. Saito, K.; Fujita, T.; Yamada, Y.; Ishida, J.I.; Kumagai, Y.; Aranami, K.; Ohmori, S.; Nagasawa, R.; Kumagai, S.; Muroi, C.; et al. The operational JMA nonhydrostatic mesoscale model. *Mon. Wea. Rev.* **2006**, *134*, 1266–1298. [CrossRef]
26. Saito, K.; Ishida, J.; Aranami, K.; Hara, T.; Segawa, T.; Narita, M.; Honda, Y. Nonhydrostatic atmospheric models and operational development at JMA. *J. Meteorol. Soc. Jpn.* **2007**, *85*, 271–304. [CrossRef]
27. JMA Website. Outline of the Operational Numerical Weather Prediction at the Japan Meteorological Agency (Outline NWP March 2013). Available online: http://www.jma.go.jp/jma/jma-eng/jma-center/nwp/outline2013-nwp/index.htm (accessed on 6 March 2019).
28. Lin, Y.-L.; Farley, R.D.; Orville, H.D. Bulk parameterization of the snow field in a cloud model. *J. Clim. Appl. Meteorol.* **1983**, *22*, 1065–1092. [CrossRef]
29. Cotton, W.R.; Tripoli, G.J.; Rauber, R.M.; Mulvihill, E.A. Numerical simulation of the effects of varying ice crystal nucleation rates and aggregation processes on orographic snowfall. *J. Clim. Appl. Meteorol.* **1986**, *25*, 1658–1680. [CrossRef]
30. Nakanishi, M.; Niino, H. Development of an improved turbulence closure model for the atmospheric boundary layer. *J. Meteorol. Soc. Jpn.* **2009**, *87*, 895–912. [CrossRef]
31. Slingo, A. A GCM Parameterization for the Shortwave Radiative Properties of Water Clouds. *J. Atmos. Sci.* **1989**, *46*, 1419–1427. [CrossRef]
32. Ebert, E.E.; Curry, J.A. A parameterization for ice optical properties for climate models. *J. Geophys. Res.* **1992**, *97*, 3831–3836. [CrossRef]
33. Hu, Y.X.; Stamnes, K. An Accurate Parameterization of the Radiative Properties of Water Clouds Suitable for Use in Climate Models. *J. Clim.* **1993**, *6*, 728–742. [CrossRef]
34. Ou, S.C.; Liou, K.-N. Ice microphysics and climatic temperature feedback. *Atmos. Res.* **1995**, *35*, 127–138. [CrossRef]
35. Sommeria, G.; Deardorff, J.W. Subgrid-scale condensation in models of nonprecipitating clouds. *J. Atmos. Sci.* **1977**, *34*, 344–355. [CrossRef]
36. Davy, R.J.; Huang, J.R.; Troccoli, A. Improving the accuracy of hourly satellite-derived solar irradiance by combining with dynamically downscaled estimates using generalised additive models. *Sol. Energy* **2016**, *135*, 854–863. [CrossRef]
37. Hara, T. Land-surface processes. JMA non-hydrostatic model II. *Suuchiyohouka Houkoku Bessatsu* **2008**, *54*, 166–186. (In Japanese)
38. Kusabiraki, H. Improvement of snow analysis using an offline land-surface model. *CAS/JSC WGNE Res. Activities Atmos. Ocean. Model.* **2015**, *45*, 1–13. Available online: http://bluebook.meteoinfo.ru/uploads/2015/documents/author-list.html (accessed on 9 April 2019).
39. Weverberg, K.V.; Morcrette, C.J.; Petch, J.; Klein, S.A.; Ma, H.-Y.; Zhang, C.; Xie, S.; Tang, Q.; Gustafson, W.I., Jr.; Qian, Y.; et al. CAUSES: Attribution of Surface Radiation Biases in NWP and Climate Models near the U.S. Southern Great Plains. *J. Geophys. Res.* **2018**, *123*, 3612–3644. [CrossRef]
40. Mocrette, J.-J. The Surface Downward Longwave Radiation in the ECMWF Forecast System. *J. Clim.* **2002**, *15*, 1875–1892. [CrossRef]

 © 2019 by the authors. Licensee MDPI, Basel, Switzerland. This article is an open access article distributed under the terms and conditions of the Creative Commons Attribution (CC BY) license (http://creativecommons.org/licenses/by/4.0/).

Article

Accuracy Enhancement for Zone Mapping of a Solar Radiation Forecasting Based Multi-Objective Model for Better Management of the Generation of Renewable Energy

Mohammad Ehteram [1], Ali Najah Ahmed [2], Chow Ming Fai [2,3], Haitham Abdulmohsin Afan [4,*] and Ahmed El-Shafie [4,*]

[1] Department of Water Engineering and Hydraulic Structures, Faculty of Civil Engineering, Semnan University, Semnan 35131-19111, Iran
[2] Institute of Energy Infrastructure (IEI), Universiti Tenaga Nasional (UNITEN), Kajang 43000, Selangor, Malaysia
[3] Department of Civil Engineering, College of Engineering, Universiti Tenaga Nasional (UNITEN), Kajang 43000, Selangor, Malaysia
[4] Department of Civil Engineering, Faculty of Engineering, University Malaya, Kuala Lumpur 50603, Malaysia
* Correspondence: haitham.afan@gmail.com (H.A.A.); elshafie@um.edu.my (A.E.-S.)

Received: 29 May 2019; Accepted: 25 June 2019; Published: 17 July 2019

Abstract: The estimation of solar radiation for planning current and future periods in different fields, such as renewable energy generation, is very important for decision makers. The current study presents a hybrid model structure based on a multi-objective shark algorithm and fuzzy method for forecasting and generating a zone map for solar radiation as an alternative solution for future renewable energy production. The multi-objective shark algorithm attempts to select the best input combination for solar radiation (SR) estimation and the optimal value of the adaptive neuro-fuzzy inference system (ANFIS) parameter, and the power parameter of the inverse distance weight (IDW) is computed. Three provinces in Iran with different climates and air quality index conditions have been considered as case studies for this research. In addition, comparative analysis has been carried out with other models, including multi-objective genetic algorithm-ANFIS and multi-objective particle swarm optimization-ANFIS. The Taguchi model is used to obtain the best value of random parameters for multi-objective algorithms. The comparison of the results shows that the relative deviation index (RDI) of the distributed solutions in the Pareto front based multi-objective shark algorithm has the lowest value in the spread index, spacing metric index, favorable distribution, and good diversity. The generated Pareto solutions based on the multi-objective shark algorithm are compared to those based on the genetic algorithm and particle swarm algorithm and found to be the optimal and near ideal solutions. In addition, the determination of the best solution based on a multi-criteria decision model enables the best input to the model to be selected based on different effective parameters. Three different performance indices have been used in this study, including the root mean square error, Nash–Sutcliffe efficiency, and mean absolute error. The generated zone map based on the multi-objective shark algorithm-ANFIS highly matches with the observed data in all zones in all case studies. Additionally, the analysis shows that the air quality index (AQI) should be considered as effective input for SR estimation. Finally, the measurement and analysis of the uncertainty based on the multi-objective shark algorithm-ANFIS were carried out. As a result, the proposed new hybrid model is highly suitable for the generation of accurate zone mapping for different renewable energy generation fields. In addition, the proposed hybrid model showed outstanding performance for the development of a forecasting model for the solar radiation value, which is essential for the decision-makers to draw a future plan for generating renewable energy based solar radiation.

Keywords: renewable energy forecasting; solar radiation; shark algorithm; particle swarm optimization; ANFIS

1. Introduction

The Sun is considered a fundamental resource for most physical and chemical processes on Earth. Thus, processes related to the Sun are important for researchers [1,2]. The application of solar energy for different aims as a renewable energy source is an important priority for researchers [3]. Renewable energy technologies are fast becoming more effective and cheaper and their application is widely growing. Solar radiation is one of the most important resources of renewable energy [4]. In fact, the wide application of solar energy as renewable energy can decrease greenhouse emissions. Renewable energy, such as solar radiation, has a low effect on the environment. In this context, decision makers should know the available solar radiation in order to be able to identify the expected generation of renewable energy. Therefore, the development of an accurate forecasting model and generation of zone mapping for solar radiation is of vital importance for decision-makers. In addition, with the world becoming warmer, renewable energies can balance the ecological conditions and the production of domestic power can be carried out based on the use of solar energy [5,6].

In fact, Solar Radiation (SR) affects climate processes, and agricultural production is governed by solar energy [7]. SR is necessary for plant growth, photosynthesis, and regulation of the growth duration. The estimation of SR has many applications in different fields, including agricultural engineering, building engineering, the energy and hydrology fields, and power and heat production [8]. For example, the accurate estimation of irrigation is considered an important issue for irrigation planning and design [9]. Additionally, SR is considered an important issue for evaporation computation [10]. Thus, access to solar data and an accurate model for the prediction of SR data are important for the leverage of the solar energy potential in particular locations. There are different empirical models and equations for SR estimation. These models are highly important because of the economic limitation and measurement complexity of SR estimation in some locations, rendering empirical models suitable for radiation estimation [11]. Additionally, remote sensing and satellite images can be considered effective tools for solar energy estimation. However, these empirical models require various parameters that might be too complex to be accurately estimated in some locations. In addition, some models cannot provide a good estimation of SR under changing climate conditions and other conditions [12]. The station height from sea level, longitude, latitude, relative humidity, and pollution accumulation in the atmosphere affect radiation estimations. Thus, the pollution content is an important and influential issue affecting SR, and the application of accurate tools for radiation estimation under the effect of different parameters is very important for decision makers [13]. Soft computing methods can be effective as powerful tools based on large data sets because these methods can accurately simulate hydrological or other variables. These methods can effectively adapt to climate and hydrological boundaries, decrease the computational time, and ensure high accuracy in hydrological predictions [14]. In addition, soft computing methods can be effective tools for estimating SR when different parameters, such as particle pollution, temperature, humidity, etc., have a significant effect on SR [15]. The present study attempts to simulate the SR in East Azarbayejan in Iran in some stations in the presence of particle pollution. The current article presents self-organizing maps (SOMs) with a multi-objective shark algorithm (MOSA), and the fuzzy method is applied to select the optimal input combinations, identify the best value of the ANFIS parameters, and generate the spatial distribution of SR. The SOM is used as a clustering method to identify impactful spatial SR values, and the results are compared with those obtained using the multi-objective genetic algorithm (MOGA) and multi-objective particle swarm optimization (MPSO).

2. Background

An ANN model based on different numbers of inputs in different cities in Turkey was previously used to simulate SR [16]. The results indicated that using a pre-selection procedure is necessary for the determination of the inputs because the elimination of some input parameters could significantly decrease the accuracy of the models. The ANN used for different climate change conditions had acceptable performance for SR based on the accuracy of the parameter selection.

Another study simulated the monthly and daily total global SR based on an artificial neural network (ANN) [17]. The soil temperature, wind speed, temperature, and mean monthly rainfall were used to estimate SR. The results indicated that the mean absolute percentage error (MAPE) based on the ANN was approximately 5.34% and that the ANN method decreased the MAPE compared with the output of the regression method with a high correlation.

The auto regressive moving average mode (ARMA) and multi-layer perceptron (MLP) have been used as hybrid models to simulate hourly SR [18]. The results indicated that the hybrid model had a lower root mean square error (RMSE) than the simple MLP model, and the necessary data for this research were obtained based on a numerical weather simulation model.

A regression model was used to estimate hourly SR values [19]. The relative humidity, atmospheric pressure, air pollution index, and mean rainfall were used as inputs in the models. The results indicated that the models incorporating the air pollution index could produce a relatively accurate estimate of SR with a high Nash–Sutcliffe efficiency value and a small RMSE value.

The daily SR has also been predicted by a support vector machine (SVM) [20]. The SVM method was used based on the sunshine ratio as an effective input, and the results were compared with those obtained using empirical models. Compared to the empirical models, the SVM models could significantly reduce the relative error and provide more accurate predictions of the winter season.

Vakili et al. [21] simulated the daily SR based on an ANN while considering the suspended particulate matter. The temperature, rainfall, humidity, and suspended particle characteristics were used to simulate SR in Tehran Province, Iran, and the results indicated that compared to the other applied methods' input structure, the ANN considering the suspended particles could simulate SR with the lowest error in terms of the RMSE and mean absolute error.

The hourly global horizontal irradiance (HGI) was estimated based on Meteosat imagery and an ANN [22]. The data were obtained from a radiometric station. The Heliosat-2 model was used to compare the results with those obtained using an ANN. The results obtained by the ANN based on different sky conditions had a significantly lower RMSE value than those obtained by the Heliosat-2 model.

Celik et al. [23] applied an optimized ANN for SR estimation over the Eastern Mediterranean. The results indicated that the daily SR could be predicted based on the optimized ANN with high accuracy such that the optimized model could accurately determine the number of hidden neurons and weights in the ANN, and the RMSE was decreased by approximately 10% to 12% compared to that obtained using regression methods.

A moderate-resolution imaging spectroradiometer (MODIS) and an ANN have also been used to obtain SR estimates [24]. Land surface temperature (LST) data were used as input data to the ANN, and the results indicated that the relative error of the ANN was 5.35%, while the error of the regression models was 10.23%.

The new daily SR model (NDSRM) using the air quality index (AQI) was applied in multiple cities [25]. The results differed according to whether the AQI was added or removed from the inputs such that the predicted SR based on the elimination of the AQI in some cities had high accuracy, whereas the model accuracy depended on the AQI value as input in other cities.

The ANN model and inverse distance weight (IDW)-based model were used to simulate SR at distances greater than 50 km [26]. The results indicated that the IDW model had an RMSE of approximately 6.4%, while the RMSE of the ANN model was 5.11%, and the IDW model was simpler and more accurate than the ANN model.

Yoe et al. [25] applied the SVM to SR estimation based on the air quality index. The results indicated that the Nash–Sutcliffe efficiency varied from 0.682 to 0.740, and the models with the AQI input provided a higher accuracy in solar estimation than those without this input.

The daily SR considering the air quality index was simulated by a support vector machine (SVM) in a large region with different climates [2]. The results indicated that the elimination of the AQI input among the other inputs could significantly decrease the accuracy of the estimation model; the SVM featured a high accuracy, and simple structures were found to have a high ability to absorb SR.

Fan et al. [27] applied an SVM for SR estimation while considering atmospheric particulate matter (PM). Daily metrological and air pollution data were used to simulate SR. The inclusion of PM with a diameter of 2.5 micrometres (PM 2.5), PM 10, and O_3 in the input combinations were considered the best combination of inputs and improved the results of the SVM.

Furthermore, many research efforts have attempted to simulate SR based on soft computing methods under different conditions, such as using air pollution data. Different air quality indexes have been used to evaluate the air quality, such as the air pollution index (API) and air quality index (AQI) [28]. The AQI is used to provide a daily evaluation of the air quality. This index is used to present the air quality to the population while focusing on the respiratory effects that can be observed some hours or days after exposure [29]. The AQI is computed based on models or air monitors considering nitrogen dioxide, ozone, carbon monoxide, and sulfur dioxide [30]. This index ranges from 0 to 500, which is divided into different classifications, and each classification is related to different levels of human health. The country of Iran experiences considerable air pollution in different cities; this air pollution is usually measured and evaluated with the index, and the classification changes accordingly [31]. For example, when the index value is within the range of 0 to 50, the air quality is good, and when the index value is greater than 300, the air quality is very dangerous. A literature review of the pollutant particle effect on SR shows that several studies have considered the effect of air quality on SR [31].

Thus, the main purpose of the current paper is to evaluate the effect of air quality on SR using soft computing models to provide a comprehensive discussion concerning the influences of the air quality on the accuracy of SR prediction. In this study, the ANFIS model was selected as the soft computing method because ANFIS is suitable for predicting stochastic nature variables, such as SR. However, in the ANFIS model procedure, there is a need to initialize a few internal parameters that are usually selected using the trial-and-error process. The selection of the optimal values of these parameters significantly affects the accuracy of the model performance. In this context, there is a need to optimize the value of the ANFIS's internal parameter to ensure an acceptable level of prediction accuracy. In fact, the shark algorithm is widely accepted and has been successfully applied in the fields of water resources and power generation, mathematical simulations, the design of trusses, and other fields [32,33]. Therefore, the shark algorithm is used as an effective optimization tool to obtain the optimal parameters for the ANFIS. The rotational movement of the shark in this algorithm improves the ability to search for the global optima of the ANFIS's internal parameters. The proposed integrated adaptive neuro-fuzzy inference system with multi-objective shark algorithm (ANFIS-MOSA) model was examined in SR prediction in three different case studies. In addition, comprehensive analyses were carried out to compare the proposed model to other models.

3. Materials and Methods

3.1. Fuzzy Method

The ANFIS model, which is based on large amounts of data, can accurately simulate different variables, such as hydrological parameters, water quality parameters, climate parameters, and other parameters. The following six layers are used in the ANFIS:

- The first layer with inputs x and y is connected to the neurons in the adjacent layer.

- The nodes in layer 2 constitute the fuzzification layer, and this layer includes an adaptive node that has a function. The fuzzy membership function is computed in this layer as follows:

$$Q_{1,i} = \mu_{A_i}(x_1), \text{for}(i = 1, 2, ..)$$
$$Q_{2,i} = \mu_{B_{i-2}}(x_1), \text{for}(i = 1, 2, ..)$$
(1)

There are different types of membership functions, and the following bell function has been widely used with successful applications in previous research:

$$\mu_{Ai} = \frac{1}{1 + \left(\frac{x-c_i}{a_i}\right)^{2b_i}},$$
(2)

where a_i, b_i, and c_i are premise parameters that can be obtained based on the optimization process.

- The firing strength (w_i) is computed in the third layer as follows:

$$Q_{2,i} = w_i = \mu_{Ai}(x) \cdot \mu_B(y).$$
(3)

Each node in the fourth layer computes the ratio of the firing strength of the membership rules to the firing strength of the total number of rules. The output is computed based on the following equation:

$$Q_{3,i} = \frac{w_i}{\sum w_i} = \frac{w_i}{w_1 + w_2} = \overline{w_i}.$$
(4)

- Each node in the fifth layer is considered an adaptive square node with a node function:

$$Q_{4,1} = \overline{w_i} f_i = \overline{w_i} \cdot (p_i x + + q_i y + r_i),$$
(5)

where p_i, q_i, and r_i are considered the consequent parameters.

- The overall output in the sixth layer is computed based on signals received from the defuzzification layer and the following equation:

$$Q_{5,1} = \frac{\sum w_i f_i}{\sum w_i} = f_{out}.$$
(6)

However, the consequent parameters and premise parameters should be determined accurately, and an optimization algorithm can obtain the accurate value of these parameters if the initial estimates of the parameter values are inserted into the algorithms as decision variables.

3.2. Shark Algorithm (SA)

The shark algorithm acts based on smell receptors in idealized sharks. The algorithm, which features a simple structure, high flexibility, resistance to trapping in local optima, and fast convergence, has been successfully applied in the fields of water resource management, reservoir operation, and power generation. The initial positions of the sharks are considered decision variables based on the following equation [34]:

$$X_i^l = \left[x_{il}^1, x_{il}^2, ..., x_{il}^{ND}\right],$$
(7)

where x_{il}^j is the jth dimension of the ith shark position, ND is the number of decision variables, and X_i^l is the initial shark position. The SA has the following three main assumptions:

- The injured fish are considered prey to the sharks, and the movement velocity of the fish is very low compared to the shark velocity due to their injuries. The sharks find the fish locations by detecting blood from the injured fish.

- Blood is regularly emitted from the fish bodies, and the water flow has a negligible effect on the blood emission and movement in the water. When the number of blood particles around the injured fish is considerable, the smell receptors of the shark can detect the blood odor, and the shark moves to the location of the fish.
- Each injured fish is considered one blood source.

Sharks moving in water have a specific velocity [34] as follows:

$$V_i^1 = \left[v_{i,1}^1, v_{i,2}^1, \ldots, v_{il}^{ND}\right], \tag{8}$$

where v_{ij}^1 is the *j*th dimension of the *i*th velocity position, ND is the number of decision variables, and V_i^1 is the initial shark velocity.

The shark velocity varies based on the detected smell intensity, and when sharks detect a higher odor concentration, they rapidly move in the direction of the target. If the variation in the odor concentrations is considered a gradient of the objective function, the velocity varies according to the gradient of the objective function based on the following equation:

$$V_i^k = \eta_k . R_1 . \nabla(OF)\big|_{x_i^k}, i = 1, \ldots NP, k = 1, \ldots, k_{max}, \tag{9}$$

where OF is the objective function, *k* is the stage number, NP is the population size, η_k is a random value between 0 and 1, and R_1 is a random value. The forward movement of the sharks is based on the *k* number stage such that the maximum number stage equals k_{max}. The sharks are subjected to an inertial limitation, and thus, they move at a specific velocity as follows:

$$v_{i,j}^k = \eta_k . R_1 . \frac{\partial(OF)}{\partial x_j}\big|_{x_{jk}} + \alpha_k R_2 v_{i,j}^{k-1}, \tag{10}$$

where α_k is a momentum coefficient between 0 and 1, and R_2 is a random value. A momentum rate with a higher value indicates greater inertia, and thus, the current velocity strongly depends on the previous velocity. The normal velocity of a shark is 20 km/hr, and the maximum velocity of a shark is 80 km/hr; thus, there is a limitation to the velocity based on the following equation:

$$\left|v_{i,j}^k\right| = \min\left[\left|\eta_k . R_1 . \frac{\partial(OF)}{\partial x} + \alpha_k . R_2 . v_{i,j}^{k-1}\right|, \left|\beta_k . v_{ij}^{k-1}\right|\right], \tag{11}$$

where β_k is the velocity limiter. The new shark position is computed based on the following equation:

$$Y_i^{k+1} = X_i^k + V_i^k \Delta t_k, \tag{12}$$

where Y_i^{k+1} is the new shark position, and Δt_k is the time interval. The rotational movement of the shark is considered a local search to find the best solution based on the following equation:

$$Z_i^{k+1,m} = Y_i^{k+1} + R_3 Y_i^{k+1}, \tag{13}$$

where R_3 is a random number, M is the total number of points in the local search and rotational point, and m is the number of each rotation level. In fact, there are M points around Y_i^{k+1}; thus, the sharks manifest rotational movement around these points. Then, the sharks select the best position based on the obtained Y_i^{k+1} positions and $Z_i^{k+1,m}$ positions.

3.3. Multi-Objective Algorithms

An optimization problem can have one or more objective functions. The framework of a multi-objective function can be defined based on the following equation [35]:

$$\text{Min } \vec{f}_j(\vec{p}), j = 1, .., m$$
$$\text{subject, to}: g_i(\vec{p}) \leq b_i \qquad (14)$$
$$\vec{lb} \leq \vec{p} \leq \vec{ub}$$

where $(\vec{p}) = (p_1, .., p_n)$ is the solution vector, f_j is the jth objective function, m is the number of objective functions, l is the number of constraints, $\vec{ub}$ and $\vec{lb}$ are the upper and lower constraints, respectively, and g_l and b_l are the ith constraints related to the right-hand side.

When the problem has one objective function, the solutions based on different methods can be easily compared, but the domination concept should be used when two objective functions can be identified for the problem. The solution, $\vec{p}$, dominates the solution, $\vec{q}$, if the following is satisfied with the problem:

$$\vec{p} > \vec{q} : [\forall_j \in \{1, \ldots, m\} | f_j(\vec{p}) \leq f_j(\vec{q})] \& [\exists_j \in \{1, \ldots, m\} | f_j(\vec{p}) < f_j(\vec{q})]. \qquad (15)$$

The Pareto front includes solutions such that any other solution cannot dominate the Pareto set in the problem space. The optimal Pareto front is used to define the objective function values based on the following equation:

$$PF = [\vec{p} \in U | \vec{q} \in U : \vec{q} > \vec{p}], \qquad (16)$$

$$POF = \{f_j(\vec{x}) | \vec{x} \in PF\}. \qquad (17)$$

The main aim of the problem is related to identifying the Pareto front with the highest diversity. A set of non-dominated solutions for the multi-objective problem is selected, and the decision maker selects one solution as the best solution based on an accurate evaluation.

3.4. Multi-Objective Shark Algorithm (MOSA)

The shark positions in the MOSA are considered solutions to the problem, and these positions are generated randomly at the initial level of the MOSA. The non-dominated solutions are saved in an archive and generate a POF. Then, the equality of the solutions is investigated based on a computation of the objective function. The non-dominated sharks in the new archive are computed and recorded in an archive. The following aspects of the archive should be considered:

- If the new solution is dominated by at least one current solution in the archive, the new solution is not inserted into the archive.
- If the new solution dominates over older solutions, the non-dominated solutions are eliminated, and the new solution is inserted into the archive.
- If no new solutions or current solutions in the archive can dominate each other, the new solutions are added to the archive.

Several hypercubes are produced by dividing the objective function spaces by a grid-based mechanism. When the archive size exceeds the capacity, additional sharks should be eliminated from the archive. Hypercubes with a relatively low density are important for the generation of a uniform distribution, and elimination can occur in areas with greater hypercube crowding.

3.5. Non-Dominated Sorting Genetic Algorithm (NSGA)

NSGA $\prod$ was generated by Deb [36]. A random population based on generation chromosomes is considered in the first level of NSGA $\prod$. The children chromosomes are generated based on crossover and mutation operators, and their objective function is computed. Then, the combination of the parent and children populations is divided over the fronts based on non-dominated sorting as described by Deb [36]. The crowding distance of each member is computed, and the population is sorted based on this index. Then, the combined population after the sorting mechanism is truncated, such as the parent population, and a new population is prepared to produce a child population. After a ranking process, the first-ranked solution represents the best solution.

3.6. Multi-Objective Particle Swarm Optimization (MPSO)

If the problem space is considered with d dimensions and particles, the ith position particle at the ith position, $X_i(x_{i1}, .., x_{id})$, has a velocity of $V_i = (v_{i1}, .., v_{id})$. The best performance of each particle in the swarm is $P_i(p_{i1}, .., p_{id})$ [37]. Each particle attempts to improve its position, velocity, and distance with respect to the best particle. The position and velocity of the particles are updated based on the following equations:

$$V_i^{t+1} = \omega V_i^t + c_1 r_1 (x_{pbest} - X_i^t) + c_2 r_2 (x_{gbest} - X_i^t)$$
$$X_i^{t+1} = X_i^t + V_i^t$$
(18)

where ω is the inertia coefficient, x_{gbest} is the global best output of the particle, x_{pbest} is the personal best output of the particle, c_1 is the cognitive acceleration coefficient, c_2 is the social acceleration coefficient, r_1, r_2 are random numbers, V_i^{t+1} is the velocity at time $t + 1$, and X_i^{t+1} is the position at time $t + 1$.

The MPSO algorithm encounters a set of solutions as a Pareto front [38]. The archive of a non-dominated solution is considered in the solutions developed at each level. First, the initial position and velocity of the particles are considered in the MPSO, and their objective function is computed [39]. Then, the leader is selected from the archive as the particle with the best objective function value such that each particle follows one leader in the archive, and the velocity and position of each particle are updated. The objective function of the new positions and velocities is computed. If the new solution can dominate over x_{pbest}, the new solution is inserted into the archive instead of x_{pbest}. An efficient mutation strategy for increasing diversity can be considered at this level and applied to the particles. One of the objective functions is selected for the level mutation, and then, the particles are sorted in descending order at this level based on the objective function computation. The crowding distance of the particles in the archive is computed, and sorting is performed in descending order. Then, elitist mutation is applied to a predefined number of available solutions in the archives. The convergence criteria are checked, and if the criteria are not satisfied, the system returns to updating the velocity and position and determining whether the algorithm has completed.

4. Case Study

The current study considers SR over Ardebil, Gilan, and East Azarbayejan. East Azarbayejan has an area of approximately 47,830 km² (Figure 1). The northern extent of East Azarbayejan is a part of the Republic of Azarbayejan and Armenia. Zanjan Province is located in the south of this province, and the Sahand Mountain, which has a height of 3707 m, is one of the highest points in the province. The annual temperature from 2008 to 2018 ranged between 25 and −15 °C. This province is located between the longitudes of 45°0′ E and 47°50′ E and latitudes of 36°50′ to 39°50′. This province is mountainous as follows: 40% of the province area has mountainous conditions. The climate in this province is cold and dry, but the different topographies cause variations in the climates. The maximum temperature, precipitation, and number of sunny hours obtained from three stations, i.e., Ahar (longitude: 47°48′ and latitude: 38°28′), Bonab (longitude: 45°70′ and latitude: 37°20′), and Sarab (longitude: 47°23′ and latitude: 37°51′), were considered. Additionally, the AQI values at these 3 stations or cities were obtained based on air quality monitoring. Ardebil Province occupies an area of approximately 3.17 km^2 between the longitudes of 47°00′ E and 48°50′ E and latitudes of 37°00′ N to 40°00′ N. Different climates, such as Mediterranean, moderate Mediterranean, and mountainous climates, can be observed in this province. The average annual precipitation in this province is 462.5 mm, and the annual temperature varied between −3 °C (minimum temperature) and 34 °C (maximum temperature) from 2008 to 2018. The following three stations in this province were used to characterize the temperature, precipitation, sunny hours, and AQI data: Ardebil station (longitude: 47°29′ and latitude: 38°00′), Ebrahimabad (longitude: 48°29′ and latitude: 38°22′), and Phyroozabad (longitude: 48°20′ and latitude: 37°59′). Gilan Province is located in the north of Iran within the boundary demarcated by these stations.

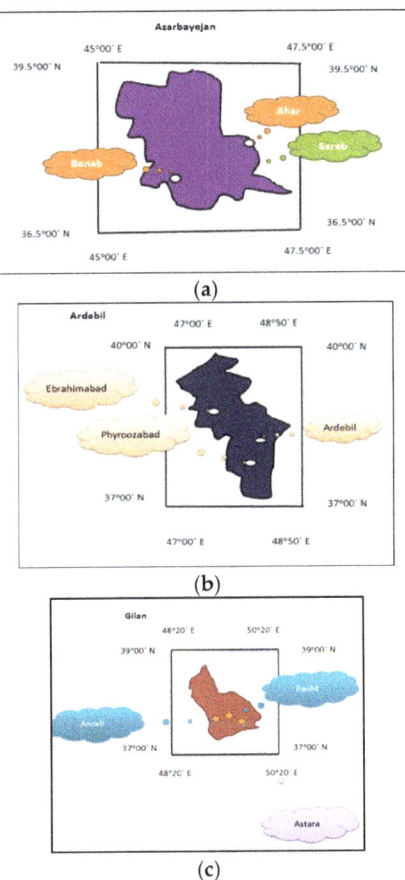

Figure 1. Location of case study (**a**) East Azarbayjan, (**b**) Ardebil, and (**c**) Gilan.

A moderate climate is observed in this basin. The area of this province is 14,044 km^2, and the average annual precipitation varied from 1200 to 1800 mm during the period from 2008 to 2018. The climatology-oriented Rasht Institute (longitude: 49°39′ and latitude: 37°12′), Anzali (longitude: 49°39′ and latitude: 38°20′) and Astara (longitude: 48°51′ and latitude: 38°21′) are used to record the different data. The AQI at the different stations is computed based on the following equation:

$$I = \frac{I_{high} - I_{low}}{C_{high} - C_{low}}(C - C_{low}) + I_{low},\tag{19}$$

45

where I is the air quality index, C is the pollutant concentration, C_{low} is the concentration extreme equal to or lower than C, C_{high} is the concentration extreme equal to or greater than C, I_{high} is the index breakpoint related to C_{high}, and I_{low} is the index breakpoint related to C_{low}. Table 1 shows the different classifications of AQI (Air Quality Index).

Table 1. Classification of the AQI index.

Air Pollution Level	AQI Index	Value
1	excellent	0–50
2	good	51–100
3	Lightly polluted	101–150
4	Moderately populated	151–200
5	Heavily polluted	201–300
6	Severely polluted	>300

Figure 2 shows the average AQI during different months in Ardebil, East Azarbayejan, and Gilan Provinces. Clearly, the greatest variation in East Azarbayejan can be observed in January as follows: The minimum AQI in January is 46, and the maximum value is 74. The lowest variation in East Azarbayejan is observed in May (minimum AQI = 60 and maximum AQI = 62). The highest AQI value is observed in March (AQI = 79.5) in East Azarbayejan, and the lowest AQI value in East Azarbayejan Province (EAZP) occurs in April. This index in EAZP shows that the first quartile and third quartile in most months exhibit an AQI > 50, and the conditions appear to differ only in April. The other values of the AQI index can be observed in the other provinces.

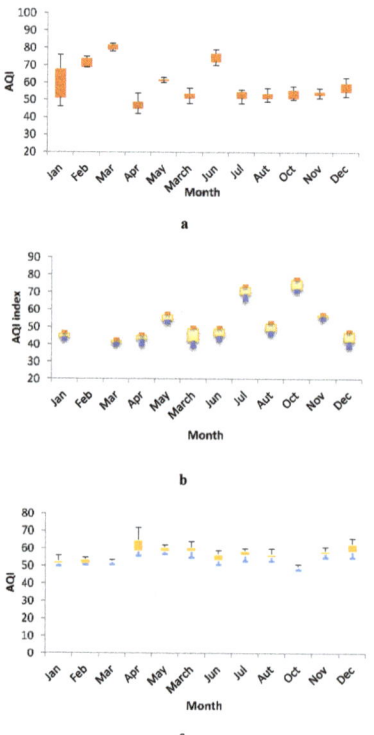

Figure 2. AQI for (a) East Azarbayjan, (b) Ardebil, and (c) Gilan.

Table 2 shows the variation in the average temperature during the different months during the period of 2008–2018. July exhibits the greatest variation based on the high value of the variation coefficient in Azarbayejan (EAZP), and the maximum temperature in this province occurs during this month. The minimum temperature in Ardebil Province (AP) occurs in January, and the greatest variation in this province can be observed in May. The other details are listed in Table 2. Figure 3 shows the precipitation values in the different provinces. For example, the highest precipitation value in AP occurs in March, and the greatest variation in precipitation in AP is observed in this month. The lowest precipitation value in AP occurs in December. The other details of the other provinces are shown in Figure 3. Additionally, the number of sunny hours at different stations is shown. For example, the number of sunny hours (NSN) in June in Gilan Province (GP) is as follows: The most variation in the NSN occurs in June, and the lowest variation can be observed in July. The other details are shown in Figure 3.

Table 2. Temperature variation for the different months (2008–2019).

Month	Jan	Feb	Mar	Apr	May	Jun	Jul	Aug	Sep	Oct	Nov	Dec
Max. Temp. °C	2	4	10	12	19	28	31	18	17	14	9	8
Min. Temp. °C	−8	−2	6	8	10	16	15	10	10	9	5	3
Variation Coeff. ×10^{-2}	0.34	022.	0.12	0.10	0.31	0.35	0.44	0.30	0.27	0.21	0.20	0.32
Ardebil												
Max. Temp. °C	1	5	12	14	23	27	32	17	18	15	7	6
Min. Temp. °C	−7	−3	7	9	11	19	23	9	12	10	2	5
Variation Coeff. ×10^{-2}	0.31	0.24	0.32	0.28	0.45	0.34	0.21	0.18	0.17	0.12	0.11	0.11
Gilan Province												
Max. Temp. °C	6	7	14	15	22	28	33	18	19	17	8	7
Min. Temp. °C	−5	−2	9	10	15	16	23	12	14	12	3	2
Variation Coeff. ×10^{-2}	0.31	0.30	0.28	0.23	0.26	0.39	0.31	0.28	0.15	0.12	0.10	0.10

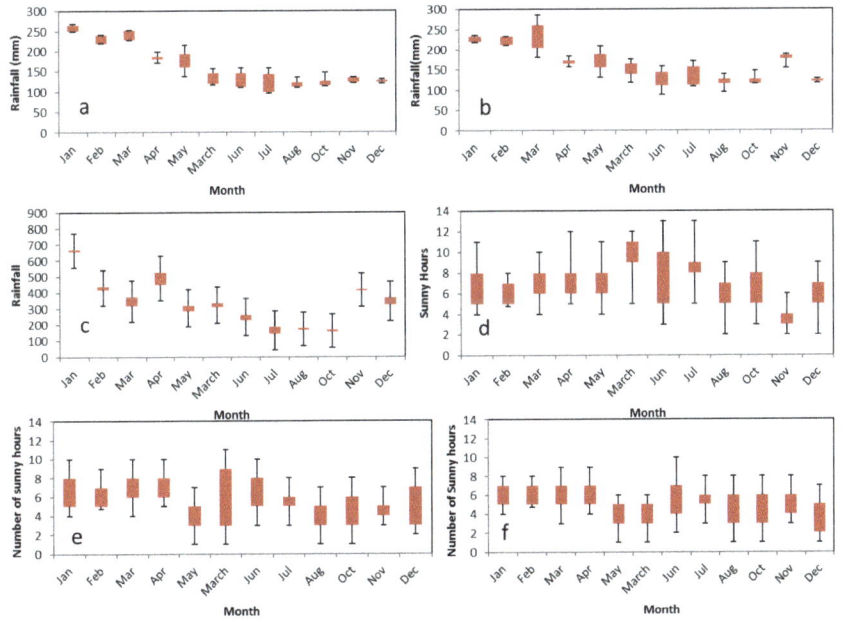

Figure 3. (a) Monthly temperature for the 2008–2018 in EAZP, (b) monthly temperature for the 2008–2018 in AP, (c) monthly temperature for the 2008–2018 in GP, (d) monthly number of sunny hours for the 2008–2018 in EAZP, (e) monthly number of sunny hours for the 2008–2018 in AP, (f) monthly number of sunny hours for the 2008–2018 in GP.

The inverse distance weight (IDW) method was used to obtain the SR in the different zones. This method has an effective feature allowing the optimal value to be obtained based on a multi-objective optimization framework as follows:

$$z^* = \frac{\sum_{i=1}^{n} \frac{1}{D_i^q} z_i}{\sum_{i=1}^{n} \frac{1}{D_i^q}}, \tag{20}$$

where z^* is the estimated precipitation at each point, D_i is the difference between the predicted and observed data, and q is the power parameter.

Three objective functions are considered in the ANFISmulti-objective optimization algorithm [40]. The *obg* function is considered to select the best input parameters for the ANFIS, and the RMSE is used as an objective function to obtain the best value of the ANFIS parameters. The general standard deviation (GSD) is used to obtain the optimal power parameter for the IDW:

$$\text{Minimize}(obg) = \left(\frac{NO_{train} - NO_{valTest}}{NO_{train} - NO_{valTest}}\right) + \frac{RMSE_{train} + MAE_{train}}{R_{train+1}} + \left(\frac{2NO_{valTest}}{NO_{Train} + NO_{Valtest}}\right) \frac{RMSE_{ValTest} + MAE_{ValTest}}{R_{valTest+1}}$$

$$RMSE = \sqrt{\frac{1}{p}\sum_{i=1}^{p}(T_i - O_i)^2}$$

$$MAE = \frac{1}{p}\sum_{i=1}^{p}|T_i - O_i| \tag{21}$$

$$R = \frac{p\sum_{i=1}^{p} T_i O_i}{\left(p\sum_{i=1}^{p} T_i^2 - \left(\sum_{i=1}^{p} T_i\right)^2\right)\left(p\sum_{i=1}^{p} O_i^2 - \left(\sum_{i=1}^{p} O_i\right)^2\right)}$$

$$GSD = \frac{RMSE}{\overline{Z}}$$

where NO_{train} is the training data number, $RMSE_{train}$ is the root mean square error of the training data, $RMSE_{ValTest}$ is the root mean square error of the test data, MAE_{train} is the mean absolute error of the training data, $MAE_{ValTest}$ is the mean absolute error of the test data, $R_{train+1}$ is the correlation coefficient of the training data, $R_{valTest+1}$ is the correlation coefficient of the test data, T_i represents the simulated data, $\overline{Z}$ is the average value of the simulated data at different points, and O_i represents the observed data. Lower RMSE, MAE, and GSD values are considered better. First, the ANFIS model is considered based on the initial estimates of the linear and nonlinear parameters, and different components (N_s: Number of sunny hours, $T_{max(t-3)}$: Maximum temperature with a three-month lag, $T_{max(t-6)}$: Maximum temperature with a 6-month lag, $T_{min(t-3)}$: Minimum temperature with a three-month lag, $T_{min(t-6)}$: Minimum temperature with a six-month lag, Rainfall$_{(t-3)}$: Precipitation value with a three-month lag, Rainfall$_{(t-6)}$: Precipitation value with a six-month lag, and the AQI indexes) are used as inputs. The ANFIS simulates the results. The IDW is used to simulate SR in the different zones, and then, the Multi-Objective Shark Algorithm (MOSA) is used based on the initial population used for the selection of inputs, the optimal determination of the adaptive neuro-fuzzy inference system (ANFIS) parameters, and the selection of the power parameters for the IDW. As shown in Figure 4, the different operators apply the candidate solutions, and then, these solutions are returned to the ANFIS subroutine for another simulation iteration. If the stopping criteria are satisfied, the process finishes with the optimal results. The modified technique for order preference by similarity by the ideal solution (M-TOPSIS) is used to select the best solution from the Pareto form based on the following equations:

$$D_j^+ = \sqrt{\sum_{j=1}^{n}\left(x_{ij} - x_j^+\right)^2}$$

$$D_j^- = \sqrt{\sum_{j=1}^{n}\left(x_{ij} - x_j^-\right)^2} \qquad (22)$$

$$R_j^* = \sqrt{\left[D_j^+ - \min\left(D_j^+\right)\right]^2 + \left[D_j^- - \max\left(D_j^-\right)\right]^2}$$

where x_j^+ is the ideal solution (largest maximization criterion value or smallest minimization criterion value), x_j^- is the negative ideal solution (largest minimization criterion value or smallest maximization criterion value), D_j^+ is the distance from the ideal solution, D_j^+ is the distance from the least ideal solution, x_{ij} represents the results of alternative i considering criterion j, and R_j^* is the similarity ratio, and this index of the solution is sorted by descending values to show the rank of each solution.

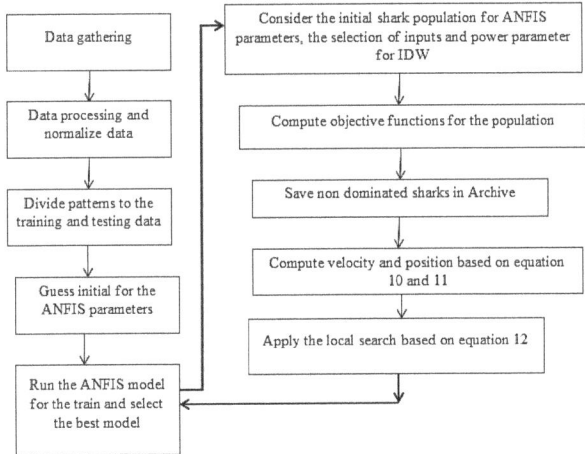

Figure 4. MOSA and ANFIS structure.

Additionally, the following indexes are used to select the best multi-objective algorithm:

- Cover Surface (CS)

This index presents the relative score of the solutions in set B that are weakly dominated by set A as follows:

$$CS(A, B) = \frac{|\{b \in B | \exists a \in A : a \leq b\}|}{|B|} \qquad (23)$$

If the index value equals 1, all solutions in set B are weakly dominated by those in set A, and if the index value equals −1, any solution in set B is dominated by the solutions in set A. However, the index value can have values other than 1 and −1, which could indicate that the number of solutions in set A is covered by those in set B.

- General Distance (GD)

This index shows the closeness value of the computed Pareto solutions to the true Pareto solution. If Q is considered a set obtained by the MOSA, the GD is computed based on the following equation:

$$GD = \frac{\left(\sum_{i=1}^{|Q|} d_i^p\right)^{\frac{1}{p}}}{|Q|},$$

$$d_i = \min_k \sqrt{\sum_{m=1}^{|P*|}\left(f_m^i - f_m^{*k}\right)^2}$$

(24)

where P^* is a reference solution (a set of all possible true Pareto solutions), d_i^p is the distance of the solution obtained by the algorithm to the best solution, and f_m^{*k} is the m-objective value of the kth member of P^*. A lower value of this index is more favorable for decision makers.

- Spread Index (SI)

The SI presents the diversity value of the obtained and archived solutions among the non-dominated solutions.

$$\Delta = \frac{\sum_{m=1}^{M} d_m^e + \sum_{i=1}^{N-1} \left|d_i - \bar{d}\right|}{\sum_{m=1}^{M} d_m^e + (N-1)\bar{d}},$$

(25)

where d_i is the Euclidean distance between successive solutions among the obtained non-dominated solutions, $\bar{d}$ is the average of all distances d_i, N is the number of solutions in the best non-dominated front, and d_m^e is the computed distance of the extreme solution between the obtained Pareto of the m^{oth} objective and the true optimal Pareto.

- Spacing Metric (SM)

This index is computed by measuring the distances of successive solutions in a non-dominated front and shows an evaluation of the spread of vectors in the total set of non-dominated solutions.

$$S = \sqrt{\frac{1}{|Q|}\sum_{i=1}^{Q}(d_i - \bar{d})^2},$$

(26)

where $d_i = \min_{k \in Q \wedge k \neq i} \sum_{m=1}^{M} \left|f_i^m - f_i^k\right|$, $\bar{d} = \sum_{i=1}^{Q} \frac{d_i}{|Q|}$ and f_i^k is the value of the ith objective function of the kth member.

5. Discussion and Results of the Algorithm Parameters

5.1. Results of the Sensitivity Analysis

Evolutionary algorithms are usually initialized using random parameters to allow accurate prediction values to be determined when these random parameters have been optimally selected. The Taguchi model is used to set the values of the random parameters. Its advantages include decreased time and cost in the selection of effective parameters with respect to the results. The selection of an orthogonal array is important for the Taguchi model. Table 3 shows the effective parameters of each algorithm. For example, the MOSA has four effective parameters with four levels, and the other details of the other algorithms are also shown. Then, the relative deviation index (RDI) is used based on the computed CS, GD, SI, and SM values.

Table 3. Level and effective parameters for MOSA, NSGAII, and MOPSO (Multi objective particle swarm algorithm.

	MOSA			
Parameter	Level 1	Level 2	Level 3	Level 4
Population size	10	30	50	70
M	20	40	60	80
α	0.2	0.4	0.60	0.80
β	1	2	3	4
Parameter	Level1	Level2	Level3	Level 4

	NSGAII		
Parameter	Level 1	Level 2	Level 3
Population size	10	30	50
Mutation probability	20	40	60
Crossover probability	0.2	0.4	0.60

	MOPSO			
Parameter	Level 1	Level 2	Level 3	Level 4
Population size	10	20	30	40
Inertia weight	0.2	0.40	0.60	0.80
Mutation probability	0.05	0.10	0.15	0.20
C1	1.6	1.8	2.0	2.2
C2	1.6	1.8	2.0	2.2

In fact, this index shows the difference among the computed solutions with the best solution in each index. Notably, the best solution in some indexes, such as the CS, is considered based on the computed largest value of this index, and the best solution in other indexes, such as the GD, SI, and SM, is computed based on the lowest value of the indexes.

$$RDI = \left| \frac{Alg_{sol} - Best_{sol}}{Best_{sol}} \right| \times 100, \qquad (27)$$

where Alg_{sol} is the computed solution of each index, and $Best_{sol}$ is the best computed solution of each index.

When the RDI is computed for the CS, GD, SI, and SM, the product of a Wight parameter (WP) and the computed RDI is obtained, and then, the products of the RDI and WP for the different indexes are summed. The WP of the SC, GD, SI, and SM is 0.25 because the four indexes have the same priority to decision makers, and the algorithm should satisfy all indexes. Then, Minitab software is used, and the L_9, L_{12}, and L_9 arrays of the MOSA, NSGAII, and MOPSO, respectively, are considered for the model. The orthogonal array and results are shown in Table 4. The lowest CMRDI value shows that the algorithm parameters can obtain the solutions with a small difference between the best solutions. For example, the size population for the MOSA with level 2 has the lowest value among the levels, and thus, the population size for level 2 equals 30. The best values of the other parameters can be computed similarity. The best CMRDI value is shown in Table 4.

Table 4. Computed results based on the computed average of the relative deviation index (CMRDI) for MOSA, MOPSOA, and NSGAII.

MOSA			
Population size			
Level 1	Level 2	Level 3	Level 4
2.8	2.4	3.1	3.5
M			
Level 1	Level 2	Level 3	Level 4
2.9	2.7	3.3	3.5
α			
Level 1	Level 2	Level 3	Level 4
3.6	3.1	2.7	3.9
β			
Level 1	Level 2	Level 3	Level 4
3.5	2.6	2.8	2.9
MOPSOA			
Population size			
Level 1	Level 2	Level 3	
20	40	60	
4.1	3.9	4.5	
Mutation probability			
Level 1	Level 2	Level 3	
4.5	3.7	3.9	
Inertia weight			
Level 1	Level 2	Level 3	
C_1			
Level 1	Level 2	Level 3	
3.9	3.7	4.00	
C_2			
Level 1	Level 2	Level 3	
4.3	3.7	3.9	
NSGAII			
Population size			
Level 1	Level 2	Level 3	Level 4
20	30	40	50
5.6	5.3	5.1	5.9
Mutation probability			
Level 1	Level 2	Level 3	Level 4
6.2	5.6	5.8	6.2
Level 1	Level 2	Level 3	Level 4
6.3	6.1	6.7	6.9

5.2. Results of the Different Multi-Objective Algorithms

The three Pareto fronts of the three algorithms are shown in Figure 5. The results and discussion were applied to Azarbayejan Province to avoid repetition. Additionally, when a large Pareto is obtained, many points act as solutions, and thus, the methods are compared using a decreased number of points for ease of presentation and to facilitate understanding (Table 5). A cluster method was used in the current article. First, the N cluster is considered, and the distance between each cluster is computed. The two clusters with the smallest distance are selected and combined to generate one cluster, and this process continues until the number of clusters reaches the lowest probable value. Linear programming is used to obtain each objective function value separately without considering the other two objective functions. B, C, and A are the best values of the first objective function (F(1) or GSD), the second objective function (F(2) or *RMSE*), and third objective function (F(3) or MAE), respectively. By comparing the nearest points of the different algorithm Pareto results to the mentioned points (A, B, and C), in contrast to the cases of the other algorithms, the nearest point in the MOPSO can obtain and match the optimal objective function separately. For the ease of comparison, the objective function values are converted to values between 0 and 1. A, B, and C have the best optimal values for the three objective functions, F(3), F(1), and F(2), respectively, based on the lowest values. Thus, the value of the objective function of the ideal points is considered the closest value to 1; consequently, the other points could be sorted and ranked. Additionally, the nearest points (NPs) to the reference points A, B, and C in the MOSA, MOPSO, and NSGAII are shown by NPMOSA, NPMOPSO, and NPNSGAII, respectively. Clearly, the NPMOSA has a good match with the best optimal objective function. For example, the best value of F(1) based on the B point is considered equal to 1, and this value of NPMOSA is 0.99; the other points are similar. In fact, when F(1) equals 1, the best objective function value of F(1) can be observed in point B, and then, F(2) and F(3) are computed for this point. Clearly, when an ideal point can satisfy one objective function, the value of the other objective functions at this point does not have the best value. In addition, the results of the other two provinces are shown in Figure 5. Figure 6 shows the RDI percentage of the different indexes in Azarbayejan, and the values in Ardebil and Gilan are not shown in this section to avoid repetition. The low RDI value shows better solutions and Pareto for the multi-objective algorithms. Notably, the variation in the values of the different indexes in the MOSA is smaller than that using the other two methods, and thus, the reliability of the generated data based on the ANFIS-MOSA is higher than that of the other algorithms. Additionally, the minimum, maximum, and median RDI in the box plots of the NFIS-SOMA have lower values than those based on the other two algorithms. Thus, these results show that the formed Pareto for the ANFIS-MOSA has good and diverse distribution in the space problem compared with that for the other two algorithms. Notably, the input data in each separate province during different months are associated with separate outputs per province.

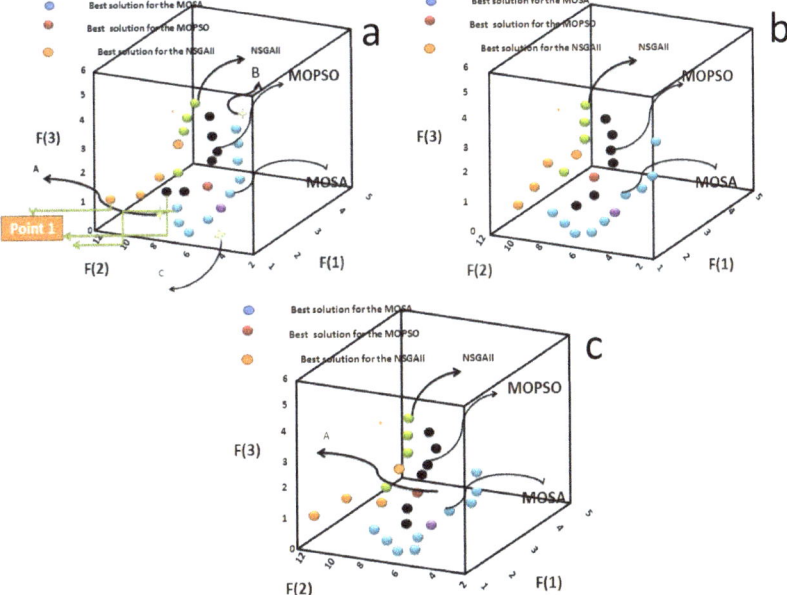

Figure 5. Pareto front for (a) Azarbayjen province, (b) Ardebil, and (C) Gilan.

Table 5. Comparison of the condition of the nearest point of each Pareto with the best values of each objective function for Azarbayejan.

Model	F(1)			F(2)			F(3)		
Lingo (NLP) (B)	0.12			1			0.0.14		
Lingo (NLP) (C)	1			0.75			0.50		
Lingo (NLP) (A)	0.75			0.50			1		
description	The value of first objective function for the nearest points in the Paretos to the ideal point of first objective function			The value of second objective function for the nearest points in the Paretos to the ideal point of first objective function			The value of the third objective function for the nearest points in the Paretos to the ideal point of first objective function		
(optimal value for F1)	0.99	0.75	0.50	0.62	0.50	0.27	0.86	0.33	0.50
	NPSA	NPPSO	NPGA	NPSA	NPPSO	NPGA	NPSA	NPPSO	NPSGA
description	The value of first objective function for the nearest points in the Paretos to the ideal point of second objective function			The value of second objective function for the nearest points in the Paretos to the ideal point of second objective function			The value of third objective function for the nearest points in the Paretos to the ideal point of second objective function		
(optimal value for F2)	0.50	0.49	0.48	0.95	0.63	0.30	0.14	0.11	0.10
description	The value of first objective function for the nearest points in the Paretos to the ideal point of third objective function			The value of second objective function for the nearest points in the Paretos to the ideal point of third objective function			The value of third objective function for the nearest points in the Paretos to the ideal point of third objective function		
(optimal value for F3)	0.85	0.75	0.50	0.37	0.33	0.12	0.99	0.50	0.50

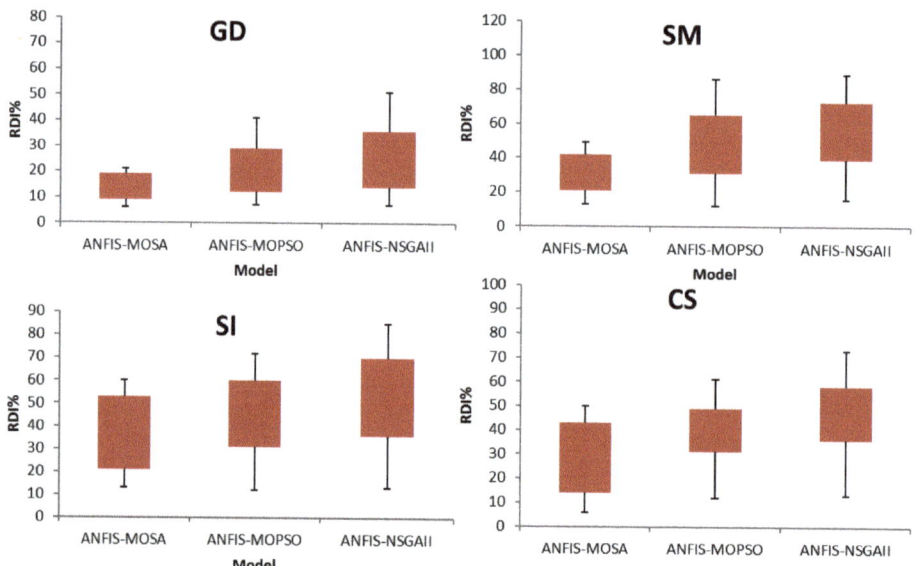

Figure 6. Computation of RDI for different indexes for Azarbayejan province.

5.3. SR Results

The previous section showed that the ANFIS-MOSA has better performance than the other two models. There are initially 10 points representing the MOSA Pareto in Figure 5. The points are labelled by a number such that point 1 is shown by the number 1, and the other numbers corresponding to the other points are allocated such that the sixth point representing the MOSA and its Pareto is based on the MTOPSIS. This index is based on the objective function value and the difference in the objective function at each point between the most ideal and least ideal points. Then, the similarity ratio is computed for each point, and the points are sorted based on the computed rank. Figure 5 shows the best solution for EAZP, and the points are computed for other similar points such that each point shows a combination of the input, the value of the q power, and the optimal values for the ANFIS model. For example, the best point for the MOSA simulates SR with inputs (N_s: Number of sunny hours, $T_{max(t-3)}$: Maximum temperature with a three-month lag, $T_{min(t-3)}$: Minimum temperature with a three-month lag, rainfall$_{(t-3)}$: Precipitation value with a three-month lag, and AQI indexes), and the other points have different input combinations. Clearly, the best model or sixth point in the MOSA Pareto and EAZP does not use all inputs, and thus, this model can obtain the best results with the fewest number of inputs. Figure 7 compares the performance of 10 points in the ANFIS-MOSA, and the Taylor diagram is based on the standard deviation, correlation, and distance from the reference point. The numbers on the radius show the RMSE values. The performance of the sixth point in different provinces shows better simulation results because it is closer to the reference point and, thus, shows the best performance.

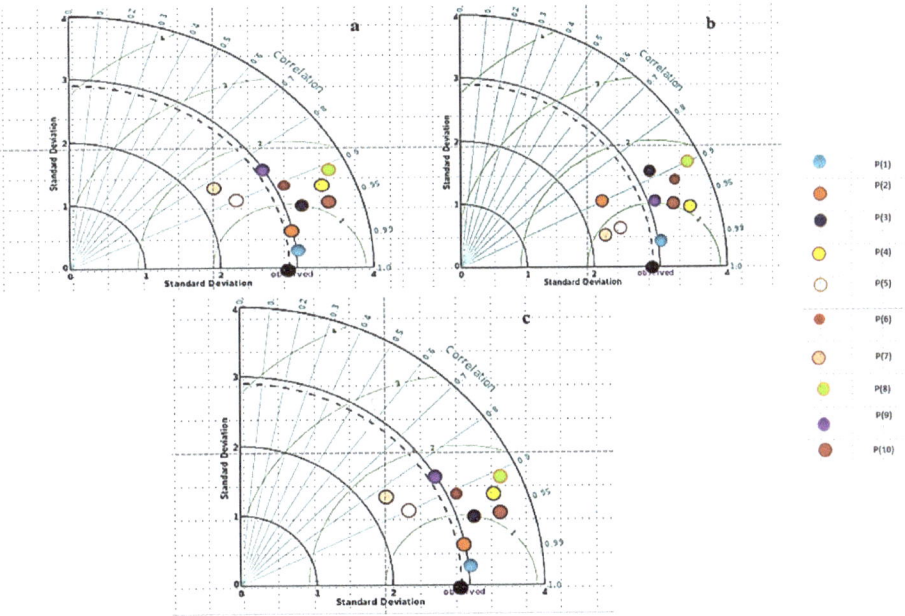

Figure 7. The results for the Pareto solutions for (**a**) EAZP, (**b**) AP, and (**c**) GP.

Figure 8 shows the process variation of the RMSE, MAE, and NSE indexes for the 10 points in the different provinces, and the locations of the points are determined on the three-dimensional graph. Clearly, the sixth point has the lowest RMSE and MAE values and the highest NSE value. In fact, the 10 points show that 10 combination inputs were generated by the MOSA and that the sixth point has the best input combination, best value of the ANFIS parameter, and best value of the power q for the IDW. The locations of the points are shown in Figure 8. Stars show the points. Table 6 shows the performance of the ideal solution of ANFIS and MOSA with and without the AQI input parameter. The results show that the elimination of the AQI input parameter significantly increases the error index and decreases the NSE because the elimination of this parameter enables the simulation of SR. Although the dependency of the performance of models to AQI is variable for different studies, it is essential to consider the AQI as one of the model inputs for solar radiation. This is due to the fact that there is a strong interaction between the solar radiation value and air pollutant and visa-versa. For example, airborne particulate and gaseous pollutants decrease the amount of solar radiation reaching the Earth's surface. In contrast, ultraviolet (UV) radiation is necessary to initiate a series of reactions that cause high urban ozone values. Understanding the interaction between solar radiation and air pollution is especially important from the viewpoint of collecting and utilizing solar energy as an alternate energy source. Aerosols in the atmosphere can alter the solar radiation incident at the ground in two ways: By depleting the total energy and by changing the relative amounts of direct and diffuse radiation. At urban sites, high aerosol concentrations reduce the total incident energy and alter the direct: diffuse ratio. At rural locales, where anthropogenic aerosol burdens are smaller, the decrease in the direct solar beam will be largely compensated by an increase in the diffuse flux. Photo-chemical pollutants, which depend on UV radiation for their formation, also affect the amount of solar energy reaching the ground. Ozone and the particles formed from photochemically induced gas to particle reactions cause absorption and scattering of incident radiation. As a result, the air pollution is one of the most effective parameters that influences on the expected value of the solar radiation. However, it is relatively difficult to understand the real interrelationship between them mathematically. In addition, in most case studies, the data for air pollution is not available. Therefore, it is necessary while developing a model for forecasting solar radiation to consider the air pollution as one of the major inputs to the model. It can be obviously observed from Table 6 that the expected errors could be increased especially if the pollution value has a high level.

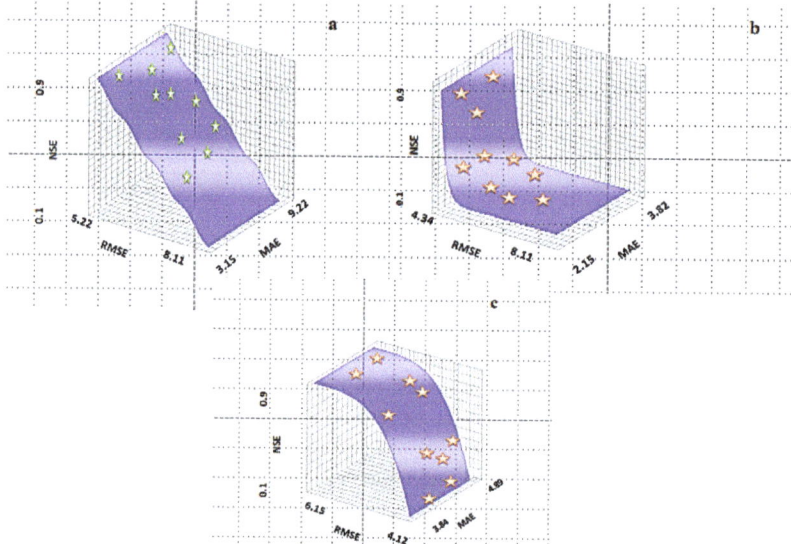

Figure 8. The tree generated three dimensional figure based on all solutions in Pareto with the determination of 10 points of the Pareto solution in the ANFIS-MOSA model for (**a**) the test level in EAZP, (**b**) test level in AP, and (**c**) test level in GP.

Table 6. Consideration of ANFIS and MOSA model with and without AQI for the best solution.

Index	RMSE (Kwh/m²)	MAE	NSE
ANIFS-MOSA with AQI input	2.12	1.25	0.95
ANFIS MOSA without AQI input	2.98	1.71	0.90

Figure 9 shows the zone map based on the IDW and AQI in EAZPs for different seasons. The Kappa coefficient is used as an index to show the degree of agreement between the observation zone map and the obtained map based on the ANFIS and IDW. The Kappa values of SR in the winter, autumn, summer, and spring are 0.85, 0.89, 0.91, and 0.92, respectively, highlighting the high accuracy of the zone map. The zone map of winter shows a lower intensity in most areas of the map, and that of summer shows a higher SR intensity.

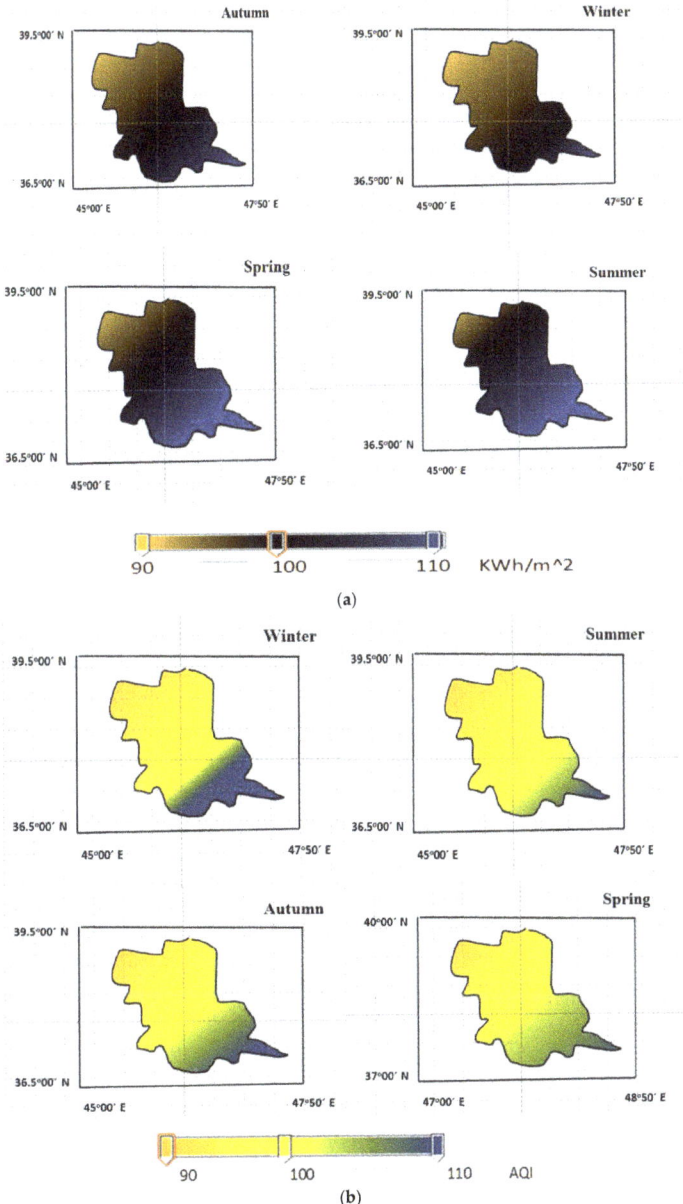

Figure 9. (a) The AQI variation for different provinces based on obtaining zoning map, (b) the zoning map for the SR.

The analysis of the results shows that the AQI value in the winter season has a greater intensity than that in the other seasons. In fact, the higher concentration of pollutant particles in the winter season significantly increases the AQI value, and these particles decrease the SR intensity. Thus, if a decision maker eliminates the AQI, the zone map of the SR in winter is drawn as a sample, and the zone map shows higher SR in the different parts of the zone map in the winter. Thus, the elimination of the AQI significantly increases the SI, highlighting the importance of considering the AQI as an input. Another point is related to the uncertainty of the simulated results because the input data and the IDW method used for the zone map contain uncertainty; thus, the computation of the uncertainty of the model is very important, and therefore, generalized likelihood uncertainty estimation (GLUE) was used in this article. When the variation domain of the input parameters of the best solution in the ANFIS and MOSA is determined, sampling is applied to the parameter space. First, the space parameter is divided into the same interval, and then, sampling is considered for each interval [41]. Thus, when the gathered parameters are obtained and compared, a series of initial parameters is prepared to be inserted into the ANFIS model. The sampling of data is repeated 10,000 times, and then, the model based on the data group (generated sample) and the computation of the probability value based on the observed data and simulated data are considered such that the input parameters are generated based on 10,000 iterations.

The objective functions are computed for 10,000 iterations, and all SR simulated data are sorted after arranging the objective function values. Then, 2.5% of the data of the upper limit and 2.5% of the data of the lower limit are considered as outlying data. Thus, the bound of 95% of the certainty level is considered, and the d factor shows the thickness bound (Figure 10).

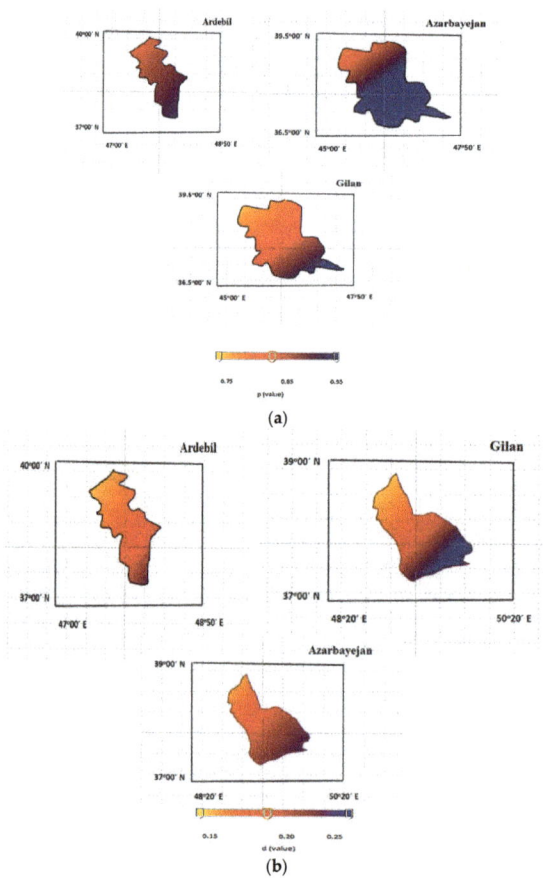

Figure 10. The zoning maps for the provinces based on uncertainty of data based on (**a**) p factor and (**b**) d value.

The p values show the percentage of data in the 95% boundary, and a higher percentage corresponds to the better performance of each method. When the zone map based on the IDW is obtained, the value, p factor, and d factor of the data from each area of the zone map are computed to show the uncertainty value of the estimated results on the complete map. Clearly, the d factor is small, and the p factor exhibits a good percentage of the maps (Figure 10).

6. Conclusions

The current paper aimed to simulate SR based on the ANFIS-MOSA, and the IDW was used to obtain zone maps of three provinces. Pareto solutions were obtained using different algorithms and the ANFIS model. The different indexes showed that ANFIS-MOSA performs better than the other models, and the low value of the RDI index showed that the Pareto obtained using the MOSA and ANFIS matched well with the ideal solution. The MTOPSIS model was used to select the optimal solutions, and different indexes, such as the RMSE, MAE, and Taylor diagram, showed that the obtained ideal solution performance was the highest for the Pareto solution with a significant difference. Then, the effect of the AQI parameter on the results was analyzed. The results showed that the elimination of the AQI parameter decreased the accuracy of the zone map. Additionally, different models with and without the AQI parameter were considered, and the results showed that the error index without the AQI parameter was significantly higher. Finally, the uncertainty of the obtained data was considered to determine its effect on the results, and the high p factor value and low d factor value illustrated the adequate performance of the proposed model. The proposed model can not only simulate SR with an acceptable level of accuracy but also add a new direction to include multi-objective functions to evaluate the performance of the prediction model. For example, to improve the performance of the proposed model, another objective function could be considered to represent the risk performance, such as experiencing ± maximum errors. In this context, an objective function that represents the risk performance could be added to address the probability occurrence of the ±maximum errors at any time during the span of the prediction time.

In fact, the current research focused on studying the performance of the proposed model considering the time period dimension. However, there is an important dimension that could be considered for further analysis, which is the importance of a certain parameter at a specific location. In this context, it is essential to recommend this direction of research to be carried out in future research.

Author Contributions: Formal analysis, M.E., A.N.A. and H.A.A.; methodology, M.E. and A.E.-S.; writing—original draft, M.E. and C.M.F.

Funding: The authors would like to appreciate the financial support received from Bold 2025 grant coded RJO 10436494 by Innovation & Research Management Center (iRMC), Universiti Tenaga Nasional, Malaysia and from research grant coded UMRG RP025A-18SUS funded by the University of Malaya.

Acknowledgments: The authors appreciate so much the facilities support by the Civil Engineering Department, Faculty of Engineering, University of Malaya, Malaysia.

Conflicts of Interest: The authors declare no conflict of interest.

Abbreviations

AQI	air quality index
a,b,c	premise parameters
C_1	cognitive coefficient
C_2	Acceleration coefficient
CS	Cover surface
d	Euclidean distance
GD	General distance
IDW	inverse distance weight
K	Number of iteration

M	the number of point
NO	number of training data
ND	Number of decision variable
OF	objective function
R_1	Random number
R_2	Random number
R^*_j	similarity ratio
SI	spread index
SR	solar radiation
Up	upper constraints
Ub	lower constraint
W_i	Wight value
X_i^{t+1}	new position
X_{gbest}	global solution
X_{pbest}	the personal best solution
v_{ij}	velocity
Z_i	new position after rotational movement
α	Momentum coefficient

References

1. Quej, V.H.; Almorox, J.; Arnaldo, J.A.; Saito, L. ANFIS, SVM and ANN soft-computing techniques to estimate daily global solar radiation in a warm sub-humid environment. *J. Atmos. Sol.-Terr. Phys.* **2017**, *155*, 62–70. [CrossRef]
2. Ma, M.; Zhao, L.; Deng, S.; Zhang, Y.; Lin, S.; Shao, Y. Estimation of horizontal direct solar radiation considering air quality index in China. *Energy Procedia* **2019**, *158*, 424–430. [CrossRef]
3. De León-Ruiz, J.; Carvajal-Mariscal, I.; De León-Ruiz, J.E.; Carvajal-Mariscal, I. Mathematical thermal modelling of a direct-expansion solar-assisted heat pump using multi-objective optimization based on the energy demand. *Energies* **2018**, *11*, 1773. [CrossRef]
4. Ghazvinian, H.; Mousavi, S.F.; Karami, H.; Farzin, S.; Ehteram, M.; Hossain, M.S.; Fai, C.M.; Hashim, H.B.; Singh, V.P.; Ros, F.C.; et al. Integrated support vector regression and an improved particle swarm optimization-based model for solar radiation prediction. *PLoS ONE* **2019**, *14*, e0217634. [CrossRef]
5. Atuahene, S.; Bao, Y.; Ziggah, Y.; Gyan, P.; Li, F.; Atuahene, S.; Bao, Y.; Ziggah, Y.Y.; Gyan, P.S.; Li, F. Short-Term Electric Power Forecasting Using Dual-Stage Hierarchical Wavelet-Particle swarm optimization-adaptive neuro-fuzzy inference system PSO-ANFIS approach based on climate change. *Energies* **2018**, *11*, 2822. [CrossRef]
6. Lotfinejad, M.; Hafezi, R.; Khanali, M.; Hosseini, S.; Mehrpooya, M.; Shamshirband, S.; Lotfinejad, M.M.; Hafezi, R.; Khanali, M.; Hosseini, S.S.; et al. Comparative assessment of predicting daily solar radiation using bat neural network (BNN), generalized regression neural network (GRNN), and neuro-fuzzy (NF) system: A Case Study. *Energies* **2018**, *11*, 1188. [CrossRef]
7. Sharifi, S.S.; Rezaverdinejad, V.; Nourani, V. Estimation of daily global solar radiation using wavelet regression, ANN, GEP and empirical models: A comparative study of selected temperature-based approaches. *J. Atmos. Sol.-Terr. Phys.* **2016**, *149*, 131–145. [CrossRef]
8. Alsina, E.F.; Bortolini, M.; Gamberi, M.; Regattieri, A. Artificial neural network optimisation for monthly average daily global solar radiation prediction. *Energy Convers. Manag.* **2016**, *120*, 320–329. [CrossRef]
9. Yaseen, Z.; Ehteram, M.; Hossain, M.; Fai, C.; Binti Koting, S.; Mohd, N.; Binti Jaafar, W.; Afan, H.; Hin, L.; Zaini, N.; et al. A novel hybrid evolutionary data-intelligence algorithm for irrigation and power production management: Application to multi-purpose reservoir systems. *Sustainability* **2019**, *11*, 1953. [CrossRef]
10. Meenal, R.; Selvakumar, A.I. Assessment of SVM, empirical and ANN based solar radiation prediction models with most influencing input parameters. *Renew. Energy* **2018**, *121*, 324–343. [CrossRef]
11. Marzo, A.; Trigo-Gonzalez, M.; Alonso-Montesinos, J.; Martínez-Durbán, M.; López, G.; Ferrada, P.; Fuentealba, E.; Cortés, M.; Batlles, F.J. Daily global solar radiation estimation in desert areas using daily extreme temperatures and extraterrestrial radiation. *Renew. Energy* **2017**, *113*, 303–311. [CrossRef]

12. Renno, C.; Petito, F.; Gatto, A. ANN model for predicting the direct normal irradiance and the global radiation for a solar application to a residential building. *J. Clean. Prod.* **2016**, *135*, 1298–1316. [CrossRef]
13. Shamshirband, S.; Mohammadi, K.; Tong, C.W.; Zamani, M.; Motamedi, S.; Ch, S. A hybrid SVM-FFA method for prediction of monthly mean global solar radiation. *Theor. Appl. Climatol.* **2015**, *125*, 53–65. [CrossRef]
14. Samadianfard, S.; Majnooni-Heris, A.; Qasem, S.N.; Kisi, O.; Shamshirband, S.; Chau, K. Daily global solar radiation modeling using data-driven techniques and empirical equations in a semi-arid climate. *Eng. Appl. Comput. Fluid Mech.* **2019**, *13*, 142–157. [CrossRef]
15. Mehdizadeh, S.; Behmanesh, J.; Khalili, K. Comparison of artificial intelligence methods and empirical equations to estimate daily solar radiation. *J. Atmos. Sol.-Terr. Phys.* **2016**, *146*, 215–227. [CrossRef]
16. Koca, A.; Oztop, H.F.; Varol, Y.; Koca, G.O. Estimation of solar radiation using artificial neural networks with different input parameters for Mediterranean region of Anatolia in Turkey. *Expert Syst. Appl.* **2011**, *38*, 8756–8762. [CrossRef]
17. Ozgoren, M.; Bilgili, M.; Sahin, B. Estimation of global solar radiation using ANN over Turkey. *Expert Syst. Appl.* **2012**, *39*, 5043–5051. [CrossRef]
18. Voyant, C.; Muselli, M.; Paoli, C.; Nivet, M.-L. Numerical weather prediction (NWP) and hybrid ARMA/ANN model to predict global radiation. *Energy* **2012**, *39*, 341–355. [CrossRef]
19. Furlan, C.; de Oliveira, A.P.; Soares, J.; Codato, G.; Escobedo, J.F. The role of clouds in improving the regression model for hourly values of diffuse solar radiation. *Appl. Energy* **2012**, *92*, 240–254. [CrossRef]
20. Chen, J.-L.; Li, G.-S.; Wu, S.-J. Assessing the potential of support vector machine for estimating daily solar radiation using sunshine duration. *Energy Convers. Manag.* **2013**, *75*, 311–318. [CrossRef]
21. Vakili, M.; Sabbagh-Yazdi, S.-R.; Kalhor, K.; Khosrojerdi, S. Using artificial neural networks for prediction of global solar radiation in Tehran considering particulate matter air pollution. *Energy Procedia* **2015**, *74*, 1205–1212. [CrossRef]
22. Quesada-Ruiz, S.; Linares-Rodríguez, A.; Ruiz-Arias, J.A.; Pozo-Vázquez, D.; Tovar-Pescador, J. An advanced ANN-based method to estimate hourly solar radiation from multi-spectral MSG imagery. *Sol. Energy* **2015**, *115*, 494–504. [CrossRef]
23. Çelik, Ö.; Teke, A.; Yıldırım, H.B. The optimized artificial neural network model with Levenberg–Marquardt algorithm for global solar radiation estimation in eastern mediterranean region of Turkey. *J. Clean. Prod.* **2016**, *116*, 1–12. [CrossRef]
24. Deo, R.C.; Şahin, M. Forecasting long-term global solar radiation with an ANN algorithm coupled with satellite-derived (MODIS) land surface temperature (LST) for regional locations in Queensland. *Renew. Sustain. Energy Rev.* **2017**, *72*, 828–848. [CrossRef]
25. Yao, W.; Zhang, C.; Wang, X.; Sheng, J.; Zhu, Y.; Zhang, S. The research of new daily diffuse solar radiation models modified by air quality index (AQI) in the region with heavy fog and haze. *Energy Convers. Manag.* **2017**, *139*, 140–150. [CrossRef]
26. Loghmari, I.; Timoumi, Y.; Messadi, A. Performance comparison of two global solar radiation models for spatial interpolation purposes. *Renew. Sustain. Energy Rev.* **2018**, *82*, 837–844. [CrossRef]
27. Fan, J.; Wu, L.; Zhang, F.; Cai, H.; Wang, X.; Lu, X.; Xiang, Y. Evaluating the effect of air pollution on global and diffuse solar radiation prediction using support vector machine modeling based on sunshine duration and air temperature. *Renew. Sustain. Energy Rev.* **2018**, *94*, 732–747. [CrossRef]
28. Khan, I.; Zhu, H.; Yao, J.; Khan, D.; Iqbal, T. Hybrid power forecasting model for photovoltaic plants based on neural network with air quality index. *Int. J. Photoenergy* **2017**, *2017*. [CrossRef]
29. Wang, H.; Sun, F.; Wang, T.; Liu, W. Estimation of daily and monthly diffuse radiation from measurements of global solar radiation a case study across China. *Renew. Energy* **2018**, *126*, 226–241. [CrossRef]
30. Sun, H.; Gui, D.; Yan, B.; Liu, Y.; Liao, W.; Zhu, Y.; Lu, C.; Zhao, N. Assessing the potential of random forest method for estimating solar radiation using air pollution index. *Energy Convers. Manag.* **2016**, *119*, 121–129. [CrossRef]
31. Vakili, M.; Sabbagh-Yazdi, S.R.; Khosrojerdi, S.; Kalhor, K. Evaluating the effect of particulate matter pollution on estimation of daily global solar radiation using artificial neural network modeling based on meteorological data. *J. Clean. Prod.* **2017**, *141*, 1275–1285. [CrossRef]
32. Mohammad-Azari, S.; Bozorg-Haddad, O.; Chu, X. Shark smell optimization (SSO) algorithm. *Adv. Optim. Nat.-Inspir. Algorithms* **2017**, 93–103.

33. Ehteram, M.; Karami, H.; Mousavi, S.-F.; El-Shafie, A.; Amini, Z. Optimizing dam and reservoirs operation based model utilizing shark algorithm approach. *Knowl.-Based Syst.* **2017**, *122*, 26–38. [CrossRef]
34. Abedinia, O.; Amjady, N.; Ghasemi, A. A new metaheuristic algorithm based on shark smell optimization. *Complexity* **2014**, *21*, 97–116. [CrossRef]
35. Santiago, A.; Dorronsoro, B.; Nebro, A.J.; Durillo, J.J.; Castillo, O.; Fraire, H.J. A novel multi-objective evolutionary algorithm with fuzzy logic based adaptive selection of operators: FAME. *Inf. Sci.* **2019**, *471*, 233–251. [CrossRef]
36. Deb, K.; Pratap, A.; Agarwal, S.; Meyarivan, T. A fast and elitist multiobjective genetic algorithm: NSGA-II. *IEEE Trans. Evol. Comput.* **2002**, *6*, 182–197. [CrossRef]
37. Mofid, H.; Jazayeri-Rad, H.; Shahbazian, M.; Fetanat, A. Enhancing the performance of a parallel nitrogen expansion liquefaction process (NELP) using the multi-objective particle swarm optimization (MOPSO) algorithm. *Energy* **2019**, *172*, 286–303. [CrossRef]
38. Antonakis, A.; Nikolaidis, T.; Pilidis, P.; Antonakis, A.; Nikolaidis, T.; Pilidis, P. Multi-objective climb path optimization for aircraft/engine integration using particle swarm optimization. *Appl. Sci.* **2017**, *7*, 469. [CrossRef]
39. Wang, Y.; Hua, Z.; Wang, L.; Wang, Y.; Hua, Z.; Wang, L. Parameter estimation of water quality models using an improved multi-objective particle swarm optimization. *Water* **2018**, *10*, 32. [CrossRef]
40. Ehteram, M.; Afan, H.A.; Dianatikhah, M.; Ahmed, A.N.; Fai, C.M.; Hossain, M.S.; Allawi, M.F.; Elshafie, A.; Ehteram, M.; Afan, H.A.; et al. Assessing the predictability of an improved ANFIS model for monthly streamflow using lagged climate indices as predictors. *Water* **2019**, *11*, 1130. [CrossRef]
41. Karimi, S.; Amiri, B.J.; Malekian, A. Similarity metrics-based uncertainty analysis of river water quality models. *Water Resour. Manag.* **2019**, *33*, 1927–1945. [CrossRef]

© 2019 by the authors. Licensee MDPI, Basel, Switzerland. This article is an open access article distributed under the terms and conditions of the Creative Commons Attribution (CC BY) license (http://creativecommons.org/licenses/by/4.0/).

Article

Development of a High-Resolution Wind Forecast System Based on the WRF Model and a Hybrid Kalman-Bayesian Filter

Carlos Otero-Casal [1,2,*], Platon Patlakas [3], Miguel A. Prósper [4], George Galanis [5] and Gonzalo Miguez-Macho [1]

1. Nonlinear Physics Group, Universidade de Santiago de Compostela, 15705 Santiago de Compostela, Spain
2. MeteoGalicia, Xunta de Galicia, 15707 Santiago de Compostela, Spain
3. School of Physics, Division of Environment and Meteorology, University of Athens, 15784 Athens, Greece
4. Siemens Gamesa, Meteorology Department, 28043 Madrid, Spain
5. Mathematical Modeling and Applications Laboratory, Hellenic Naval Academy, Hatzikiriakion, 18539 Piraeus, Greece
* Correspondence: carlos.otero.casal@usc.es

Received: 2 June 2019; Accepted: 29 July 2019; Published: 8 August 2019

Abstract: Regional microscale meteorological models have become a critical tool for wind farm production forecasting due to their capacity for resolving local flow dynamics. The high demand for reliable forecasting tools in the energy industry is the motivation for the development of an integrated system that combines the Weather Research and Forecasting (WRF) atmospheric model with an optimization obtained by the conjunction of a Kalman filter and a Bayesian model. This study focuses on the development and validation of this combined system in a very dense wind farm cluster located in Galicia (Northwest of Spain). A period of one year is simulated at 333 m horizontal resolution, with a daily operational forecasting set-up. The Kalman-Bayesian filter was tested both directly on wind speed and on the U-V (zonal and meridional) components for nowcasting periods from 10 min to 6 h periods, all of them with important applications in the wind industry. The results are quite promising, as the main statistical error indices are significantly improved in a 6 h forecasting horizon and even more in shorter horizon cases. The Mean Annual Error (MAE) for 1 h nowcasting horizon is 1.03 m/s for wind speed and 12.16° for wind direction. Moreover, the successful utilization of the integrated system in test cases with different characteristics demonstrates the potential utility that this tool may have for a variety of applications in wind farm operations and energy markets.

Keywords: nowcasting; Kalman-Bayesian filter; WRF; high-resolution; complex terrain; wind

1. Introduction

The global expansion of wind energy is a well-known reality, especially in recent years. In fact, in 2017 52.5 GW of new wind power installations were added across the globe, bringing the total worldwide capacity up to 539 GW [1]. This was a record year for Europe as well, reaching record numbers with 15.6 GW of new wind power capacity installed [2]. Part of this success has derived from offshore wind farms and their new 3.1 GW in 2017. The market forecast predicts a continuation of this tendency for the next few years [1,3], establishing wind energy as one of the main ingredients of the worldwide energy mix, which translates into an increasing number of operational wind farms, composed of more powerful wind turbines.

Considering the aforementioned scenarios, an improvement of wind energy exploitation seems necessary, employing new strategies in future wind farm projections and better daily management of current installations. The latter is where high-resolution meteorological modeling becomes increasingly prominent. Local short-term forecasts can supply valuable information in an operational wind farm, as wind speed and direction predictions can lead to a better running of the installation [4,5] or even provide alert for any extreme weather situation such as a downslope windstorm [6] or a severe frost [7].

The Weather Research and Forecasting (WRF) meso and microscale model has been used extensively during the last few years in the wind energy field for a variety of applications [8–10]. It includes a wind turbine parameterization [11] enabling the study of the wake effect from wind turbines [5,12]. However, Numerical Weather Prediction (NWP) simulations by themselves, present some limitations regarding subgrid scale phenomena forecasting. Onshore farms can be affected by local climatology and microscale events with significant wind direction and intensity changes in short periods of time. It is well known that such variations cannot be well reproduced by NWP simulations with resolution lower than that of the small-scale events commented above. To address this problem, WRF Large Eddy Simulations (LES) have been shown to be useful modeling techniques within this field. Their high resolution (<100 m) allows obtaining a good characterization of the turbulent intensity and an accurate representation of the wind fields [13–15]. Despite these capabilities, the computational demands of this modeling methodology and its numerical instability over complex terrain [14,16] make it unfeasible for operational forecasting at the present time.

The combination of NWP and statistical post-processes, like Model Output Statistics (MOS) or dynamically adjustable filters, can be an effective technique to minimize errors induced by sub-grid phenomena and local effects. In this direction, several strategies have been developed. Kariniotakis and Pinson [17], Kariniotakis et al. [18] proposed a neural network approach while Vanem [19], Giebel [20], Resconi [21] and others introduce methodologies based on heavy statistical models. Another method adopted for the reduction of systematic biases that numerical weather prediction models face, is Kalman filtering based algorithms (Resconi [21], Pelland et al. [22], Galanis et al. [23,24], Crochet [25], Galanis and Anadranistakis [26], Kalnay [27], Kalman and Bucy [28], Kalman [29]). Stathopoulos et al. [30] used the combination of an ETA-coordinate weather prediction model (running at a resolution of 0.05°) and a Kalman Filtering postprocess for wind power prediction. This showed an improved performance and favorable results. Hua et al. [31] use 3 km × 3 km WRF model simulations and a Kalman Filter to obtain more skillful wind speed forecasts. Che et al. [32], Che and Xiao [33] developed a similar forecasting system employing 0.5 km resolution WRF simulations with a Kalman filter for a wind farm in Japan. Despite the overall good performance, in many cases, such filters only reduce the systematic mean bias [34], especially in wind speed, due to the nature and variability of the parameter. A proposed solution for this drawback is described analytically in Galanis et al. [35]. The suggested optimization technique is emerging from the conjunction of a non-linear Kalman filter and a Bayesian model (K-B model). This new hybrid model led to promising results eliminating systematic biases in the model outputs and reducing the variability of the remaining white noise.

In this study, we develop and validate a combined system of WRF high-resolution simulations and the aforementioned K-B filter. The main novelty proposed is the application of the hybrid filter to the two-dimensional field of horizontal wind speed components instead of the 1-dimensional filter in Galanis et al. [35]. In this way, an improved prediction of wind speed and direction is produced for a very dense wind farm cluster located in Galicia (NW Spain). A period of one year, broken into 365 daily integrations, is simulated with the WRF atmospheric model at high resolution (333 m), and results are post-processed with the K-B filter.

The obtained integrated system is validated for different short-term forecast periods employing in-situ data from meteorological stations on the top of the wind turbines provided by the farm's operator. In addition, the new wind direction nowcasting tool is applied to the wind energy field as a backup for the current yaw orientation systems, which are based on Supervisory Control And Data Acquisition (SCADA) data obtained from each wind turbine [36]. These sensors are sometimes affected by short temporal errors due to technical problems or severe meteorological situations. Apart from the wind farm operational issue, this kind of tool can be used by electric grid operators to prevent ramp event effects. These extreme short-term situations produce significant increases or drops in the power production [37], delivering a major impact on the grid.

The article is structured as follows: In Section 2 the methodology is explained in detail, with particular attention to the WRF model configuration and the K-B filter description. Section 3 discusses the results obtained for wind speed and wind direction nowcasting as well as applications on wind energy forecasting. Finally, in Section 4 conclusions are presented.

2. Methodology and Data

2.1. Wind Farm Location and WRF Configuration

Coruxeiras wind farm is located in Serra do Xistral mountains in Galicia, in northwest Spain (Figure 1a). The prevailing winds in this area are in the southwest/northeast direction (Figure 1c). In winter, frequent passing by cyclones along the North Atlantic storm track often produce strong southwest flows, whereas in summer, the poleward displacement of the subtropical Azores high, results in winds with a northeast component. Southeast/northeast flows are further accelerated due to the packing of the isobars by interaction with the high terrain of the northwestern Iberian Peninsula; hence the position of Serra do Xistral, exposed to the ocean in both prevailing directions, results in a very high wind energy potential. Numerous wind farms exist in the area, which is one of the most productive in Europe [38,39]. The developer and current operator of Coruxerias wind farm is Norvento S.L.U. company. It was installed in 2006 and is composed of 31 turbines with 60 m hub height and 74 m rotor diameter. They are separated by a mean distance of 300 m and placed along smooth hill tops of around 800 m elevation above sea level (Figure 1b).

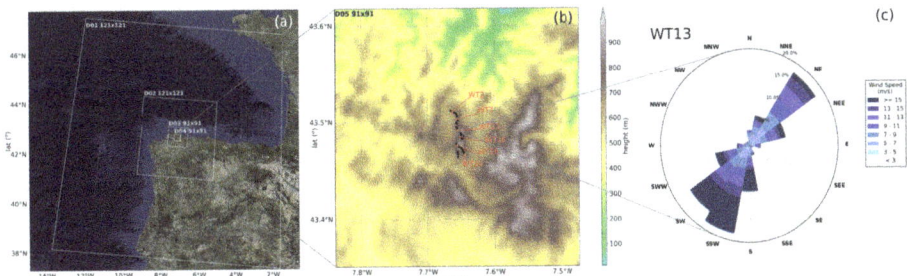

Figure 1. (a) WRF nested grid configuration, with the number of points for each domain indicated. (b) D04 is expanded showing its topography. Coruxeiras wind farm is located in the central area of D04, on top of a hill. Observational data for the study is obtained from meteorological stations at the hub of the wind turbines plotted in red. (c) Annual wind rose from observational data at hub height from a wind turbine in the center of the farm (WT13).

For the atmospheric simulations, we use the Advanced Research WRF (ARW) model version 3.6 [40] (WRFV3.6), designed for operational forecasting, as well as research. WRFV3.6 is a limited-area mesoscale and microscale model based on a fully compressible and non-hydrostatic dynamic core [41] that uses a terrain-following hydrostatic-pressure vertical coordinate. Figure 1a shows the domain's configuration where D01 is centered at 43.29 N and 7.75 W (Figure 1a) with 121 × 121 grid points of 9 km of horizontal resolution. The horizontal resolutions of D02, D03, and D04 are 3 km (121 × 121 grid points), 1 km (91 × 91 grid points) and 333 m (91 × 91 grid points) respectively. As demonstrated in Prósper et al. [5], the high horizontal resolution used over this complex terrain increases the accuracy of the wind forecast. The innermost domain has 67 vertical levels, 7 of which lie within the first 200 m above ground, at about 14, 41, 70, 99, 127, 156 and 184 m, a distribution that better captures wind and temperature variations in the surface layer [42] and improves the performance of the wind turbine parameterization [11]. Terrain elevation data is obtained from the ASTER Global Digital Elevation

Map (GDEM) from USGS (United States Geological Survey) [43] with a resolution of 30 m, and land use information from the Corine (Coordination of Information on the Environment) database [44] with 250 m resolution.

A thorough validation of this model configuration for Coruxeiras wind farm and more information about the model physics options can be found in [5]. In the next table (Table 1), we present the main physics parameterizations used.

Table 1. The principal physical schemes used in the atmospheric model.

Microphysics	Single-moment six-class scheme [45]
Cumulus Parameterization	Kain-Fritsch scheme [46] disabled in d04 and d05
Long wave radiation physics	RRTM Longwave model [47]
Short wave radiation physics	Dudhia shortwave radiation schemes [48]
Planetary boundary layer	Mellor–Yamada Nakanishi Niino Level 2.5 [49]
Surface layer option	Revised MM5 Monin-Obukhov scheme [50]
Land-surface physics	Noah land-surface model [51]
Wind Turbine Parameterization	Fitch, A. C. 2012 [11]

One year (2015/02/01–2016/02/01) of GFS forecast data from the National Centers for Environmental Prediction (NCEP) are used as initial and boundary conditions for the WRF model, with a 3-h update interval. The horizontal resolution of this dataset for all variables is 0.25 × 0.25 deg, with 32 levels ranging from 1000 to 10 hPa. The use of forecast data as boundary conditions should be particularly highlighted, as this study is performed employing a daily operational forecasting set-up, therefore including all the potential drawbacks of such an application.

Moreover, the A.C. Fitch wind turbine parameterization [11,52] is used in order to account for the wake effects of the turbines. Given the wind turbine dimensions, power curve, and their positions, the Fitch scheme represents them as momentum sinks, transferring the flow's kinetic energy into turbulent kinetic energy and electricity.

2.2. Hybrid Bayesian Kalman Filter

In this subsection, the statistical optimization postprocess adopted is presented. Beginning with the polynomial Kalman filtering local adaptation model, the main goal is the estimation of the atmospheric model bias y_t as a function of the model output m_t.

$$y_t = x_{0,t} + x_{1,t} \cdot m_t + v_t \tag{1}$$

where $x_{i,t} (i = 0, 1)$ are the parameters to be estimated and v_t is the Gaussian nonsystematic error.

The above equation can be expressed as $y_t = H_t \cdot x_t + v_t$ where $x_t = \begin{bmatrix} x_{0,t} & x_{1,t} \end{bmatrix}$ is the state vector and $H_t = \begin{bmatrix} 1 & m_t \end{bmatrix}$ the observation matrix. As a result, it can be written in a more analytical way as follows:

$$y_t = \begin{bmatrix} 1 & m_t \end{bmatrix} \cdot \begin{bmatrix} x_{0,t} \\ x_{1,t} \end{bmatrix} + v_t \tag{2}$$

Similarly, the evolution in time of x_t is described by the equation

$$x_t = x_{t-1} + w_t \tag{3}$$

where x_{t-1} is the parameter in $t-1$ and w_t is the Gaussian nonsystematic error.

In the present study, a linear approach is followed, since it was found to be sufficient both in terms of reliability and computer resource needs. The estimation of the variables v_t and w_t is based on a training period. For this case and after multiple sensitivity tests, the training period was set to 140 include the last 12 values of the sample for each forecast. We note, however, that options such as the order of the polynomial in Equation (1) used or the training period, are case sensitive and depend on local features and time period characteristics.

In particular, the covariance matrices V_t and W_t of the error terms v_t and w_t are calculated as follows:

$$V_{t_i} = \frac{1}{11} \sum_{i=0}^{11} (((y_{t_i} - H_{t_i} x_{t_i}) - (\frac{\sum_{i=0}^{11}(y_{t_i} - H_{t_i} x_{t_i})}{12})))^2 \qquad (4)$$

$$W_{t_i} = \frac{1}{11} \sum_{i=0}^{11} (((x_{t_i+1} - x_{t_i}) - (\frac{\sum_{i=0}^{11}(x_{t_i+1} - x_{t_i})}{12})))^2 \qquad (5)$$

The main steps of the Kalman algorithm for the estimation of the state vector x_t are summarized as follows:

- A first estimation of x_t is given by $x_{t/t-1} = x_{t-1}$
- The corresponding covariance matrix P_t is calculated by $P_{t/t-1} = P_{t-1} + W_t$
- With a new observation y_t available the calculation of x at time t takes the form: $x_t = x_{t/t-1} + K_t - (y_t - H_t - x_{t/t-1})$

Here, $K_t = P_{t/t-1} H_t^T (H_t P_{t/t-1} H_t^T + V_t)^{-1}$ is the *Kalman gain*, a parameter that controls the flexibility of the filter.

More details on the Kalman filtering theory can be found in [27–29]. The filter has been tested successfully in wind speed and wind gust prediction Stathopoulos et al. [30], Louka et al. [34], Patlakas et al. [53], Galanis et al. [35] with results that ensure the reduction of the systematic errors induced by numerical weather models.

However, in most of the previous studies, the variation of the remaining nonsystematic part of the error is not reduced. For a further improvement of the Kalman filter output, we incorporate the following linear Bayesian model:

$$k_t = o_t + n_t \qquad (6)$$

where, k_t represents the Kalman filtered output at time t (estimated by the Kalman filter as the model initial forecast corrected by y_t), o_t is the corresponding observation value, and n_t the remaining Gaussian nonsystematic white noise.

For the application of the Bayesian model, an approach based on Normal distribution is utilized based on the assumption that the probability density functions of the filter parameters have as follows: $P(o_t) \sim N(o_\mu, \sigma_o^2)$, $P(v_t) \sim N(0, \sigma_v^2)$. Then, the conditional pdf takes the form $P(k_t|o_t) \sim N(o_t, \sigma_v^2)$. The final estimation of the parameter o_t takes advantage of the posteriori estimator given by $\hat{o}_t = argmax P(o_t|k_t)$.

Further details concerning Bayesian theory and models can be found in the studies of Box [54] and Bernardo and Smith [55], while the combination of the two approaches is fully described by Galanis et al. [35].

The new approach proposed in this work consists in the application of the K-B model for both wind speed and U-V components for different forecasting horizons, covering in this way both wind speed and direction with quite satisfactory results as presented in the following sections.

2.3. Observational Data and Nowcasting Experiments

The observational data used in this work is provided by Norvento S.L.U., collected from anemometers located on top of each turbine's nacelle. Specifically, we use wind speed and wind direction (WS and WD hereafter) observations from six wind turbines, two in the north of the farm (WT2 and WT7), two in the center (WT13 and WT16) and two in the south (WT24 and WT26) (Figure 1b). In this way, the WRF K-B nowcasting tool is tested all over the wind farm, with each turbine differently affected by wake effects and local topography. One year of 10-min wind data for each turbine is available. We note that a priori quality control on the raw observational data is performed, filtering out discontinuities such as automatic starts after stops or preventive maintenance.

This leads to three different applications of the integrated forecasting model. More precisely, in the first scenario, the forecasting bias in WRF wind speed prediction is minimized by the Kalman filter described in the previous section based on the polynomial Equation (1). In a second step, the Bayesian component of the optimization procedure reduces the variability of the error, and therefore the forecasting uncertainty. Following the same concept, two different K-B filters are utilized separately for the U, V wind speed components. This scenario aims at the improvement of both the wind speed and direction simultaneously.

The period chosen for the present study is one year (1 February 2015–1 February 2016), divided into 365 daily simulations. Each run starts at 18 UTC the day before the one considered, and the initial 6 h are excluded from the forecast data series used in the analysis of the results, thus, only outputs after this spin-up time are taken into account.

By definition, nowcasting refers to short lead time weather forecasts. Some organizations like the U.S. National Weather Service stipulate that it pertains to lead times from zero to three hours. However, forecasts up to six hours are also considered nowcasts by different agencies [56]. To analyze the capabilities of our methodology, we evaluate the K-B filter with different nowcasting horizon periods, from 6 h to 10 min (Table 2) for each one of the daily simulations. For example, the wind speed nowcasting obtained in K-B 1 h is the result of correcting RAW data every 10 min by using model output and real data from one hour before. Results are compared with both observations and original non-post-processed WRF outputs.

Table 2. Short explanation of the experiments tested in this study. The bold names on the left are the identifiers used hereafter for each case.

Experiment	Description
RAW	WRF results without post-proccess
K-B 6 h	K-B filter nowcasting used for 6 h time horizon
K-B 1 h	K-B filter nowcasting used for 1 h time horizon
K-B 30 min	K-B filter nowcasting used for 30 min time horizon
K-B 10 min	K-B filter nowcasting used for 10 min time horizon

3. Results and Discussion

The next section is divided into three parts; the first one analyzes the results obtained with the nowcasting wind speed postprocess at different time horizons, from 6 h to 10 min in advance. The analysis is developed for different temporal periods, from annual mean results to the performance of the experiments at individual simulation timesteps. The second subsection shows the results of the K-B filter used to measure wind direction correction, same as with wind speed, with different nowcasting horizons and mean errors for several periods. Finally, Section 3.3 shows an example of the improvement that K-B 1 h applied in wind direction could induce in the daily management of the studied wind farm.

3.1. Wind Speed Nowcasting

This subsection employs several error measures at different temporal scales to provide a general view of the skill of all the experiments. The next equations show the statistical measures used in different temporal scales throughout the chapter.

$$MAE = \frac{1}{n}\sum_{i=1}^{n}|f_i - ob_i| \qquad (7)$$

$$ME = \frac{1}{n}\sum_{i=1}^{n}(f_i - ob_i) \qquad (8)$$

$$RMSE = \sqrt{\frac{\sum_{i=1}^{n}(f_i - ob_i)^2}{n}} \tag{9}$$

$$\sigma = \sqrt{\frac{\sum_{i=1}^{n}|x_i - \bar{x}|^2}{n}} \tag{10}$$

Equation (7) gives the Mean Absolute Error (MAE) between forecasts (f_i) and observations (ob_i), (8) the Mean Error (ME) estimating possible systematic biases, (9) the Root Mean Square Error (RMSE), where n is the sample size, and (10) the standard deviation (σ), where x_i is the value, x the mean value of a study period.

Annual mean results are firstly shown in Figure 2, and Table 3 for each wind turbine studied. Figure 2 displays a bar chart with the annual WS MAE and standard deviation for each turbine and experiment performed. Table 3a,b present the annual WS ME and RMSE, again, for each turbine and experiment.

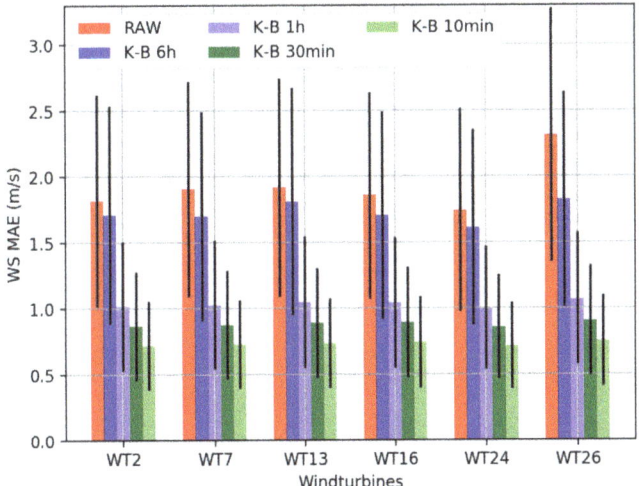

Figure 2. Barchart with annual WS MAE for all experiments and wind turbines analyzed. For each case, the standard deviation (σ) is represented by a black line on top of each bar.

Overall, a similar improvement concerning WS MAE for all the wind turbines and cases is recorded. Raw WRF forecast MAEs are between 1.91 and 1.74 m/s, except for WT26, which has a RAW MAE of 2.31 m/s. All of these results, regardless of wind farm area, present a small reduction of the WS MAE (of around 10%) in the longest nowcast period, K-B 6 h (blue). It is important to highlight that the specific K-B approach employed in this experience is the result of training the tool for the nowcast time horizons analyzed. For this reason, apply this K-B tool configuration for a longer range, as for example K-B 24 h, would generate even worse than RAW case (K-B 24 h WS MAE = 2.37 m/s). Longer forecasting time ranges, as 24 h, are associated with different temporal-spatial scale wind phenomena and the sources of the errors for this temporal resolution are also different. If we want to use this K-B tool for 24 h or a longer horizon, it is necessary to perform a different training of the K-B for the points of interest. Continuing with Figure 2's discussion, the accuracy of the K-B filter is significantly enhanced for shorter lead times, with K-B 1 h (pale blue) reaching a WS MAE of around 1 m/s in all cases. The shorter nowcasting periods, K-B 30 m (green) and K-B 10 m (pale green) show a further reduction of error, with values around 0.86 and 0.72 m/s respectively. The standard deviations plotted on top of each bar (σ) are also reduced for all turbines, in parallel with the shortening of the nowcasting periods. RAW cases have a mean σ of 0.82 m/s, and K-B 10 min present a mean value for all of the machines of 0.33 m/s.

Table 3. WS ME and WS RMSE for all experiments and wind turbines.

(a) WS ME (m/s)					
	RAW	K-B 6 h	K-B 1 h	K-B 30 min	K-B 10 min
WT2	−0.419	0.026	0.017	0.009	0.004
WT7	−0.446	0.011	0.016	0.009	0.004
WT13	−0.331	0.102	0.014	0.008	0.006
WT16	−0.590	0.069	0.025	0.013	0.006
WT24	−0.449	0.019	0.021	0.013	0.006
WT26	−1.400	0.048	0.022	0.012	0.004
TOTAL	−0.605	0.046	0.019	0.011	0.005
(b) WS RMSE (m/s)					
	RAW	K-B 6 h	K-B 1 h	K-B 30 min	K-B 10 min
WT2	2.426	2.376	1.411	1.194	0.986
WT7	2.508	2.325	1.413	1.199	0.990
WT13	2.528	2.498	1.445	1.219	1.001
WT16	2.425	2.322	1.435	1.226	1.019
WT24	2.327	2.189	1.370	1.170	0.969
WT26	3.009	2.449	1.464	1.237	1.022
TOTAL	2.537	2.359	1.423	1.207	0.998

The behavior of the WS ME is different as compared with MAE. All of the RAW WS MEs for the six wind turbines are negative, ranging from −0.33 m/s in WT13 to −1.40 in WT26. Unlike in WS MAEs, there is a clear difference between RAW and K-B 6 h MEs. In K-B 6 h cases, the ME practically disappears, with a value of 0.046 m/s. In the rest of the experiments, the ME is reduced even more; reaching a total ME of 0.005 m/s for the K-B 10 min case. WS RMSEs show a similar tendency to that of WS MAEs, with close mean values of the RAW and K-B 6 h experiments (2.54 and 2.36 m/s respectively) and a definite improvement between the three shorter nowcasting periods (K-B 1 h-30 min-10 min) and K-B 6 h.

Figure 3 analyzes the relationship between the observed and simulated wind field in terms of module and direction. As also discussed previously, there are similar error values and patterns in all of the wind turbines, which allows us to group them by areas while maintaining the rigor of the validation. In Figure 3a,c,e the wind speed distribution is displayed for the observations and the RAW, K-B 1 h and K-B 6 h experiments, for the North, Center and South areas respectively. Radar charts in Figure 3b,d,e depict the WS MAE as a function of direction for all experiments and areas.

The distribution plots in Figure 3a,c,e show a significant improvement from the K-B filter in the three areas. K-B 6 h mitigates the main wind speed deviations in all the range of the distributions. In the north area (Figure 3a) K-B filter cases correct a general overestimation from 6 to 10 m/s and an underestimation from 11 to 22 m/s. This behavior is repeated in the Central and South areas. K-B experiments (even K-B 6 h) obtain an important amelioration of the entire distribution shape, which is in agreement with the low ME errors shown in Table 3a. This demonstrates the reliability of the K-B filter in different wind situations, correcting under- and over-estimations indistinctively, which is vital for wind energy applications. Charts in Figure 3b,d,f show the skill of the K-B tool depending on wind direction. In RAW cases, the three areas present variations, as the general orientation of each zone and their position relative to the rest of the farm change forecast performance. K-B 6 h decreases mainly the higher directional errors as, for example, with southerly winds in the South area (Figure 3f). However, the K-B tool for 1 h, 30 min, and 10 min tends to smooth out errors from all sources, improving results in all the prevailing wind directions. Apart from the low MAE values observed, the symmetry in the figures in the shorter K-B time ranges leads to the conclusion that the tool is robust, rectifying wind errors in any regime. It presents the same good accuracy correcting northeast flows, principally produced in summer, as it does with southwesterly situations, characteristic of winter months.

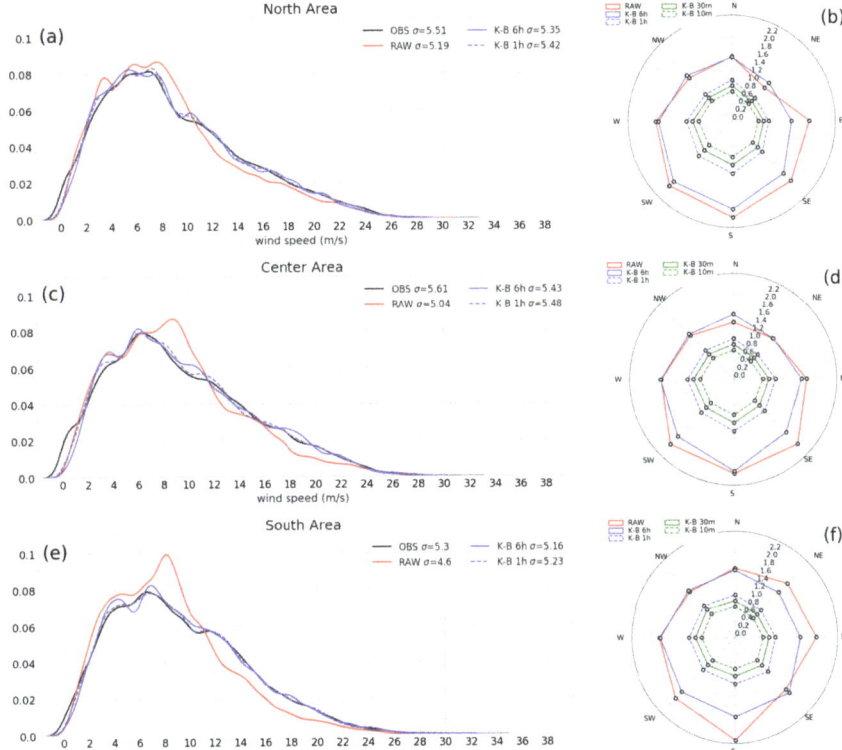

Figure 3. (a) Wind speed distribution plot for observations, RAW, K-B 6 h and K-B 1 h time experiments in the north area of the wind farm. The standard deviation of each distribution is indicated in the legend. (b) Radar charts comparing WS errors by observed wind direction in the north area. (c) Same as (a) for the center area. (d) Same as (b) for the center area. (e) Same as (a) for the south area. (f) Same as (b) for the southern area.

The seasonal cycle of errors averaged for all the turbines is shown in Figure 4, comparing wind speed monthly MAE from RAW results with the same calculation from K-B experiments. Observed monthly mean wind speeds are also shown for reference in the table below (Table 4).

Table 4. Observed mean monthly wind speed normalized with observed mean annual wind speed.

Mean Monthly WS /Mean Annual WS											
January	February	March	April	May	June	July	August	September	October	November	December
1.4	1.16	0.97	0.85	1.07	0.77	0.8	0.76	0.83	1.02	0.88	1.48

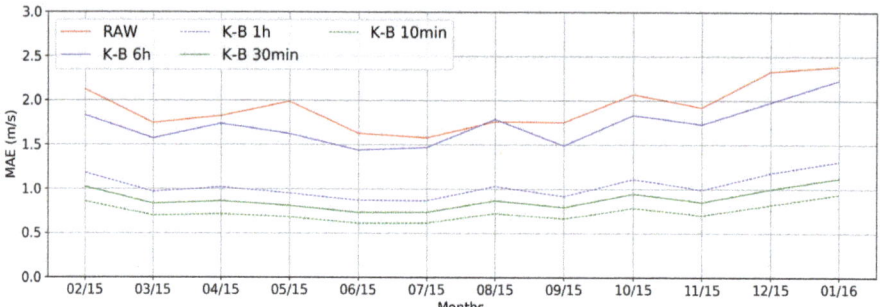

Figure 4. Monthly wind speed MAE for the wind turbines with K-B model (6 h (blue), 1 h (dotted blue), 30 min (green), and 10 min (dotted green)) and RAW (red).

RAW MAEs are relatively constant throughout the year, with values around 2 m/s in all months. Slightly larger errors are observed in the winter period, when mean wind speeds are higher, and lower biases occur in summer months, when mean wind speeds are also lower. K-B 6 h obtains the biggest improvements in the months when the RAW error is largest, such as December or February and minor improvements in the summertime. This tendency to better improve bigger RAW MAE is also reflected in previous comparisons such as for WT26 in Figure 2 or the south radar chart in Figure 3f. In K-B experiments for shorter lead times, the pattern shown in Figure 3 is repeated, with a significant error decrease relative to K-B 6 h results and very similar outcomes in all K-B cases and months. The K-B 1 h MAEs are all very close to 1 m/s, February registers the higher K-B 1 h MAE, with 1.19 m/s and June the lowest, with 0.87 m/s. This equality among monthly errors is maintained in K-B 30 m and 10 m with mean values all year long around 0.88 and 0.73 m/s respectively.

After investigating the intra-annual behavior and monthly error patterns of the experiments, we examine their accuracy on a finer timescale. Figure 5 displays the mean hourly wind speed MAE during the simulated year for all the wind turbines analyzed in each experiment.

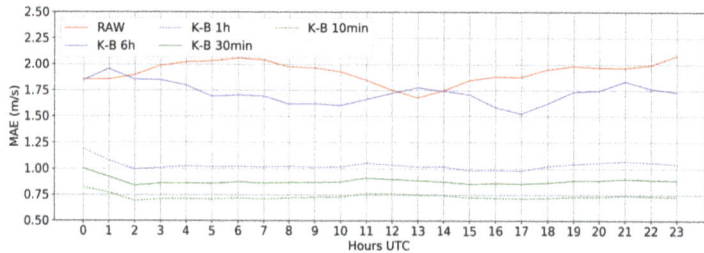

Figure 5. Wind speed MAE for all the turbines, hourly (moving average) for the entire 24 h simulation period with RAW outputs (red) and K-B model cases: K-B 6 h (blue), K-B 1 h (dotted blue), K-B 30 min (green) and K-B 10 min (dotted green).

RAW error results present a daily cycle, with values of the MAE around noon lower than during the night. These differences among the hourly WS MAEs are, however, not very significant, and the lower absolute errors at noon seem to be related with the better representation of turbulent fluxes during that part of the day [5]. In contrast, K-B 6 h errors do not follow a daily cycle, attaining values below 1.75 m/s mostly during daytime, but obtaining a worse result than RAW at 01 and 13 UTC. K-B 1 h, 30 min, and 10 min present a very flat error pattern all day long. There are practically no differences among hourly results in these experiments. The situation is very similar to that in the monthly error plot (Figure 4); shorter term K-B cases represent a significant improvement with respect to K-B 6 h and RAW, and there is homogeneity in the entire error series in K-B 1 h, 30 min, and 10 min.

The skill limit for the shorter term cases corresponds to a non-systematic part of the error, which is unavoidable for the filter

The general conclusion reached from the different time scale and wind regime comparisons presented above is that the K-B filter at short-term erases any source of error. It achieves lower errors in all conditions, regardless of the intensity and direction of the wind or the time of day or season of the year. This uniformity in the corrections toward the elimination of the ME (Table 1) from short-term K-B cases, highlights the utility of this optimization module for wind energy purposes.

To illustrate the day-to-day results of the K-B nowcasting tool regarding wind speed at hub height, we show next comparisons between RAW, K-B 1 h, and observations for the Center area (WT13 and WT16 mean values) during the entire months of May and December. These months have the lowest and highest K-B 1 h MAE in this area respectively.

Disregarding some occasional large errors, the original WRF output (RAW) generally yields a good performance during the months shown, achieving MAEs below 2 m/s in both cases and PBIAS [57] of −10.82% in May and −6.62% in December (−6.59% for the whole year). Even considering these low RAW errors, which are extensive to all the yearly series, the K-B 1 h results represent an important improvement nonetheless. This is clearly apparent in these December and May plots (Figure 6), when the procedure successfully corrects forecast error by a factor of around 50%. KB-1 h PBIAS indicators, 0.39% in May and 0.06% in December, show a virtual disappearance of the original RAW substestimation, (0.21% for the whole year). Focusing on the correlation coefficient (CC), we can see that RAW case obtains a CC = 0.90 with respect to the observation. This value shows a high relationship between the observation and the RAW case. However, K-B 1 h improves the original RAW result reaching a CC = 0.97, which indicates an extremely high correlation between K-B 1 h nowcasting and observation data series. This indicator, combined with the MAE value for this same period (0.86 m/s), reflects an accurate performance of the K-B 1 h case. Total mean values for the wind farm along the whole year displays similar behavior as the example shown in the figure above. The mean CC for all wind turbines in WS RAW case is 0.91, and WS K-B 1 h obtains a CC = 0.97.

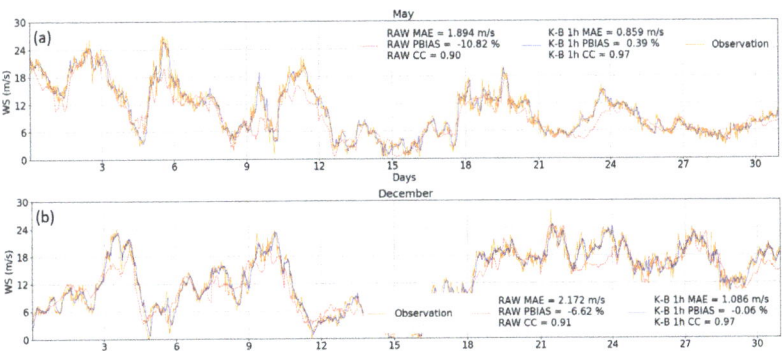

Figure 6. (a) Wind speed at hub height in the center area with observations (orange), RAW (dotted red), and K-B 1 h (blue) in May. (b) Same as (a) in December. Wind speed MAE, PBIAS and Correlation Coefficient (CC) for each series is presented in the legend for both figures.

The nowcasting tool is able to produce better results at different time ranges regardless of the origin of the error, eliminating the most substantial deviations in any situation. Overall, the K-B filter exhibits a useful short-term operational forecasting performance, offering a stable improvement of the original WRF outputs during the whole year and for all the wind turbines.

3.2. Wind Direction Nowcasting

Following the discussion on the capabilities of the K-B filter regarding WS (wind speed) nowcasting, we review the statistical tool predicting wind direction (WD) following a similar structure to Section 3.1. To perform the WD nowcasting post-process we obtain the zonal and meridional wind components (U and V) from WS and WD observations, and we use them to correct, with the K-B filter, RAW U and V variables directly extracted from the model output. With the corrected U and V, we recalculate the post-processed WD. Using the same process as in the WS analysis, we test the WD nowcasting for different short-term forecast time periods (Table 2).

Figure 7 displays the annual WD MAE and standard deviation for each turbine and experiment.

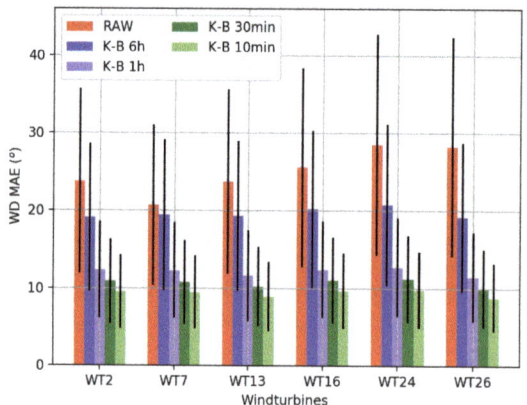

Figure 7. Barchart with annual WD MAE for all experiments and wind turbines analyzed. For each case, the standard deviation is represented by a black line (σ) on top of each bar.

The bar chart depicts in red the results of WD MAE for each RAW wind turbine. All of them are above 20°; WT7 with the lowest value, 20.69° and WT24 with the highest, 28.51°. The mean value of the MAEs for all the wind turbines in the RAW case is 25.10°. Comparing K-B 6 h with the unfiltered WRF results, we can see an important amelioration in most of the wind turbines; all of them are now below or practically at 20° MAE, reducing the mean total K-B 6 h MAE to 19.67°. This value translates into a 22% improvement with respect to RAW results, a correction impact that doubles that obtained in wind speed nowcasting for this same comparison (WS RAW vs. K-B 6 h). This difference can be attributed to the origin of the observational data used. In WS nowcasting we correct one output data series, with another single data series (observed wind speed). However, in WD nowcasting we use WS and WD data from sonic anemometers to obtain the U and V wind components and calculate the post-processed WD. We are thus introducing more sources of error in this last step that the K-B filter seems nevertheless capable of offsetting.

In the shorter lead times of wind direction nowcasting (K-B 1 h, 30 min, and 10 min) the behavior presents a similarity to that of wind speed nowcasting (Figure 2). All the turbines attain analogous error values in each experiment, with a total mean WD MAE of 12.16°, 10.69° and 9.37° for K-B 1 h, K-B 30 min and K-B 10 min respectively. Aside from these low errors, there is also a corresponding reduction of the standard deviation. The K-B filter at short-term nowcasting increases prediction accuracy significantly and eliminates more substantial punctual errors from the original WRF forecasting, which is quite important for the use of these kinds of combined nowcasting systems in wind farm applications. The reason lies in the fact that during daily operations of these installations, wind turbines are continually being orientated depending on wind direction, and potential errors in this manoeuvre lead to machine overstress and production decrease.

As in Section 3.1 for wind speed, the monthly errors for wind direction are shown in Figure 8a, and on a finer timescale, Figure 8b displays the mean hourly wind direction MAE, averaged for all the turbines in the farm. In both cases, RAW WRF outputs are compared with the results of the K-B experiments.

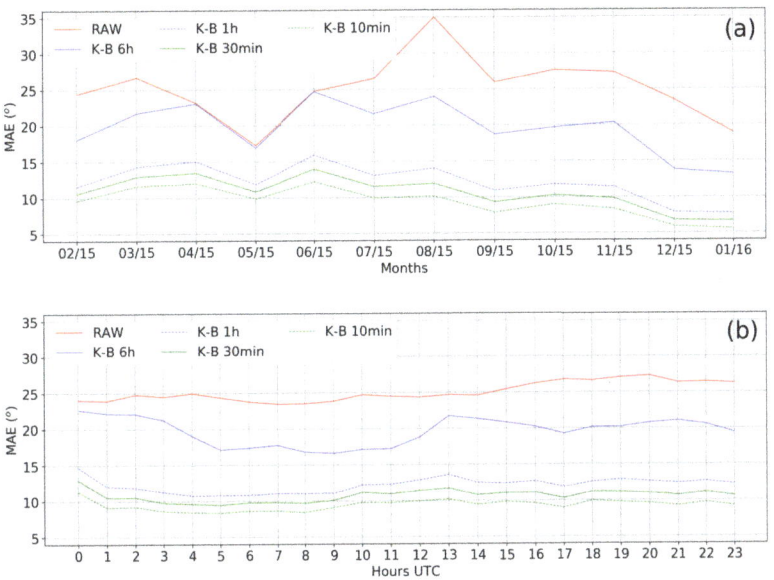

Figure 8. (a) Monthly wind direction MAE for the wind farm with K-B model experiments and RAW. (b) Wind direction MAE, hourly (moving average) for the entire 24 h simulation period. In both cases the comparison is between RAW outputs (red) and K-B model cases: K-B 6 h (blue), K-B 1 h (dotted blue), K-B 30 min (green) and K-B 10 min (dotted green).

Monthly results in Figure 8a provide insights on the origin of the 22% improvement in forecast skill of K-B 6 h with respect to RAW, commented in Figure 7. From 15 February to 15 March and from 15 July to 16 January, the effect of the K-B 6 h is noticeable, with a decrease in errors of around 7.5° during those periods. Nevertheless, the scenario is entirely different in spring, when K-B 6 h practically does not refine the original RAW result. This lack of improvement can be partially explained by the fact that, during spring months, the original WRF outputs register the lowest U and V errors of all the year, and, similarly to previous results, the K-B filter has more difficulties correcting RAW forecast results with low errors.

K-B 1 h, 30 min, and 10 min follow the general tendency of the rest of the comparisons, presenting a substantial error decrease throughout the year. K-B 30 min skill score is always below K-B 1 h and the same for K-B 10 min with respect to K-B 30 min, with all three time series showing a parallel behavior throughout the whole period. As opposed to the case of wind speed (Figure 4), winter months, particularly December and January, have the lowest errors of the series in all wind direction K-B results. For example, K-B 30 min error is around 6.6° during these winter months, but over 12° in springtime. The lower WD MAE in winter months, on the one hand, seems to be related with the higher mean wind speed registered during that part of the year (Table 3). This means that there are fewer periods of weak winds, which are associated with increased variability in wind direction, presenting, therefore, more difficulties for the K-B filter to handle. On the other hand, these good results during that period can also be related to the clear unimodal wind regime which affects the region during that period.

Changing the timescale of the analysis, Figure 8b shows the evolution of the WD MAE throughout the day. The RAW series (red line) does not present significant hourly changes; all the values are close to 25° with a slight increase in the late hours of the day. K-B 6 h differs from this homogeneity, displaying a slight daily cycle with more accurate predictions from 04 to 12 UTC, with a minimum MAE = 16.64° at 09 UTC increasing to a maximum of 21.76° at 13 UTC. Same as in the daily analysis of wind speed (Figure 5), in the shortest nowcasting periods (1 h, 30 min, and 10 min), the variability is practically eliminated. The K-B filter decreases errors and smooths out the series, with rather constant MAEs around 10° throughout the day. The comparisons at different time scales among all the cases confirm that the K-B filter yields a valuable improvement in wind direction prediction. The shortest term nowcasting cases, starting from K-B 1 h, lead to a significant correction of WRF outputs independently of the meteorological situation of the moment.

To conclude this section, in Figure 9 we exemplify the capabilities of wind direction nowcasting with K-B filter. Specifically, a two-day WD MAE comparison is displayed for WT13 between RAW (red) and K-B 1 h (blue). The arrows inside the circles show the instantaneous observed wind direction every four hours (black) and its respective RAW (red) and K-B 1 h (blue) prediction.

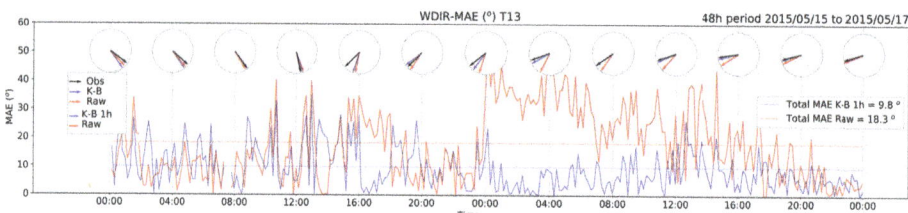

Figure 9. Wind direction MAE with K-B 1 h filter and RAW in W13 during a 48 h period. Wind direction arrows with observations, K-B 1 h, and RAW are displayed on top every four hours.

The two-day results presented, from 15 May 2015 to 17 May 2015, show different forecast skill patterns. In the first day, RAW and K-B 1 h error series are close to each other in their first 12 h. In the next hours, RAW maintains a more constant higher error (around 25°) for 3–4 h, which K-B 1 h drastically corrects. During the second day, this difference between the time series increases in magnitude; RAW predictions go over MAE = 30° for 12 h, reaching up to 40° in different moments. Throughout this entire time period, the K-B correction sharply eliminates these big MAEs to values below 10°.

Given these results, we can affirm that K-B 1 h provides a good skill improvement in all the situations where the original RAW forecast presents large errors. As in the case of wind speed nowcasting, the correction of big WRF wind direction biases is crucial for wind farm operation applications.

3.3. Application of the K-B Filter for Wind Power Forecasting

In this last section, we test the capabilities of K-B 1 h wind direction nowcasting tool on a real scenario of Coruxeiras wind farm in the studied year, with special attention to critical issues in power prediction, such as wind ramps. For this purpose, we focus on the results corresponding to the wind turbine power ramp, which is the ascending part of the wind power curve before nominal power (Figure 10). Wind speeds within this curve (in this case from 4 to 13 m/s) have the most significant effect in the forecast skill of any nowcasting tool in the wind energy field. In Figure 10, we present the relation between the wind power curve of the wind farm's turbines (ECOTECNIA74 with 1.670 MW of nominal power [58]) and the WD MAE associated with each wind speed bin, both for RAW and K-B 1 h cases. Table 5 displays the WD MAE in power ramp wind ranges for all the experiments and for each area of the wind farm. It also compares these results with persistence at 10 min, 30 min, and 1 h.

The persistence forecast is the assumption that the next timestep value in a prediction is going to be the same as the last measured value [59].

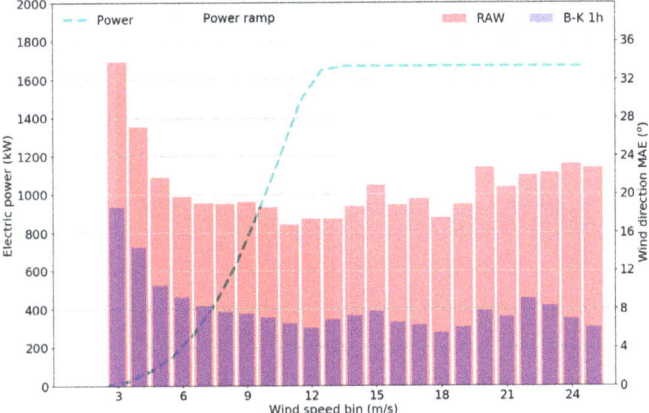

Figure 10. ECOTECNIA 74 wind power curve (density = 1.112 kg/m^3, height = 975 m, temperature = 9 °C), cyan line. Each bar of the figure represents the wind direction MAE associated with wind speed bins with a bandwidth of 1 m/s both for RAW (red) and K-B 1 h (blue).

Table 5. WD MAE for all the experiments and different persistence periods in power ramp winds.

	Power Ramp Winds (49.7%) Wind Dir. MAE (°)			
	North	Center	South	Total
Per. 10 min	6.54	6.32	6.96	6.61
Per. 30 min	8.49	8.31	8.84	8.55
Per. 1 h	10.07	9.92	10.47	10.15
K-B 10 min	6.65	6.52	6.91	6.70
K-B 30 min	7.45	7.35	7.71	7.51
K-B 1 h	8.33	8.26	8.56	8.38
K-B 6 h	13.59	13.71	14.10	13.80
RAW	16.07	19.54	22.81	19.47

The error bar chart in Figure 10 clearly illustrates the significant improvement of K-B 1 h over RAW results in power ramp winds. At the beginning of the ramp, K-B 1 h obtains a 50% of improvement over RAW. This percentage increases with wind speed, reaching practically a 75% error reduction in the last part of the ramp (12–13 m/s). This is a desirable characteristic of the nowcasting tool because it results in more accurate direction predictions in wind speed situations where energy production values are higher.

Table 5 shows no significant differences among farm areas in power ramp wind direction MAEs. In general, all the experiments achieve better results for power ramp winds than in general for all cases (Figure 7). This is mainly because power ramp situations exclude lower wind occurrences, associated with more variability. Moreover, K-B 1 h and K-B 30 min lead to better results than 1 h and 30 min persistence, respectively. K-B 1 h also reaches lower WD MAE than 30 min persistence.

A K-B nowcasting tool can be useful in the daily management of a wind farm. For instance, an operational K-B 1 h can back up current orientation methods, which are based on SCADA systems [36]. There are also new wind farm control techniques such as yaw-misalignment [60,61], in which upstream wind turbines redirect wakes to increase the production of downstream turbines, that could also benefit from the WRF K-B filter and its capacity to nowcast wind direction with MAEs below 10° with a 1 h time horizon.

4. Summary and Conclusions

In the present work, we studied the ability of a new Kalman-Bayesian postprocessing technique to improve the wind speed and wind direction forecasts derived from high-resolution simulations for an operational wind farm over complex terrain. For the needs of the study, a period of one year is simulated utilizing the WRF atmospheric model at a 333 m horizontal resolution, and its outputs post-processed with the K-B filter. We validate the method for different short-term forecast periods, from 6 h to 10 min, with in-situ real data from meteorological stations on the hub of wind turbines.

The results show that the K-B filter in the very short-term (from 1 h to 10 min) is capable of improving the initial wind speed forecasts of the atmospheric model significantly, although the latter are already considerably good (RAW annual WS MAE = 1.87 m/s). The K-B filter achieves a decrease in the WS MAE to 1 m/s for 1 h horizon nowcasting and to 0.72 m/s for the 10 min case. The ME practically disappears when using the K-B 6 h. Analyzing all the experiments in different period ranges (months, hours, instants), we demonstrated that for shorter-term cases, K-B postprocessing eliminates any source of error, improving the prediction in all conditions regardless of the intensity and direction of the wind or the moment of the day and the season of the year.

The effectiveness of the hybrid filter in correcting forecasted wind directions was also a key point of the present study. This was achieved by applying the hybrid post process system to the two-dimensional field of horizontal wind speed components (U and V). The results were quite satisfactory as the K-B 6 h presents a 22% of amelioration with respect to RAW forecasts. Shorter-term cases decrease their WD MAEs even below 10°, with an important degree of efficiency in any meteorological situation throughout the year.

It should be noted at this point that several methodologies have been proposed in the literature for the reduction of atmospheric or wave model errors (see for example Landberg and Watson [62] and Joensen et al. [63] for MOS applications, Galanis et al. [23,24]; Kalman [29]; Kalman and Bucy [28]; Kalnay [27]; Louka et al. [34] and Pelland et al. [22] for Kalman filter optimization tools. The proposed methodology in this work that combines Kalman-Bayesin filter applied to single or multiple dimension arrays provides additional advantages and better results in the improvement of wind speed and direction forecasts.

Emphasis was also given to the behavior of wind direction K-B 1 h in power ramp wind cases. The mean results for these are better than for general cases due to the filtering of low winds. K-B 1 h WD improves the persistence forecast for 1 h and 30 min.

In general, the combination of high-resolution WRF simulations and the K-B hybrid filter has shown effectivity and reliability obtaining accurate short-term wind speed and direction predictions, which are of critical importance for an operational wind farm.

Author Contributions: C.O.-C., P.P., G.G. and G.M.-M. designed the research. C.O.-C. and M.A.P. performed the simulations with the WRF model in CESGA computing facilities. P.P. and G.G. implemented the Kalman-Filter tool in the RAW results. C.O.-C., P.P. and M.A.P. analyzed the data. C.O.-C., P.P., G.G. and G.M.-M. drafted the paper. C.O.-C. and M.A.P. designed all the figures. All authors contributed to the interpretation of the data and revision of the paper.

Funding: This research received no external funding.

Acknowledgments: We would like to thank Norvento (http://www.norvento.com/en) for the valuable observational data provided and for their continuing support, both of them were fundamental to accomplishing this project. The model forecast simulations and development of the data analysis were performed at the Centro de Supercomputacion de Galicia (CESGA) (http://www.cesga.es/). Their computer facilities and support have been indispensable during the study.

Conflicts of Interest: The authors declare no conflict of interest.

References

1. Global Wind Energy Council. *Global Wind Report*; Technical Report; Global Wind Energy Council: Brussels, Belgium, 2017.

2. Wind Europe. *Wind in Power 2017. Annual Combined Onshore and Offshore Wind Energy Statistics*; Technical Report; Wind Europe: Brussels, Belgium, 2018.
3. Fried, L.; Shukla, S.; Sawyer, S. *Growth Trends and the Future of Wind Energy*; Elsevier Inc.: Cambridge, MA, USA, 2017; pp. 559–586.
4. Zhao, J.; Guo, Y.; Xiao, X.; Wang, J.; Chi, D.; Guo, Z. Multi-step wind speed and power forecasts based on a WRF simulation and an optimized association method. *Appl. Energy* **2017**, *197*, 183–202. [CrossRef]
5. Prósper, M.A.; Otero-Casal, C.; Fernández, F.C.; Miguez-Macho, G. Wind power forecasting for a real onshore wind farm on complex terrain using WRF high resolution simulations. *Renew. Energy* **2019**, *135*, 674–686. [CrossRef]
6. Pokharel, B.; Geerts, B.; Chu, X.; Bergmaier, P. Profiling radar observations and numerical simulations of a downslopewind storm and rotor on the lee of the Medicine Bow mountains in Wyoming. *Atmosphere* **2017**, *8*, 39. [CrossRef]
7. Wang, G.M.; Liu, D.F.; Xu, Y.P.; Meng, T.; Zhu, F. PET/CT imaging in diagnosing lymph node metastasis of esophageal carcinoma and its comparison with pathological findings. *Eur. Rev. Med. Pharmacol. Sci.* **2016**, *20*, 1495–1500. [CrossRef] [PubMed]
8. Mughal, M.O.; Lynch, M.; Yu, F.; Sutton, J. Forecasting and verification of winds in an East African complex terrain using coupled mesoscale—And micro-scale models. *J. Wind Eng. Ind. Aerodyn.* **2018**, *176*, 13–20. [CrossRef]
9. Archer, C.L.; Simão, H.P.; Kempton, W.; Powell, W.B.; Dvorak, M.J. The challenge of integrating offshore wind power in the U.S. electric grid. Part I: Wind forecast error. *Renew. Energy* **2017**, *103*, 346–360. [CrossRef]
10. Vanderwende, B.; Lundquist, J.K. Could Crop Height Affect the Wind Resource at Agriculturally Productive Wind Farm Sites? *Bound.-Layer Meteorol.* **2016**, *158*, 409–428. [CrossRef]
11. Division, M.; Sciences, O.; Renewable, N. Mesoscale Influences of Wind Farms throughout a Diurnal Cycle *Mon. Weather Rev.* **2013**, *141*, 2173–2198. [CrossRef]
12. Jiménez, P.A.; Navarro, J.; Palomares, A.M.; Dudhia, J. Mesoscale modeling of offshore wind turbine wakes at the wind farm resolving scale: A composite-based analysis with the Weather Research and Forecasting model over Horns Rev. *Wind Energy* **2015**, *18*, 559–566. [CrossRef]
13. Muñoz-Esparza, D.; Lundquist, J.K.; Sauer, J.A.; Kosović, B.; Linn, R.R. Coupled mesoscale-LES modeling of a diurnal cycle during the CWEX-13 field campaign: From weather to boundary-layer eddies. *J. Adv. Model. Earth Syst.* **2017**, *9*, 1572–1594. [CrossRef]
14. Rai, R.K.; Berg, L.K.; Kosović, B.; Mirocha, J.D.; Pekour, M.S.; Shaw, W.J. Comparison of Measured and Numerically Simulated Turbulence Statistics in a Convective Boundary Layer Over Complex Terrain. *Bound.-Layer Meteorol.* **2017**, *163*, 69–89. [CrossRef]
15. Muñoz-Esparza, D.; Kosović, B.; Mirocha, J.; van Beeck, J. Bridging the Transition from Mesoscale to Microscale Turbulence in Numerical Weather Prediction Models. *Bound.-Layer Meteorol.* **2014**, *153*, 409–440. [CrossRef]
16. Huang, M.; Wang, Y.; Lou, W.; Cao, S. Multi-scale simulation of time-varying wind fields for Hangzhou Jiubao Bridge during Typhoon Chan-hom. *J. Wind Eng. Ind. Aerodyn.* **2018**, *179*, 419–437. [CrossRef]
17. Kariniotakis, G.N.; Pinson, P. Evaluation of the MORE-CARE wind power prediction platform. Performance of the fuzzy logic based models. In Proceedings of the EWEC 2003—European Wind Energy Conference, Madrid, Spain, 16–19 June 2003.
18. Kariniotakis, G.; Martí, I.; Casas, D.; Pinson, P.; Nielsen, T.S.; Madsen, H.; Giebel, G.; Usaola, J.; Sanchez, I. What performance can be expected by short-term wind power prediction models depending on site characteristics? In Proceedings of the EWC 2004 Conference, Tokyo, Japan, 2–4 August 2004; pp. 22–25.
19. Vanem, E. Long-term time-dependent stochastic modelling of extreme waves. *Stoch. Environ. Res. Risk Assess.* **2011**, *25*, 185–209. [CrossRef]
20. Giebel, G. On the Benefits of Distributed Generation of Wind Energy in Europe. Ph.D. Thesis, Carl von Ossietzky Universitat Oldenburg, Oldenburg, Germany, August 2001.
21. Resconi, G. Geometry of risk analysis (morphogenetic system). *Stoch. Environ. Res. Risk Assess.* **2009**, *23*, 425–432. [CrossRef]
22. Pelland, S.; Galanis, G.; Kallos, G. Solar and photovoltaic forecasting through post-processing of the global environmental multiscale numerical weather prediction model. *Prog. Photovolt. Res. Appl.* **2011**, *21*, 284–296. [CrossRef]

23. Galanis, G.; Louka, P.; Katsafados, P.; Pytharoulis, I.; Kallos, G. Applications of Kalman filters based on non-linear functions to numerical weather predictions. *Ann. Geophys.* **2006**, *24*, 2451–2460. [CrossRef]
24. Galanis, G.; Emmanouil, G.; Chu, P.C.; Kallos, G. A new methodology for the extension of the impact of data assimilation on ocean wave prediction. *Ocean Dyn.* **2009**, *59*, 523–535. [CrossRef]
25. Crochet, P. Adaptive Kalman filtering of 2-metre temperature and 10-metre wind-speed forecasts in Iceland. *Meteorol. Appl.* **2004**, *11*, 173–187. [CrossRef]
26. Galanis, G.; Anadranistakis, M. A one-dimensional Kalman filter for the correction of near surface temperature forecasts. *Meteorol. Appl.* **2002**, *9*, 437–441. [CrossRef]
27. Kalnay, E. *Atmospheric Modeling, Data Assimilation, and Predictability*; Cambridge University Press: Cambridge, UK, 2003.
28. Kalman, R.E.; Bucy, R.S. New Results in Linear Filtering and Prediction Theory. *J. Basic Eng.* **1961**, *83*, 95. [CrossRef]
29. Kalman, R.E. A New Approach to Linear Filtering and Prediction Problems. *J. Basic Eng.* **1960**, *82*, 35. [CrossRef]
30. Stathopoulos, C.; Kaperoni, A.; Galanis, G.; Kallos, G. Wind power prediction based on numerical and statistical models. *J. Wind Eng. Ind. Aerodyn.* **2013**, *112*, 25–38. [CrossRef]
31. Hua, S.; Wang, S.; Jin, S.; Feng, S.; Wang, B. Wind speed optimisation method of numerical prediction for wind farm based on Kalman filter method. *J. Eng.* **2017**, *2017*, 1146–1149. [CrossRef]
32. Che, Y.; Peng, X.; Monache, L.D.; Kawaguchi, T.; Xiao, F. A wind power forecasting system based on the weather research and forecasting model and Kalman filtering over a wind-farm in Japan. *J. Renew. Sustain. Energy* **2016**, *8*, 013302. [CrossRef]
33. Che, Y.; Xiao, F. An integrated wind-forecast system based on the weather research and forecasting model, Kalman filter, and data assimilation with nacelle-wind observation. *J. Renew. Sustain. Energy* **2016**, *8*, 053308. [CrossRef]
34. Louka, P.; Galanis, G.; Siebert, N.; Kariniotakis, G.; Katsafados, P.; Pytharoulis, I.; Kallos, G. Improvements in wind speed forecasts for wind power prediction purposes using Kalman filtering. *J. Wind Eng. Ind. Aerodyn.* **2008**, *96*, 2348–2362. [CrossRef]
35. Galanis, G.; Papageorgiou, E.; Liakatas, A. A hybrid Bayesian Kalman filter and applications to numerical wind speed modeling. *J. Wind Eng. Ind. Aerodyn.* **2017**, *167*, 1–22. [CrossRef]
36. Zaher, A.; McArthur, S.D.; Infield, D.G.; Patel, Y. Online wind turbine fault detection through automated SCADA data analysis. *Wind Energy* **2009**, *12*, 574–593. [CrossRef]
37. Drew, D.R.; Cannon, D.J.; Barlow, J.F.; Coker, P.J.; Frame, T.H. The importance of forecasting regional wind power ramping: A case study for the UK. *Renew. Energy* **2017**, *114*, 1201–1208. [CrossRef]
38. Lorente-Plazas, R.; Montávez, J.P.; Jimenez, P.A.; Jerez, S.; Gómez-Navarro, J.J.; García-Valero, J.A.; Jimenez-Guerrero, P. Characterization of surface winds over the Iberian Peninsula. *Int. J. Climatol.* **2015**, *35*, 1007–1026. [CrossRef]
39. Santos, J.A.; Rochinha, C.; Liberato, M.L.; Reyers, M.; Pinto, J.G. Projected changes in wind energy potentials over Iberia. *Renew. Energy* **2015**, *75*, 68–80. [CrossRef]
40. Skamarock, W.; Klemp, J.; Dudhi, J.; Gill, D.; Barker, D.; Duda, M.; Huang, X.Y.; Wang, W.; Powers, J. *A Description of the Advanced Research WRF Version 3*; Technical Report; NCAR: Boulder, CO, USA, 2008.
41. Wang, W.; Bruyère, C.; Duda, M.; Dudhia, J.; Gill, D.; Kavulich, M.; Keene, K.; Lin, H.C.; Michalakes, J.; Rizvi, S.; Zhang, X.; Berner, J.; Smith, K. ARW Version 3 Modeling System User's Guide. *J. Palest. Stud.* **2007**, *37*, 204–205. [CrossRef]
42. Shin, H.H.; Hong, S.Y.; Dudhia, J. Impacts of the Lowest Model Level Height on the Performance of Planetary Boundary Layer Parameterizations. *Mon. Weather Rev.* **2012**, *140*, 664–682. [CrossRef]
43. Slater, J.A.; Heady, B.; Kroenung, G.; Curtis, W.; Haase, J.; Hoegemann, D.; Shockley, C.; Tracy, K. *Evaluation of the New ASTER Global Digital Elevation Model*; National Geospatial-Intelligence Agency: Reston, VA, USA, 2009; pp. 335–349.
44. European Environment Agency. *CLC2006 Technical Guidelines*; European Environment Agency: Copenhagen, Denmark, 2007; Volume 2010, p. 70. [CrossRef]
45. Hong, S.; Lim, J. The WRF single-moment 6-class microphysics scheme (WSM6). *Asia-Pac. J. Atmos. Sci.* **2006**, *42*, 129–151.

46. Kain, J.S. The Kain–Fritsch Convective Parameterization: An Update. *J. Appl. Meteorol.* **2004**, *43*, 170–181. [CrossRef]
47. Mlawer, E.J.; Taubman, S.J.; Brown, P.D.; Iacono, M.J.; Clough, S.A. Radiative transfer for inhomogeneous atmospheres: RRTM, a validated correlated-k model for the longwave. *J. Geophys. Res. Atmos.* **1997**, *102*, 16663–16682. [CrossRef]
48. Dudhia, J. Numerical Study of Convection Observed during the Winter Monsoon Experiment Using a Mesoscale Two-Dimensional Model. *J. Atmos. Sci.* **1989**, *46*, 3077–3107. [CrossRef]
49. Nakanishi, M.; Niino, H. An improved Mellor-Yamada Level-3 model: Its numerical stability and application to a regional prediction of advection fog. *Bound.-Layer Meteorol.* **2006**, *119*, 397–407. [CrossRef]
50. Beljaars, A.C. The parametrization of surface fluxes in large-scale models under free convection. *Q. J. R. Meteorol. Soc.* **1995**, *121*, 255–270. [CrossRef]
51. Tewari, M.; Chen, F.; Wang, W.; Dudhia, J.; Lemone, M.A.; Mitchell, K.; Cuenca, R.H.; Springs, C.; Force, A.; Agency, W. Implementation and Verification of The Unified NOAH Land Surface Model In The WRF Model. In Proceedings of the 20th Conference on Weather Analysis and Forecasting/16th Conference on Numerical Weather Prediction, Seattle, WA, USA, 12–16 January 2004.
52. Fitch, A.C.; Olson, J.B.; Lundquist, J.K.; Dudhia, J.; Gupta, A.K.; Michalakes, J.; Barstad, I. Local and Mesoscale Impacts of Wind Farms as Parameterized in a Mesoscale NWP Model. *Mon. Weather Rev.* **2012**, *140*, 3017–3038. [CrossRef]
53. Patlakas, P.; Drakaki, E.; Galanis, G.; Spyrou, C.; Kallos, G. Wind gust estimation by combining a numerical weather prediction model and statistical post-processing. *Energy Procedia* **2017**, *125*, 190–198. [CrossRef]
54. Box, G., T.G. *Bayesian Inference in Statistical Analysis*; Wiley: Hoboken, NJ, USA, 1992. [CrossRef]
55. Bernardo, J.; Smith, A. *Bayesian Theory*; Wiley: Hoboken, NJ, USA, 2000.
56. Kuikka, I. *Wind Nowcasting: Optimizing Runway in Use*; Technical Report; Helsinki University of Technology Systems Analysis Laboratory: Espoo, Finland, 2009.
57. Gupta, H.V.; Sorooshian, S.; Yapo, P.O. Status of automatic calibration for hydrologic models: Comparison with multilevel expert calibration. *J. Hydrol. Eng.* **1999**, *4*, 135–143. [CrossRef]
58. Ecotècnia. *Technical Report 74 1.6*; Technical Report; ECOTÈCNIA: Barcelona, Spain, 2007.
59. Utsumi, H.; Misaka, T.; Moteki, W. Prediction Model of Internal Relative Humidity During Self-desiccation in Hardened Cement Pastes. *Concr. Res. Technol.* **2015**, *26*, 11–19. [CrossRef]
60. Van Dijk, M.T.; Van Wingerden, J.W.; Ashuri, T.; Li, Y.; Rotea, M.A. Yaw-Misalignment and its Impact on Wind Turbine Loads and Wind Farm Power Output. *J. Phys. Conf. Ser.* **2016**, *753*, 062013. [CrossRef]
61. Churchfield, M.; Fleming, P. Wind Turbine Wake-Redirection Control at the Fishermen's Atlantic City Windfarm. *Offshore Technol.* **2015**. [CrossRef]
62. Landberg, L.; Watson, S.J. Short-term prediction of local wind conditions. *Bound.-Layer Meteorol.* **1994**, *70*, 171–195. [CrossRef]
63. Joensen, A.; Giebel, G.; Landberg, L.; Madsen, H.; Neilsen, H. Model output statistics applied to wind power prediction. In Proceedings of the EWEC-Conference, Nice, France, 1–5 March 1999; pp. 1177–1180.

© 2019 by the authors. Licensee MDPI, Basel, Switzerland. This article is an open access article distributed under the terms and conditions of the Creative Commons Attribution (CC BY) license (http://creativecommons.org/licenses/by/4.0/).

Article

Evaluation of Photovoltaic Power Generation by Using Deep Learning in Solar Panels Installed in Buildings

Chih-Chiang Wei

Department of Marine Environmental Informatics & Center of Excellence for Ocean Engineering, National Taiwan Ocean University, No.2, Beining Rd., Jhongjheng District, Keelung City 20224, Taiwan; ccwei@ntou.edu.tw; Tel.: +886-2-24622192

Received: 14 August 2019; Accepted: 15 September 2019; Published: 17 September 2019

Abstract: Southern Taiwan has excellent solar energy resources that remain largely unused. This study incorporated a measure that aids in providing simple and effective power generation efficiency assessments of solar panel brands in the planning stage of installing these panels on roofs. The proposed methodology can be applied to evaluate photovoltaic (PV) power generation panels installed on building rooftops in Southern Taiwan. In the first phase, this study selected panels of the BP3 series, including BP350, BP365, BP380, and BP3125, to assess their PV output efficiency. BP Solar is a manufacturer and installer of photovoltaic solar cells. This study first derived ideal PV power generation and then determined the suitable tilt angle for the PV panels leading to direct sunlight that could be acquired to increase power output by panels installed on building rooftops. The potential annual power outputs for these solar panels were calculated. Climate data of 2016 were used to estimate the annual solar power output of the BP3 series per unit area. The results indicated that BP380 was the most efficient model for power generation (183.5 KWh/m^2-y), followed by BP3125 (182.2 KWh/m^2-y); by contrast, BP350 was the least efficient (164.2 KWh/m^2-y). In the second phase, to simulate meteorological uncertainty during hourly PV power generation, a surface solar radiation prediction model was developed. This study used a deep learning–based deep neural network (DNN) for predicting hourly irradiation. The simulation results of the DNN were compared with those of a backpropagation neural network (BPN) and a linear regression (LR) model. In the final phase, the panel of module BP3125 was used as an example and demonstrated the hourly PV power output prediction at different lead times on a solar panel. The results demonstrated that the proposed method is useful for evaluating the power generation efficiency of the solar panels.

Keywords: solar irradiation; photovoltaic solar energy; deep learning; prediction

1. Introduction

Solar photovoltaic (PV) energy systems generate electricity without causing pollution. Moreover, grid-connected PV cells can easily be installed on residential building roofs and commercial building walls [1]. In Taiwan, the solar energy industry has become popular; various PV power generation systems can be installed in buildings for improving energy-use efficiency. Because Taiwan is located in a subtropical region, it receives abundant sunlight and is suitable for developing solar energy. Solar cells, also known as PV cells, directly convert solar energy to electricity. Solar energy has become an alternative energy used in Taiwan because it does not cause environmental pollution or noise.

Solar cells convert sunlight to electricity through the PV effect, in which an appropriate energy-level design is employed to effectively absorb sunlight and convert it to electric voltage and currents; this conversion process is known as photovoltaics. Numerous semiconductor materials are available for solar power generation. Silicon, a prominent raw material for solar cells, is commercially divided into monocrystalline, polycrystalline, and noncrystalline silicon. Monocrystalline and polycrystalline silicon materials exhibit the same power generation mechanism despite differences in their crystal structures. The cell temperature mainly depends on the irradiance intensity and ambient temperature [2]. Semiconductors are temperature-sensitive; after the cell temperature exceeds 25 °C, each increase of 1 °C reduces the overall efficiency of monocrystalline, polycrystalline, and noncrystalline silicon cells by 0.71%, 0.5–0.66%, and 0.2–0.3%, respectively [3]. Currently, polycrystalline silicon solar cells are dominating the solar cell market, representing 90% of the market sales, with a 20–30% market growth. In 2017, silicon-based solar cells could generate PV power of up to 55 GW; this value is estimated to reach 100 GW by 2020 [4].

Solar power generation is a form of environmentally friendly power generation method, in which no greenhouse gases, such as carbon dioxide, are generated. However, solar radiation is absorbed, reflected, or refracted by clouds of varying thicknesses after entering the atmosphere, which cause an inconsistency in solar radiant energy sources. When the density of energy collected by a set of solar panels is low, several additional solar panels must be installed, which increases their investment costs. Approximately 45% of the cost of a silicon cell solar module is determined by the cost of the silicon wafer. Thus, efforts are being made to use less silicon in the manufacture of solar cells [5]. Therefore, appropriate solar panels must be selected while developing a rooftop PV system in Taiwan. Previous studies have evaluated PV power generation panels of various geometries for different buildings [6–8]. For example, Jeong et al. [9] used amorphous silicon PV panels to develop prototype models of blinds with smart photovoltaic systems. Mahmud et al. [10] presented an environmental lifecycle assessment of a solar PV system by using single crystalline Si solar cells and a solar thermal system that used evacuated glass tube collectors. Kouhestani et al. [11] used a multi-criteria approach based on geographic information systems and light detection and ranging (LiDAR) to estimate rooftop PV electricity potential of buildings in an urban environment. Additionally, for evaluating the solar panel suppliers, several studies have used multi-criteria decision making (MCDM) approaches in various fields of science and engineering [12,13]. For example, Wang and Tsai [14] presented a fuzzy MCDM approach using a fuzzy analytical hierarchy process model and data envelopment analysis for selecting solar panel suppliers for a photovoltaic system design in Taiwan.

Seasons, daytime length, the Earth's revolution and rotation, and climate changes affect the reception of solar energy [15]. The Earth's revolution and rotation are regular and can be accurately calculated using mathematical models, but the atmospheric conditions (e.g., clouds, temperature, and wind velocity) of Taiwan, which has an island climate, change rapidly and are difficult to forecast. Therefore, solar radiation must be accurately predicted in advance to accurately evaluate the total power generated through rooftop PV systems and their overall efficiency. A surface solar radiation prediction model should be established for this purpose.

The aforementioned problems indicate that efficient and prompt evaluation of the power generation efficiency of specific solar panel brands is essential in planning the installation of rooftop PV systems in Taiwan. Therefore, this study has three objectives: (1) evaluation of the potential annual power outputs of selected solar panels, (2) prediction of future hour-based solar radiation levels, and (3) assessment of the annual power for a specific solar panel when the forecasting horizon increases. The study site was Tainan (Figure 1), where the average annual solar irradiance was 1.65 MWh/m^2-y according to the statistical data from 2010 to 2016. Tainan, which has a stable climate and frequent sunshine, is suitable for solar power generation in all seasons.

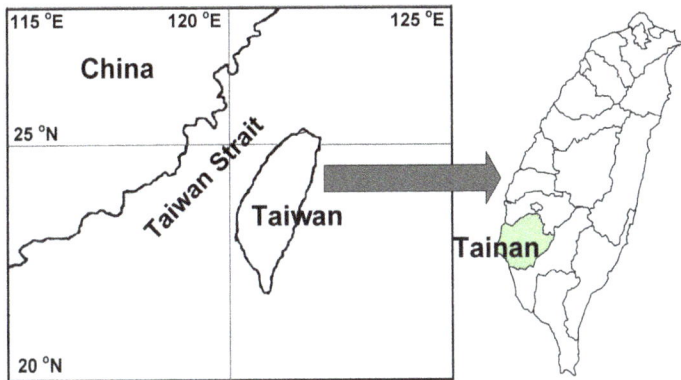

Figure 1. Map of Tainan, Taiwan.

Most rooftop PV systems in Taiwan are fixed. Therefore, the solar radiation received by solar panels at different inclination angles must be estimated. The amount of solar radiation incident on a solar thermal collector or a PV panel is strongly affected by its installation angle and orientation [16]. Over the past decade, various models have been proposed for predicting solar radiation on inclined surfaces [17–20]. Maleki et al. [21] reviewed several models for estimating solar radiation components on horizontal and inclined surfaces. As indicated by [16], all these models require hourly global irradiations and hourly horizontal diffuse solar irradiations.

Moreover, relevant literature on surface solar irradiation prediction, PV power estimation, and current state-of-the-art studies were reviewed [22–34]. Studies have used machine learning algorithms, such as k-nearest neighbor (kNN) [35,36], multilayer perceptron [37–40], and wavelet neural network [41]; some compared or combined multiple machine learning models in the prediction results. For instance, Urraca et al. [42] compared the prediction results of support vector regression with those of random forests, linear regression (LR) and kNN. Yousif et al. [43] compared a self-organizing feature map with multilayer perceptron and support vector machine for forecasting energy production in PV panels. The aforementioned studies have successfully applied their methods in either forecasting future solar resource or estimating solar resource.

Deep learning is a specific subfield of machine learning intended to enable machines to simulate the manner in which the human brain thinks, and its operational model is based on neuroscience [44]. Deep learning is designed to use a neural network structure to represent input and target data. These models use multiple feature extraction layers and learn the complex relationships within the data more efficiently [45]. Recent studies have successfully employed deep learning models in predicting energy efficiency. For instance, Li et al. [46] developed an extreme deep learning approach to improve building-energy consumption–prediction accuracy. Ryu et al. [47] applied deep neural network (DNN)-based load forecasting models and applied them to a demand-side empirical load database. Ghimire et al. [45] used the DNN and deep belief network, the two fundamental categories of DL algorithms, coupled with satellite-derived data to predict monthly global solar radiation. Because deep learning models are applicable for predicting time series, DNN was adopted herein to predict hourly solar radiation to effectively determine the amount of power generated by solar cells.

2. Methodology

In this section, methodology used for developing a usable scheme for evaluating the annual power produced by various solar panels installed on the rooftop of buildings and developing a surface solar radiation prediction model for PV power generation panels is described. Next, we described the following methods: (1) calculation of the potential annual power outputs for the selected solar panels,

(2) derivation of solar radiation prediction models, and (3) evaluation of power prediction errors on the future of solar panels.

Figure 2 illustrates a flow, which comprises a series of analysis steps. The methodology can be grouped three phases: In phase I, several solar panels were selected for comparison. Solar radiation, module temperature, and power conversion efficiency affect solar cell power output. Therefore, these three parameters were used to derive a formula for ideal PV power generation. Conventional rooftop solar panels can be installed at an inclined angle to maximize the irradiance absorption according to the locations of their installation. Thus, the suitable tilt angle for the PV power generation panels leading to direct sunlight can be achieved to increase power output of the panels installed on building rooftops. Therefore, the potential annual power outputs of these solar panels can be calculated.

In phase II, the surface solar radiation prediction model, which was developed for predicting hourly solar irradiation in the future, was created. First, the model input and output attributes from the ground weather data and solar position parameters are preprocess. A traditional training–validation–testing procedure is adopted for formulating the surface solar radiation prediction model. A DNN is used to create a solar radiation prediction model, and a backpropagation neural network (BPN) and a LR model are implemented as benchmark models. Finally, the testing data set is simulated using the optimal trained model, and the forecast results are evaluated according to the performance measures.

In phase III, the hourly PV power outputs on a solar panel were simulated. The solar radiation prediction model was used to predict the hourly solar radiation, and the hourly amount of power generated by the solar panels was calculated using the process in phase I. Finally, the solar panels were evaluated according to the errors made by the predicted power outputs on various lead times.

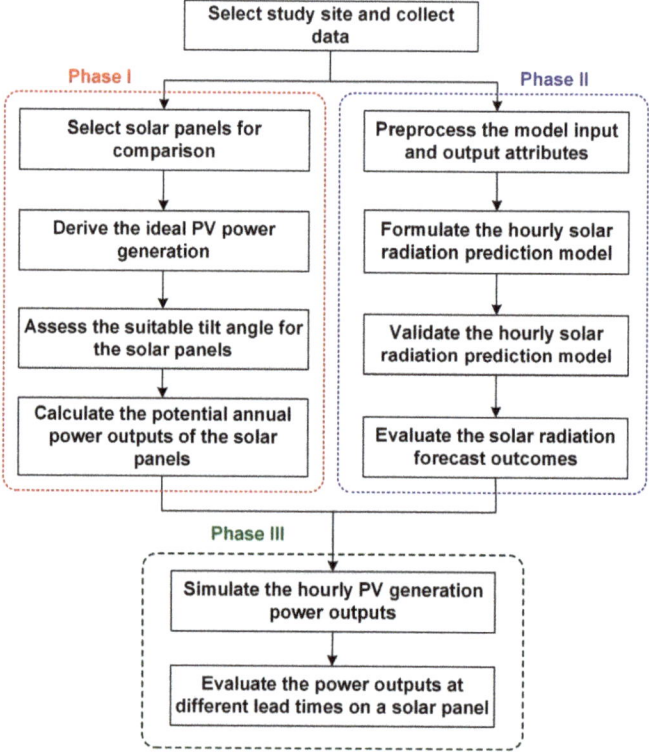

Figure 2. Flow of evaluation of power outputs of a solar panel.

3. Selection of Solar Panels

The PV system is a well-recognized system and is widely used to convert the solar energy for electric power generation applications [48]. A cell is defined as a semiconductor device that converts sunlight into electricity. A PV module refers to numerous cells connected in series, and in a PV array, modules are connected in series and in parallel [49]. PV cells represent the fundamental conversion unit of a PV power generation system. Solar insolation, PV cell temperature and operating voltage strongly influence the PV output current and power characteristics [50].

The most critical components of PV power generation systems are PV cells. The quality of these cells directly affects the efficiency and lifetime of a PV system. This study focused on the solar panels manufactured by BP Solar. BP has been involved in solar power since 1973. Its subsidiary, BP Solar, is a manufacturer and installer of photovoltaic solar cells headquartered in Madrid, Spain, with production facilities in United States, Spain, India, China and Australia [51]. The BP3 series solar panel is an advanced PV module that incorporates polycrystalline cells by using SiN coating to provide high efficiency. For evaluation, different BP3 series solar panel types, namely BP350, BP365, BP380, and BP3125, were selected. Table 1 lists the characteristics of the BP3 series solar panels; in particular, BP3125 generated the highest amount of power (125 W).

Table 1. Characteristics of BP3 series solar panel products.

Module	Maximum Power (Pmax)	Voltage at Pmax	Current at Pmax	Temperature Coefficient of Power	Dimension of Module
BP350	50 W	17.3 V	2.89 A	−0.5%/°C	839 mm × 537 mm
BP365	65 W	17.6 V	3.69 A	−0.5%/°C	1111 mm × 502 mm
BP380	80 W	17.6 V	4.55 A	−0.5%/°C	1204 mm × 537 mm
BP3125	125 W	17.6 V	7.1 A	−0.5%/°C	1510 mm × 674 mm

3.1. Deriving Ideal PV Power Generation

Irradiance is a major factor affecting the amount of solar cell–generated power. The higher the irradiance, the higher the amount of power a solar cell generates. Accordingly, the relationship between irradiance and amount of solar cell–generated power is as follows [52]:

$$P = G \times A \times \eta \quad (1)$$

where P indicates the amount of power generated by a solar cell, G indicates the clear-sky global horizontal irradiance (W/m^2), A indicates the area of the solar cell (m^2), and η indicates the power conversion efficiency (%). The standard environmental parameters for solar panels are G_0 = 1000 W/m^2 and T_{c0} (module temperature of the solar panel) = 25 °C.

By using Equation (1), the power conversion efficiency of a solar panel at the module temperature of 25 °C is calculated as follows:

$$\eta = \frac{P_{max}}{G_0 \times A_0} \quad (2)$$

where P_{max} indicates the maximal power output by a solar panel module.

Equation (2) yields the η value at the module temperature of 25 °C. P_{max} and A_0 can be referenced from the characteristic data of the four BP3 series solar panels as listed in Table 1. The value of η for BP350, BP365, BP380, and BP3125 was 11.1%, 11.7%, 12.4%, and 12.3%, respectively.

Higher solar cell module temperature results in lower power generation efficiency. Specifically, for BP3 series solar panels, when the module temperature exceeds 25 °C, each additional 1 °C reduces the overall efficiency by λ = −0.5%/°C. The efficiency change ε in this condition is accordingly expressed as follows:

$$\varepsilon = \lambda \cdot (T_c - 25) \cdot \eta \quad (3)$$

This formula is used to further evaluate the amount of power output after temperature changes:

$$P = G \cdot A \cdot (\eta - \varepsilon) \tag{4}$$

Applying Equation (3) to Equation (4) we obtain the following:

$$P = G \cdot A \cdot [1 - \lambda \cdot (T_c - 25)] \cdot \eta \tag{5}$$

As indicated, the operating cell temperatures of PV modules directly affect the performance of the PV system. Temperature of PV cells is one of the most important parameters for assessing the long term performance of PV module systems and their annual amounts of electrical energy production [53]. Therefore, estimating the operating temperature of a PV module is required. Several mathematical equations for PV module temperature have been found in the literature (e.g., [54–58]). These proposed approaches used empirical formulas to derive the PV cell temperature from the environmental variables, such as ambient temperature, irradiance, and wind speed [5]. A detailed comparison among these empirical formulas was reported by [59,60]. Because Taiwan has an island climate, wind velocity must be factored in. For simplicity, a mathematical model proposed by [61] was employed to account for ambient temperature, irradiance, and wind speed:

$$T_c = T_a + 0.0138 \cdot (1 + 0.031 \cdot T_a)(1 - 0.042 \cdot V_w) \cdot G \tag{6}$$

where V_w is the wind speed.

3.2. Equation for Irradiance Received by Inclined Solar Panels

Most available solar radiation data around the world are global solar radiations on a horizontal surface. In practice, solar collectors (flat plate thermal or PV collectors) are tilted; thus, computing the solar radiation incident on such tilted planes is necessary [16]. As calculated in Equation (5), G represents clear-sky global horizontal irradiance. The amount of solar radiation received by solar panels at an inclination must be estimated to accurately calculate their power output.

According to [62], the theoretical global irradiance with the solar panels at a tilted position (G_{tilt}) can be estimated using the following expression:

$$G_{tilt} = D_C + I_C \cdot \cos \Theta \tag{7}$$

where D_C is the diffuse horizontal irradiance; I_C is the direct irradiance; and Θ the solar incident angle, defined as the angle between the sun and the normal line of the solar panels.

In Equation (7), D_C is calculated as follows [63]:

$$D_C = G_C - I_C \cdot \cos \theta \tag{8}$$

where G_C is the clear-sky solar irradiance and θ is the zenith angle.

In Equation (7), the solar incident angle Θ on the inclined solar panel should also be estimated; Θ is expressed according to the latitude of the panel (λ), declination angle (δ), and hour angle (ω) as shown in the following equation:

$$\Theta = \cos^{-1}(\cos(\beta - \lambda) \cdot \cos \delta \cdot \cos \omega - \sin(\beta - \lambda) \cdot \sin \delta) \tag{9}$$

where β indicates the horizontal inclination angle of the panel.

3.3. Estimating the Power Output of the Inclined Solar Panels

This study focused on the weather station in Tainan (22°99′ N and 120°20′ E). The station is under the jurisdiction by the Central Weather Bureau of Taiwan and records hourly surface climate data, including clear-sky global horizontal irradiance, ambient temperature, and wind velocity. The year 2016 was designated as the simulation year. The hourly clear-sky global horizontal irradiance data recorded in the station were applied in the formulas specified in Section 3.2 to calculate the hourly irradiance received the inclined solar panels. According to [39], the maximal annual global irradiance at Southern Taiwan occurred at β ranging from 20° to 22°. Thus, this study selected $\beta = 21°$ as the tilt angle of solar panels for the BP3 series.

The process of evaluating the selected BP3 series solar panels started with the simulation of hourly irradiance captured by the inclined solar panels. Figure 3a illustrates the simulation results of the hourly solar radiation at the spring and autumn equinoxes and summer and winter solstices in 2016. The formula for the ideal PV power generation was subsequently implemented to estimate the power generated by the solar panels. Figure 3b depicts the power output of the four solar panel modules when $\beta = 21°$. Figure 4 illustrates the simulated total power output by the panels in 2016. Figure 4a displays the single module power output; in particular, BP3125 exhibited the highest power generation efficiency (185.4 KWh-y), whereas BP350 was the least efficient module in power generation (74.2 KWh-y). To estimate the PV power output per unit area in a reasonable manner, the power output by each module was divided by the its area to identify its unit area PV power output (Figure 4b). In particular, BP380 exhibited the highest unit area power output (183.5 KWh/m²-y), followed by BP3125 (182.2 KWh/m²-y); BP350 exhibited the lowest unit area power output (164.2 KWh/m²-y).

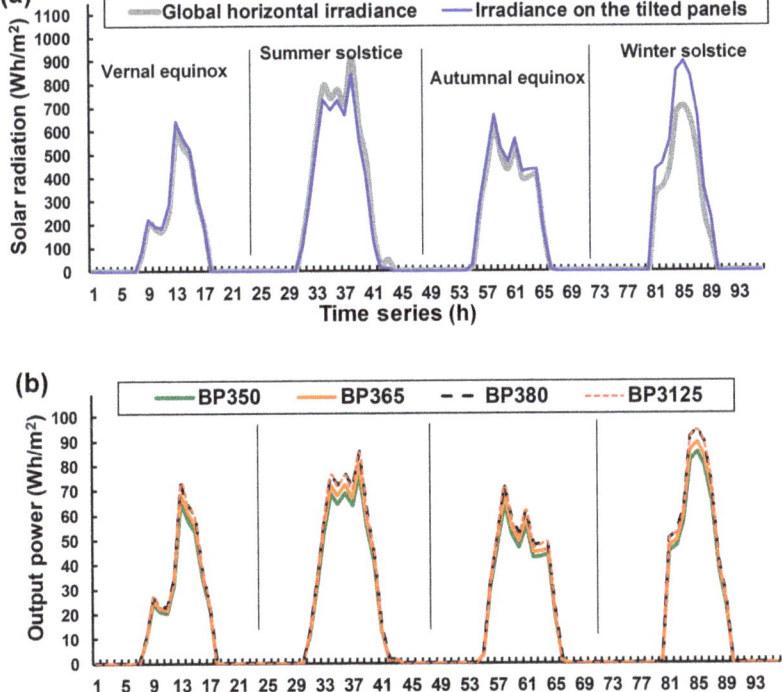

Figure 3. Spring and autumn equinoxes and summer and winter solstices in 2016: (**a**) Hourly solar radiation; (**b**) Estimated hourly output power of BP3 series modules.

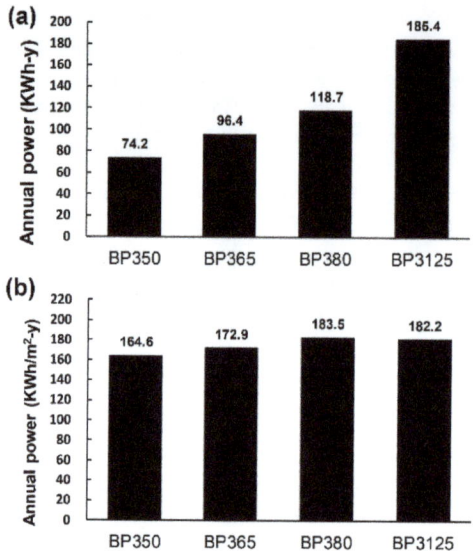

Figure 4. Simulated annual power output by the BP3 series: (**a**) Single module; (**b**) Unit area.

4. Hourly Solar Radiation Prediction

As indicated by [16], the efficiency or productivity of such a system depends on the temporal fluctuations of energy input and output. Therefore, an hourly solar radiation prediction model was established to further simulate the hourly power output of the solar panels in future for subsequent analysis.

4.1. Experimental Data

Data on surface climate and sun positions were applied in establishing the solar radiation prediction model. In addition to the 2016 data, the 2010–2015 hourly climate data from the weather station were implemented. Seven surface climate parameters related to solar radiation were selected: air pressure on the ground (hPa), ground temperature (°C), relative humidity (%), and surface wind velocity (maximum 10 min mean, 10 m above the surface) (m/s), precipitation within 1 h (mm), sunshine duration (h), and surface solar radiation (Wh/m^2); all data were recorded hourly. Sunshine duration, a measure of the time interval for which sunshine is observed in 1 h (at the study location), was used as a climatological indicator of cloudiness. In total, 61,368 data points were organized. Table 2 presents the seven attributes selected and their statistics with mean, standard deviation, maximum, and minimum values; all variables in the database were measured hourly.

Table 2. Weather attributes and statistics at the Tainan station.

Attribute	Air Pressure on the Ground	Ground Temperature	Relative Humidity	Surface Wind Velocity	Precipitation	Sunshine Duration	Surface Solar Radiation
Unit	hPa	°C	%	m/s	mm	h	Wh/m^2
Mean	1009.8	24.60	74.38	2.97	0.21	0.23	188.87
Std. dev.	5.663	5.234	10.34	1.689	1.829	0.37	276.52
Maximum	1029.6	35.8	100	19.5	90.5	1	1161.1
Minimum	973.4	5.8	22	0	0	0	0

Five sun position parameters were selected to illustrate the sun position data relative to the station over time, namely declination, hour, zenith, elevation, and azimuth angles. These parameters are based on the formula proposed by [62,63]. Table 3 lists the statistical data of the hourly sun positions from 2010 to 2016.

Table 3. Solar position attributes and statistics at the Tainan station.

Title Attribute	Declination Angle	Hour Angle	Zenith Angle	Elevation Angle	Azimuth Angle
Unit	Degree	Degree	degree	Degree	degree
Mean	−0.01	7.50	90	0	0
Std. dev.	16.58	103.83	43.84	43.84	65.10
Maximum	23.45	180	179.98	89.98	90.00
Minimum	−23.45	−165	0.02	−89.98	−90.00

4.2. DNNs and Modeling

The solar radiation prediction model was based on a DNN, which is a machine learning model that automatically identifies the representative features through linear or nonlinear transforms in multiple layers. A neural network, which is a mathematical model mimicking the neural system of an organism, features several layers of neurons, which sum the data inputted from the neurons of previous layers and convert them to output data through an activation function. Each neuron is linked to the neurons of the follow-up layer in a unique manner; the data output produced by the neurons from a layer are weighted and transmitted to the neurons of the follow-up layer [64].

DNNs consist of at least one hidden layer. Similar to shallow neural networks, DNNs establish models based on complex nonlinear systems; however, multiple hidden layers are included to enhance the learning efficacy of these models and thereby their prediction and categorization capability. Most DNNs are constructed as feedforward neural networks [65]. DNNs are trained through backpropagation. The weight updates between layers are calculated through stochastic gradient descent:

$$\Delta w_{ij}(t+1) = \beta \Delta w_{ij}(t) + \eta \frac{\partial E}{\partial w_{ij}} \quad (10)$$

where $w_{ij}(t)$ is the weight set connecting the layers i and j at time t; Δw is the weight correction; η is the learning rate; β is a momentum coefficient; and E is a cost function, which indicates the difference between the target and predicted values. In particular, η and β are hyperparameters for the adjustment of the spacing of weight correction.

In the DNN training process, the weight set in the model can be optimized through a numerical method for minimizing learning target values. This is typically achieved through stochastic gradient descent, in which all weights in high-dimensional spaces descend by one dimension per step; the iteration is repeated multiple times to identify the optimal weight set.

4.3. Modeling and Parameter Calibration

A DNN was used to establish an accurate solar radiation prediction model. The model was established according to the data attributes mentioned in Section 4.1. The employed data were divided into two datasets: 2010–2015 data defined as the training set for model training and verification and 2016 data were designated as the testing set. Model training and verification were performed through 10-fold cross-validation, in which the training set was divided into 10 subsamples, one of which was retained for model verification and the other nine were used for model training; in the verification process, each subsample must be verified.

Building the prediction model requires setting up the neural network structure; in particular, the number of DNN layers, the number of neurons per layer, and the activation function must be defined. Parameter settings affect the efficacy of a DNN model; therefore, an optimized weight set is required for DNN training. Most activation functions are designed for nonlinear transformation for information transfer in a complicated neural network. Conventional activation functions are designed as sigmoid functions or hyperbolic tangent functions. In DNNs, rectified linear unit function has been recently used to replace the sigmoid function, resulting in high performance and short training times, as reported by [66]. Moreover, the number of neurons in the hidden layer was determined according to [67] method: summing numbers of neurons in the input and output layers, subtracting 1 from this sum, and dividing this number by 2.

The hyperparameters η and β were verified. Because both parameters range from 0 to 1, intervals of 0.1 were adopted in the verification; subsequently, the root mean square errors (RMSEs) between their output and target values were calculated. Figure 5 illustrates the verification results, in which minimal RMSEs for η and β (77.195 and 76.612 Wh/m^2, respectively) were obtained when $\eta = 0.2$ and $\beta = 0.2$. In addition, the number of DNN layers was determined as 1–15. Figure 6a depicts the RMSEs of the model in the follow-up hour according to the number of layers. When the number of layers was 6, the RMSE decreased (74.846 Wh/m^2). Figure 6b–d depicts the RMSEs at lead times of 3, 6, and 12 h, respectively, which were minimized when the numbers of layers were 12, 6, and 11, respectively.

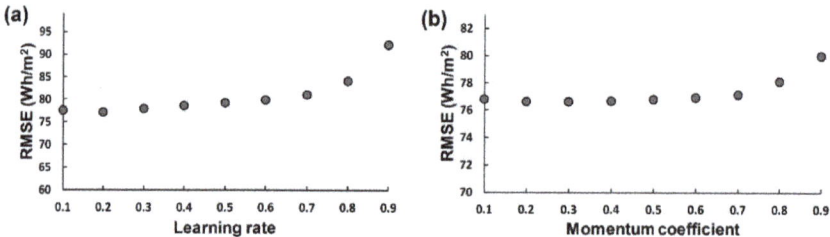

Figure 5. Calibration of hyperparameters: (**a**) Learning rate; (**b**) Momentum coefficient.

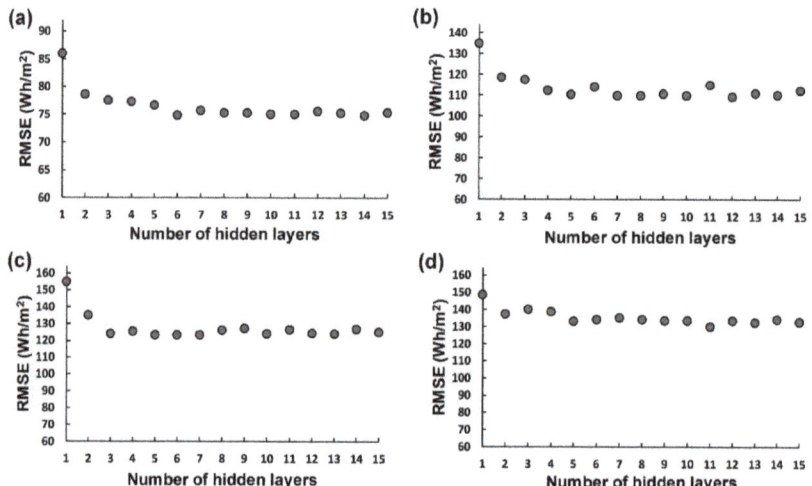

Figure 6. Calibration of hidden layer amount at lead times of (**a**) 1; (**b**) 3; (**c**) 6 and (**d**) 12 h.

4.4. Forecast of Solar Radiation Prediction

After the model parameters were verified, the testing set was used to predict solar radiation. The simulation results of the DNN model were compared with those of the benchmark models (i.e., BPN and LR models) to verify its quality. A typical BPN, which is a shallow neural network, with three layers (i.e., an input layer, a hidden layer, and an output layer) is a feedforward neural network trained with the standard backpropagation algorithm [68].

Figure 7 illustrates the scatter diagrams of the predicted and observed values of DNN, BPN, and LR. In particular, Figure 7a–d depict the results at the lead times of 1, 3, 6, and 12 h, respectively. As the lead time increased, the predicted values of DNN became closer to the observed values than did those of BPN and LR. Regarding data correlation, all four figures depicts that DNN exhibited the highest coefficient of determination (R^2), followed sequentially by those of BPN and LR.

Regarding prediction errors, Figure 8 depicts the calculation results of the three mentioned models. At lead times of 1, 3, 6, and 12 h, DNN exhibited the lowest mean absolute errors (MAEs), indicating that its prediction errors were the lowest overall. Furthermore, DNN displayed the smallest RMSE.

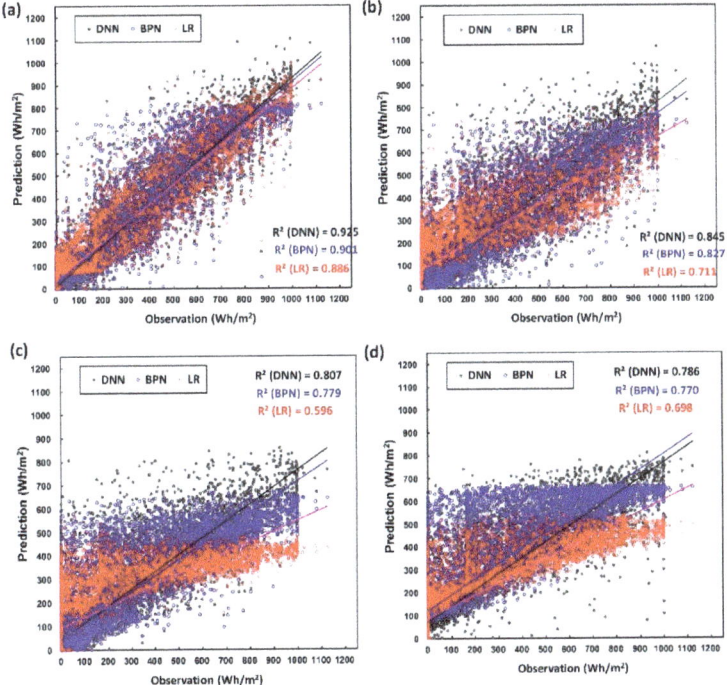

Figure 7. Scatter diagram of deep neural network (DNN), backpropagation neural network (BPN) and linear regression (LR) at the lead times of (**a**) 1; (**b**) 3; (**c**) 6 and (**d**) 12 h.

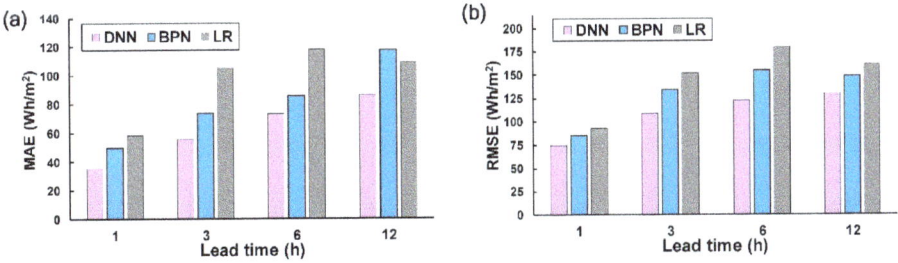

Figure 8. Solar radiation prediction performance: (**a**) mean absolute error (MAE) and (**b**) root mean square error (RMSE).

5. Simulation of PV Power Generation

BP3125 was used as the sample solar panel for estimating the PV power output at lead times of 1–12 h. As mentioned previously, after the hourly solar radiation captured by the inclined BP3 series solar panels was simulated using the clear-sky global horizontal irradiance data in 2016, it was used to estimate the hourly power output by the panels (Figure 3). The DNN-based solar radiation prediction model was subsequently employed to replace the predicted values with the observed values to calculate the hourly power output.

Figure 9 depicts the time sequences of the simulation and predicted values of the power output by BP3125 at the spring and autumn equinoxes and summer and winter solstices in 2016. Figure 9a–d depict the results at the lead times of 1, 3, 6, and 12 h, respectively. The simulation values indicated in Figure 9a represent the simulated power output of BP3125 from Figure 3b; the DNN, BPN, and LR values are the power output values calculated according to the clear-sky global horizontal irradiance estimated through the DNN, BPN, and LR models. Because solar radiation changes substantially each day, the results as depicted in Figure 9 do not represent the daily prediction results in the entire year. Therefore, the evaluation indices for the hourly predicted and simulation values from each model in 2016 were calculated. As shown in Figure 10, DNN exhibited the lowest MAEs and RMSEs at the lead times of 1, 3, 6, and 12 h; this revealed that DNN was the most satisfactory of the three models in predicting solar radiation.

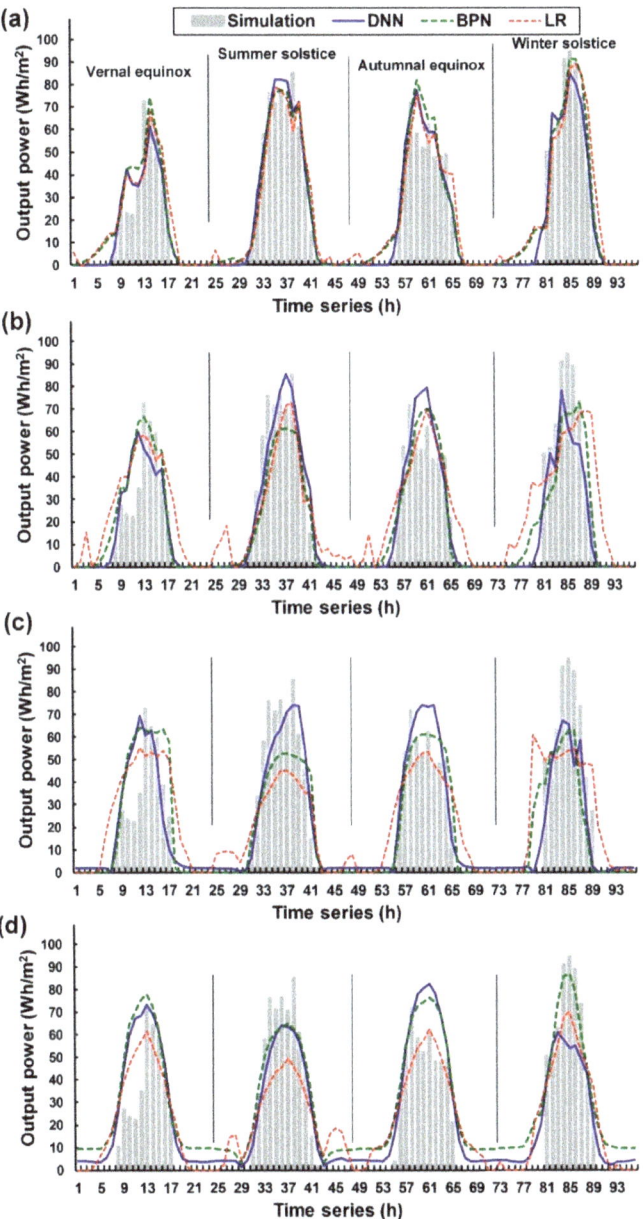

Figure 9. Simulation and predicted values on the hourly power output of BP3125 at spring and autumn equinoxes and summer and winter solstices in 2016 at lead times of (**a**) 1; (**b**) 3; (**c**) 6 and (**d**) 12 h.

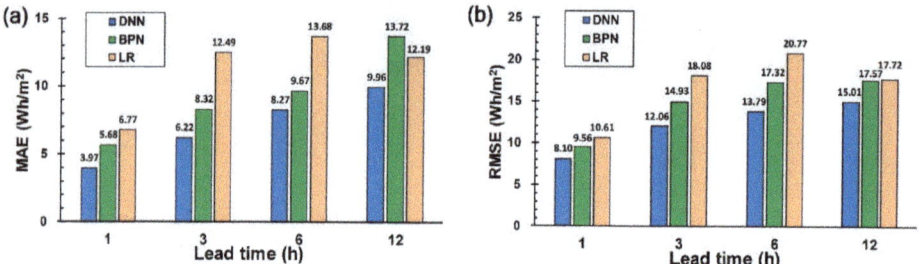

Figure 10. Annual performance of predicted power outputs of module BP3125: (a) MAE and (b) RMSE.

6. Usage and Limitations of the Methodology

In this study, we demonstrated a methodology for developing a usable scheme for evaluating the annual power produced by various solar panels installed on the rooftops of buildings. Tainan, Taiwan, a subtropical region, was used as the experimental site. In practice, the proposed methodology can be used in regions that are suitable for generating and using solar energy. However, the different regions might not have the same climatic conditions as Tainan. Therefore, when using the proposed methodology in such regions, some parameters used in the PV module should be carefully adjusted to suit the local climatic conditions. For instance, when determining the PV cell temperature in these regions, the empirical formulas of PV cell temperature should be reselected because the PV cell temperature is affected by climatic variables, such as irradiance and wind speed.

For examining PV panel brands (e.g., BP Solar), data of the panel manufacturer were obtained. The module products of this brand, such as BP350, BP365, BP380, and BP3125, were assessed for determining their PV output efficiency. When assessing the module products created by different manufacturers, the evaluation process also can be reproduced by the proposed approach. Thus, when the characteristics of the solar panel products are known (i.e., maximum power, voltage and current at maximum power, temperature coefficient of power, and dimension of module), the potential annual power outputs of solar panels can be calculated.

When we developed a surface solar radiation prediction model for PV power generation panels, the solar radiation was estimated by using DNN and BPN, which are data-driven prediction models based on machine learning. A data-driven model is based on the analysis of the data regarding a specific system [69]. Thus, in the modeling process, we used data that were collected from ground-based meteorological stations. However, for regions without ground-based meteorological stations, machine learning-based solar radiation prediction models cannot be developed. Hence, the use of data from satellite measurements is suggested for constructing machine learning-based prediction models in these regions, without ground-based meteorological stations.

7. Conclusions

This study proposed a simple and effective model for evaluating the PV power generation efficiency of each brand of solar panels when planning the installation of rooftop PV systems. Tainan, which has a stable climate and constant sunshine and is suitable for solar power generation in all seasons, was selected as the experimental site. Four BP3 series solar panel types were selected for evaluation: BP350, BP365, BP380, and BP3125.

In phase I, a formula for ideal PV power generation was derived for calculating the power conversion efficiency of each BP series module (11.1%, 11.7%, 12.4%, and 12.3% for BP350, BP365, BP380, and BP3125, respectively). Solar panels are installed at an inclined angle to maximize their reception of solar radiation. Thus, we determined the suitable tilt angle for the PV system leading to direct sunlight that could be acquired to increase power output installed on building rooftop. Subsequently, the potential annual power outputs for these solar panels were calculated. The annual power output of the BP3 series solar panels per unit area was calculated according to the 2016 climate data. The results indicated that BP380 was the most efficient module for annual power output (183.5 KWh/m^2), followed by BP3125 (182.2 KWh/m^2); BP350 was the least efficient module (164.2 KWh/m^2).

In phase II, to simulate hourly PV power generation with regard to meteorological uncertainty, the surface solar radiation prediction model was developed. The model inputs employed the solar position and meteorological information inputs. This study employed the deep learning–based DNN for predicting hourly irradiation. The prediction results of the DNN model were then compared with those of the BPN and LR models. A traditional training–validation–testing procedure was adopted for formulating the surface solar radiation prediction model. The results indicated that the DNN exhibited the lowest MAEs and RMSEs among all three models at the lead times of 1, 3, 6, and 12 h, highlighting its satisfactory prediction accuracy. In phase III, we used the panel of module BP3125 as an example and predicted hourly PV power outputs at different lead times on a solar panel. The evaluation indices for the hourly predicted and simulation values of each model in 2016 were calculated, revealing that DNN exhibited the lowest MAEs and RMSEs among all the three models at the lead times of 1, 3, 6, and 12 h.

The approach proposed in this study was confirmed to be applicable to the evaluation of the power generation efficiency of solar panels and prediction of their hourly power output in an entire year. This approach is suitable for assessing the power generation efficiency of a rooftop PV system.

Author Contributions: C.-C.W. devised the experimental strategy and carried out this experiment and wrote the manuscript and contributed to the revisions.

Funding: This study was supported by the Ministry of Science and Technology, Taiwan, under Grant No. MOST107-2111-M-019-001-CC3. The authors would like to express their sincere appreciation for this grant.

Acknowledgments: The author acknowledges the experimental data provided by the Central Weather Bureau of Taiwan.

Conflicts of Interest: The authors declare no conflict of interest.

References

1. Zahedi, A. Solar photovoltaic (PV) energy; latest developments in the building integrated and hybrid PV systems. *Renew. Energy* **2006**, *31*, 711–718. [CrossRef]
2. Messenger, R.; Ventre, J. *Photovoltaic Systems Engineering*; CRC Press: Boca Raton, FL, USA, 2000; pp. 11–12, 41–51.
3. Chen, F.C. A Meteorology Assessment for Photovoltaic Generation. Master's Thesis, Southern Taiwan University of Science and Technology, Tainan, China, 2006. (In Chinese).
4. Sopian, K.; Cheow, S.L.; Zaidi, S.H. An overview of crystalline silicon solar cell technology: Past, present, and future. *AIP Conf. Proc.* **2017**, *1877*, 020004.
5. Muzathik, A.M. Photovoltaic modules operating temperature estimation using a simple correlation. *Int. J. Energy Eng.* **2014**, *4*, 151–158.
6. Kim, B.C.; Kim, J.; Kim, K. Evaluation model for investment in solar photovoltaic power generation using fuzzy analytic hierarchy process. *Sustainability* **2019**, *11*, 2905. [CrossRef]
7. Kovač, M.; Stegnar, G.; Al-Mansour, F.; Merše, S.; Pečjak, A. Assessing solar potential and battery instalment for self-sufficient buildings with simplified model. *Energy* **2019**, *173*, 1182–1195. [CrossRef]
8. Le, N.T.; Benjapolakul, W. Evaluation of contribution of PV array and inverter configurations to rooftop PV system energy yield using machine learning techniques. *Energies* **2019**, *12*, 3158. [CrossRef]
9. Jeong, K.; Hong, T.; Koo, C.; Oh, J.; Lee, M.; Kim, J. A prototype design and development of the smart photovoltaic system blind considering the photovoltaic panel, tracking system, and monitoring system. *Appl. Sci.* **2017**, *7*, 1077. [CrossRef]
10. Mahmud, M.A.P.; Huda, N.; Farjana, S.H.; Lang, C. Environmental impacts of solar-photovoltaic and solar-thermal systems with life-cycle assessment. *Energies* **2018**, *11*, 2346. [CrossRef]

11. Kouhestani, F.M.; Byrne, J.; Johnson, D.; Spencer, L.; Hazendonk, P.; Brown, B. Evaluating solar energy technical and economic potential on rooftops in an urban setting: The city of Lethbridge, Canada. *Int. J. Energy Environ. Eng.* **2019**, *10*, 13–32. [CrossRef]
12. Balo, F.; Şağbanşua, L. The selection of the best solar panel for the photovoltaic system design by using AHP. *Energy Procedia* **2016**, *100*, 50–53. [CrossRef]
13. Nfaoui, M.; El-Hami, K. Extracting the maximum energy from solar panels. *Energy Rep.* **2018**, *4*, 536–545. [CrossRef]
14. Wang, T.C.; Tsai, S.Y. Solar panel supplier selection for the photovoltaic system design by using fuzzy multi-criteria decision making (MCDM) approaches. *Energies* **2018**, *11*, 1989. [CrossRef]
15. Guechi, A.; Chegaar, M. Effects of diffuse spectral illumination on microcrystalline solar cells. *J. Electron Devices Soc.* **2007**, *5*, 116–121.
16. Notton, G.; Cristofari, C.; Muselli, M.; Poggi, P. Calculation on an hourly basis of solar diffuse irradiations from global data for horizontal surfaces in Ajaccio. *Energy Convers. Manag.* **2004**, *45*, 2849–2866. [CrossRef]
17. Gopinathan, K.K. Solar radiation on variously oriented sloping surfaces. *Sol. Energy* **1991**, *47*, 173–179. [CrossRef]
18. Kambezidis, H.D.; Psiloglou, B.E.; Synodinou, B.M. Comparison between measurements and models for daily solar irradiation on tilted surfaces in Athens, Greece. *Renew. Energy* **1997**, *10*, 505–518. [CrossRef]
19. Robledo, L.; Soler, A. Modelling daylight on inclined surfaces for applications to daylight conscious architecture. *Renew. Energy* **1997**, *11*, 149–152. [CrossRef]
20. Nijmeh, S.; Mamlook, R. Testing of two models for computing global solar radiation on tilted surfaces. *Renew. Energy* **2000**, *20*, 75–81. [CrossRef]
21. Maleki, S.A.M.; Hizam, H.; Gomes, C. Estimation of hourly, daily and monthly global solar radiation on inclined surfaces: Models re-visited. *Energies* **2017**, *10*, 134. [CrossRef]
22. Chen, S.X.; Gooi, H.B.; Wang, M.Q. Solar radiation forecast based on fuzzy logic and neural networks. *Renew. Energy* **2013**, *60*, 195–201. [CrossRef]
23. Ding, M.; Wang, L.; Bi, R. An ANN-based approach for forecasting the power output of photovoltaic system. *Procedia Environ. Sci.* **2011**, *11 Pt C*, 1308–1315. [CrossRef]
24. Dorvlo, A.S.S.; Jervaseb, J.; Al-Lawatib, A. Solar radiation estimation using artificial neural networks. *Appl. Energy* **2015**, *71*, 307–319. [CrossRef]
25. Inman, R.H.; Pedro, H.T.C.; Coimbra, C.F.M. Solar forecasting methods for renewable energy integration. *Prog. Energy Combust. Sci.* **2013**, *39*, 535–576. [CrossRef]
26. Khandakar, A.; Chowdhury, M.E.H.; Kazi, M.K.; Benhmed, K.; Touati, F.; Al-Hitmi, M.; Gonzales, S.P., Jr. Machine learning based photovoltaics (PV) power prediction using different environmental parameters of Qatar. *Energies* **2019**, *12*, 2782. [CrossRef]
27. Li, Z.; Rahman, S.M.; Vega, R.; Dong, B. A hierarchical approach using machine learning methods in solar photovoltaic energy production forecasting. *Energies* **2016**, *9*, 55. [CrossRef]
28. Monteiro, C.; Santos, T.; Fernandez-Jimenez, L.A.; Ramirez-Rosado, I.J.; Terreros-Olarte, M.S. Short-term power forecasting model for photovoltaic plants based on historical similarity. *Energies* **2013**, *6*, 2624–2643. [CrossRef]
29. Notton, G.; Paoli, C.; Diaf, S. Estimation of tilted solar irradiation using Artificial Neural Networks. *Energy Procedia* **2013**, *42*, 33–42. [CrossRef]
30. Persson, C.; Bacher, P.; Shiga, T.; Madsen, H. Multi-site solar power forecasting using gradient boosted regression trees. *Sol. Energy* **2017**, *150*, 423–436. [CrossRef]
31. Reda, I.; Andreas, A. Solar position algorithm for solar radiation applications. *Sol. Energy* **2004**, *76*, 577–589. [CrossRef]
32. Shen, C.; He, Y.L.; Liu, Y.W.; Tao, W.Q. Modelling and simulation of solar radiation data processing with Simulink. *Simul. Model. Pract. Theory* **2008**, *16*, 721–735. [CrossRef]
33. Yang, D.; Gu, C.; Dong, Z.; Jirutitijaroen, P.; Chen, N.; Walsh, W.M. Solar irradiance forecasting using spatial-temporal covariance structures and time-forward kriging. *Renew. Energy* **2013**, *60*, 235–245. [CrossRef]
34. Yeom, J.M.; Han, K.S. Improved estimation of surface solar insolation using a neural network and MTSAT-1R data. *Comput. Geosci.* **2010**, *36*, 590–597. [CrossRef]
35. Hocaoglu, F.O. Stochastic approach for daily solar radiation modeling. *Sol. Energy* **2011**, *85*, 278–287. [CrossRef]

36. Pedro, H.T.C.; Coimbra, C.F.M. Nearest-neighbor methodology for prediction of intra-hour global horizontal and direct normal irradiances. *Renew. Energy* **2015**, *80*, 770–782. [CrossRef]
37. Bosch, J.L.; Lopez, G.; Batlles, F.J. Daily solar irradiation estimation over a mountainous area using artificial neural network. *Renew. Energy* **2008**, *33*, 1622–1628. [CrossRef]
38. Wang, F.; Mi, Z.; Su, S.; Zhao, H. Short-term solar irradiance forecasting model based on artificial neural network using statistical feature parameters. *Energies* **2012**, *5*, 1355–1370. [CrossRef]
39. Wei, C.C. Predictions of surface solar radiation on tilted solar panels using machine learning models: Case study of Tainan City, Taiwan. *Energies* **2017**, *10*, 1660. [CrossRef]
40. Zhu, H.; Li, X.; Sun, Q.; Nie, L.; Yao, J.; Zhao, G. A power prediction method for photovoltaic power plant based on wavelet decomposition and artificial neural networks. *Energies* **2016**, *9*, 11. [CrossRef]
41. Cao, J.; Lin, X. Application of the diagonal recurrent wavelet neural network to solar irradiation forecast assisted with fuzzy technique. *Eng. Appl. Artif. Intell.* **2008**, *21*, 1255–1263. [CrossRef]
42. Urraca, R.; Antonanzas, J.; Alia-Martinez, M.; Martinez-de-Pison, F.J.; Antonanzas-Torres, F. Smart baseline models for solar irradiation forecasting. *Energy Convers. Manag.* **2016**, *108*, 539–548. [CrossRef]
43. Yousif, J.H.; Kazem, H.A.; Boland, J. Predictive models for photovoltaic electricity production in hot weather conditions. *Energies* **2017**, *10*, 971. [CrossRef]
44. Hinton, G.E.; Osindero, S.; Teh, Y.W. A fast learning algorithm for deep belief nets. *Neural Comput.* **2014**, *18*, 1527–1554. [CrossRef] [PubMed]
45. Ghimire, S.; Deo, R.C.; Raj, N.; Mi, J. Deep learning neural networks trained with MODIS satellite-derived predictors for long-term global solar radiation prediction. *Energies* **2019**, *12*, 2407. [CrossRef]
46. Li, C.; Ding, Z.; Zhao, D.; Yi, J.; Zhang, G. Building energy consumption prediction: An extreme deep learning approach. *Energies* **2017**, *10*, 1525. [CrossRef]
47. Ryu, S.; Noh, J.; Kim, H. Deep neural network based demand side short term load forecasting. *Energies* **2017**, *10*, 3. [CrossRef]
48. Tsai, H.L. Insolation-oriented model of photovoltaic module using Matlab/Simulink. *Sol. Energy* **2010**, *84*, 1318–1326. [CrossRef]
49. Tian, H.; Mancilla-David, F.; Ellis, K.; Muljadi, E.; Jenkins, P. A cell-to-module-to-array detailed model for photovoltaic panels. *Sol. Energy* **2012**, *86*, 2695–2706. [CrossRef]
50. Angrist, S.W. *Direct Energy Conversion*, 4th ed.; Allyn and Bacon, Inc.: Boston, MA, USA, 1982; pp. 177–227.
51. BP Official Website. Available online: https://www.bp.com/ (accessed on 10 July 2019).
52. Shih, H. Cost and Benefit Analysis of Community-based Solar Power System. Master's Thesis, National Chiao Tung University, Taiwan, China, 2009. (In Chinese).
53. Piyatida, T.; Chumnong, S.; Dhirayut, C. Estimating operating cell temperature of BIPV modules in Thailand. *Renew. Energy* **2009**, *4*, 2515–2523.
54. Rauschenbach, H.S. *Solar Cell Array Design Handbook*; Van Nostrand Reinhold: New York, NY, USA, 1980; pp. 390–391.
55. Ross, R.G. Interface Design Considerations for Terrestrial Solar Cells Modules. In Proceedings of the 12th IEEE Photovoltaic Specialist's Conference, Baton Rouge, LA, USA, 7–10 January 1976; pp. 801–806.
56. Risser, V.V.; Fuentes, M.K. Linear Regression Analysis of Flat-plate Photovoltaic System Performance Data. In Proceedings of the 5th Photovoltaic Solar Energy Conference, Athens, Greece, 17–21 October 1983.
57. Schott, T. Operation Temperatures of PV Modules: A Theoretical and Experimental Spproach. In Proceedings of the Sixth EC Photovoltaic Solar Energy Conference, London, UK, 15–19 April 1985.
58. Koehl, M.; Heck, M.; Wiesmeier, S.; Wirth, J. Modeling of the nominal operating cell temperature based on outdoor weathering. *Sol. Energy Mater. Sol. Cells* **2011**, *95*, 1638–1646. [CrossRef]
59. Jakhrani, A.Q.; Othman, A.K.; Rigit, A.R.H.; Samo, S.R. Comparison of solar photovoltaic module temperature models. *World Appl. Sci. J.* **2011**, *14*, 01–08.
60. Kamuyu, W.C.L.; Lim, J.R.; Won, C.S.; Ahn, H.K. Prediction model of photovoltaic module temperature for power performance of floating PVs. *Energies* **2018**, *11*, 447. [CrossRef]
61. Servant, J.M. Calculation of the Cell Temperature for Photovoltaic Modules from Climatic Data. In Proceedings of the 9th biennial congress of ISES- Intersol 85, Montreal, QC, Canada; 1985; p. 370.
62. Exell, R.H.B. A mathematical model for solar radiation in South-East Asia (Thailand). *Sol. Energy* **1981**, *26*, 161–168. [CrossRef]

63. Markvart, T. *Solar Electricity*; John Wiley & Sons Ltd.: New York, NY, USA, 1994.
64. Lo, D.C.; Wei, C.C.; Tsai, N.P. Parameter automatic calibration approach for neural-network-based cyclonic precipitation forecast models. *Water* **2015**, *7*, 3963–3977. [CrossRef]
65. Wei, C.C.; Hsieh, C.J. Using adjacent buoy information to predict wave heights of typhoons offshore of northeastern Taiwan. *Water* **2018**, *10*, 1800. [CrossRef]
66. Nair, V.; Hinton, G. Rectified Linear Units Improve Restricted Boltzmann machines. In Proceedings of the 27th International Conference on Machine Learning, Haifa, Israel; 2010; pp. 807–814.
67. Trenn, S. Multilayer perceptrons: Approximation order and necessary number of hidden units. *IEEE Trans. Neural Netw.* **2008**, *19*, 836–844. [CrossRef] [PubMed]
68. Tsai, C.C.; Lu, M.C.; Wei, C.C. Decision tree-based classifier combined with neural-based predictor for water-stage forecasts in a river basin during typhoons: A case study in Taiwan. *Environ. Eng. Sci.* **2012**, *29*, 108–116. [CrossRef]
69. Solomatine, D.P.; Ostfeld, A. Data-driven modelling: Some past experiences and new approaches. *J. Hydroinform.* **2008**, *10*, 3–22. [CrossRef]

© 2019 by the author. Licensee MDPI, Basel, Switzerland. This article is an open access article distributed under the terms and conditions of the Creative Commons Attribution (CC BY) license (http://creativecommons.org/licenses/by/4.0/).

Article

Biofuel Application as a Factor of Sustainable Development Ensuring: The Case of Russia

Ekaterina S. Titova

Federal Research Centre «Fundamentals of Biotechnology» of the Russian Academy of Sciences, Leninsky Prospect, 33, bld. 2, 119071 Moscow, Russia; es_titova@inbox.ru

Received: 24 September 2019; Accepted: 15 October 2019; Published: 17 October 2019

Abstract: Diffusion of the biofuels (BF) using is justified by opening up the opportunities for obtaining fuel and energy from previously inaccessible sources and by the existence of energy-deficient regions, in particular in Russia. Works of different scientists on the problems of creating and using BF were the methodological basis of this study. Information on the state and prospects of the development of renewable energy sources in Russian regions was collected from regulatory documents and was obtained by employing a questionnaire survey. For the study of the collected materials, the different methods of comparative analysis, and the methods of expert assessments were used. The results of the Status-Quo analysis of BF production in Russia have shown that the creation of BF performed relatively successfully. However, there are many more perspectives, connected with expanding the utilization of the different raw materials. Also, the analysis of organizational and economic mechanisms applied for production of BF and the obtained data on several organizations-producers allowed for proposing six indexes for the assessment of the BF production effectiveness. It is suggested that BF production in Russia will contribute to the sustainable development of a number of the country's regions in the near future.

Keywords: biofuel; risk analysis; sustainable development; renewable energy; biomass; biotechnology; anthropogenic waste processing; energy resource assessment

1. Introduction

The power supply is one of the most critical factors for sustainable development both for particular countries and for the global economy as a whole. In this vein, the UN sets seventeen main targets for the provision of sustainable development of all countries of the world and "cheap and clean energy" is number seven in this list [1]. Among the problems requiring a solution to achieve this target is an increase the share of clean energy and accessibility of technologies of its production using renewable energy (REn). Intensive application of conventional energy over decades had a negative impact on the environment and those linked with it have established additional conditions for comprehensive development of REn in different regions of the world. At the same time, the concept of sustainable state development comprises continuous and complete work for the improvement of the existing social and economic setup. That is why the application of REn is promising for an increase in the share of clean energy in the economy that produced employing innovative technologies, which will assist in solving this problem too. For instance, in 2018, considered to be the first in the world, the REn sector of Germany produced more energy than coal and more than 40% of the electricity generated throughout the year was created using of REn [2]. In Russia, the energy sector operation is historically based on hydrocarbons crude, as well as on hydro- and nuclear power stations.

Moreover, Russia is one of the leading suppliers of hydrocarbons in the global market. These circumstances significantly influence the perception of green energy production and general use of REn in the country. However, in the second decade of the 21st century, the environmental impact of permanent anthropogenic factors has required the taking of appropriate measures in this field both at the federal and regional levels. In particular, the accumulation of industrial, agricultural, and household wastes is becoming an essential incentive for the development of innovative activities aiming at the production of biofuel (BF). According to the existing data [3], the resource potential of recyclable organic waste biomass, including solid household wastes and sediments of wastewaters reaches 13,490.26 tons of coal equivalent (TCE) per year in Russia. This study led both at the federal level and in particular regions which have significant differences in terms of economic, geographic, and climatic conditions.

According to pessimistic forecasts, the share of green energy in the Russian economy will arrive at 3–5% by 2035, according to positive ones, it will be equal to 8–10%. The proposed share of BF in the production of electric and heat energy with targeted management could be about 2–3% [4].

Different geographic, climatic, economic, and other conditions in regions of Russia necessitate solving several scientific problems for optimization of the existing regional programs for the provision of sustainable development and determination of the role of REn and BF production in them.

Therefore, the progress in BF production achieved in some leading countries has become a significant factor which causes a positive impact on the sustainable development of their economies. This gives rise to the suggestion that Russia has some conditions for a more extensive application of REn, in particular, for BF production and use. Correspondingly, this article characterizes some general trends of BF production development and presents the results of the analysis of organizational and economic indexes of BF actual production in Russia as well as some data about planned production. The potential risks of BF production in the Russian Federation are also analyzed.

2. Materials and Methods

The global trends of BF production development, as well as stimulating institutional measures, were mainly analyzed using the works conducted in the leading countries in 2015–2019. BF production capabilities in Russia were analyzed using the published official documents of ministries and agencies, official reports of the Government and ministries, and business enterprises. Information on 74 Russian state and private enterprises involved in BF production or planning to start such production were gathered. Also, the information gathering was conducted by using a questionnaire survey—the questionnaires compiled by recommendations and with consideration of data of other authors [5–8]. The questionnaires contained 20 questions including (i) form of entity; (ii) date of commencement (planned commencement) of BF production; (iii) types of BF produced by the enterprise; (iv) raw materials used for BF production; (v) education level of employees; (vi) type of equipment used for BF production; (vii) sales of produced BF, etc. The questionnaire survey was conducted via e-mail as well as online and offline [9]. Also, the materials of publicly available databases accumulating information on the application of REn in production, such as IRENA.org [10–12], REN21.com [13], GIS RESR [3], etc. were used in work.

Russia possesses ample natural resources, including conventional energy sources. Hence, forecasting of BF production requires the development of different approaches and a set of tools for an unbiased assessment. It is worth noting that geographic, climatic, and economic conditions of Russian regions are substantially different. Due to this, a situational analysis was conducted in this work using the Russian regulatory documents, which are quoted below.

Based on obtained results, possible tools for assessment of BF production development were set out with the determination of several indexes and calculation manners. Relevant management measures for the provision of long-term BF production were elaborated. For setting out of the tools, the "before and after" conditions comparison method was used [14], as well as the contemporary methodology of Global Innovation Index [15–17], National Report on Innovations in Russia [18], some materials from publications on innovation management [19–23] were taken into account. Based on proposed tools with the using of grouping and determination methods [24–26], potential risks have been estimated. Risk assessment was performed using the method of E. Kulikova [27]. Based on the performed analysis, using the method of expert estimations, tendency function, and least square method, the estimated forecast of BF production development in Russia was formulated. Through the exploratory forecast method and using the theoretical basis of Russian and foreign scientists [28,29], possible conditions and future continuations of the found trends of BF production determined. Generally, the study algorithm included six main steps presented in Figure 1.

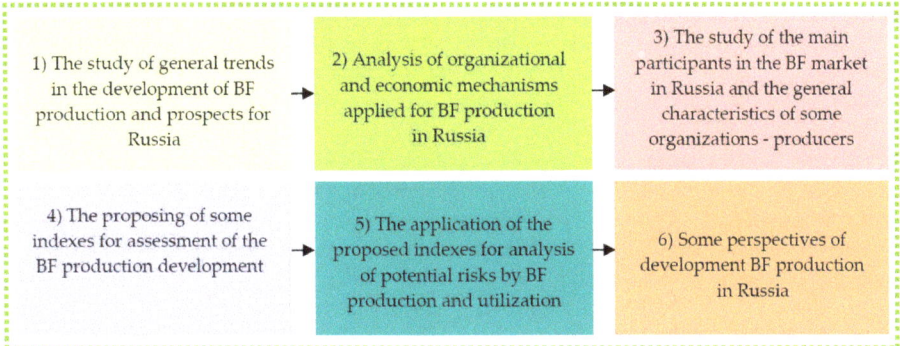

Figure 1. General study algorithm.

3. General Trends of Biofuel Production Development and Some Prospectives for Russia (Research Background)

Over several decades of development, some of the leading BF producers (USA, Brazil, Germany, China, India, etc.) [10] have covered a lot of ground in development and improvement of the used technologies. The first-generation technologies created in the beginning provided industrial manufacturing of the three main types of BF: Bioethanol, biodiesel, and biogas [30–32]. At this stage, starch-containing agriculture products were mainly used as raw materials. At the beginning of the 21st century, several new approaches and technologies were developed, which led to the expansion of BF production and appearing of the products named the second- and the third-generation BF.

For instance, biotechnologies allowing the transformation of carbohydrates not applied in the food industry but widely spread in plants (cellulose, hemicellulose, lignocellulose, xylogen, etc.) [33–35] has been created. In particular, the problem of efficient destruction of beta-glycosidic bonds which bond monomers in such biopolymers was solved. At the same time, wild-growing and cultivating oil plants were found. These plants began to serve as a more efficient raw material in the production of BF than various conventional plants (sunflower, soybean, etc.). As a result, this practice allowed the preservation of the purpose of traditional crops that are used in the food industry [35–37]. Pyrolysis started to be applied in BF production. In the course of this process, quick thermochemical decomposition of lignocellulose occurs at high temperatures and without oxygen [38,39]. The main product of it is levoglucosan, hexose anhydride (six-carbon monosaccharide) which is considered the most suitable substrate for further transformation into the second-generation BF [39]. Then the products of pyrolysis are exposed to, for instance, specific microbic conversion for production of ready-to-use BF, in particular, bioethanol [39].

Moreover, pyrolysis and other particular technologies allowed the use of some prokaryotic (cyanobacterial) and eukaryotic algae as raw materials for BF production [40–42]. Based on appropriate criteria, some authors categorize them as the third generation of technologies [36,42].

Among the reasons for the use of cyanobacterial algal, it is noted that these microorganisms grow fast, and do not compete for agricultural lands and resources. At the same time, they can efficiently transform CO_2 into different organic substances, including lipids and carbohydrates, in their cells [40]. With consideration of the high content of lipids and carbohydrates in biomass of Cyanobacteria and eukaryotic microalgae, technologies for the production of the third-generation BF were created and tested [43,44]. For instance, three main techniques are categorized as the third generation of bioethanol production technologies: (i) fermentation of processed biomass; (ii) dark fermentation of reserved carbohydrates and (iii) direct photofermentation from carbon dioxide to bioethanol using light energy. However, the conventional technologies of fermentation of biomass by using these microorganisms are still of great interest [43].

Considerable attention is also paid to the creation of technologies for the production of bio-oil and biodiesel from cyanobacterial algal [40,45]. Although the cost of algae bio-oil is still rather high (about $2 L^{-1}), as compared to the costs of similar fossil fuel, the cost of such bio-oil may become competitive and occupy a significant market share in the following decades [46].

As can be seen from the given above information, among the characteristics of the three generations of BF, an important place is given to the raw material principle, that is, the ability to process certain raw materials. Alas, this area is also undergoing evolution transformation. Recently there are reports surrounding the development of the new, fourth generation of technologies. Among the features of which is the application of technological principles involving the creation of specialized genetic engineered constructions, modifications of microorganism complexes, etc. [45,47]. The main steps of technological evolution in BF production are presented in Figure 2.

	USED RAW MATERIALS	RAW MATERIALS PROCESSING
G1	Starchy agricultural plants (sugar beet, wheat, corn) and oil plants (rape etc.), palm oil	Thermochemical (pyrolysis, acid hydrolysis, gasification, refining, enrichment); biochemical (enzyme conversion, distillation); microbiological (using of microorganisms for anaerobic fermentation)
G2	Cellulose containing biomass (wood, straw, grass, various industrial and household wastes)	
G3	Wild oil plants, saffron, jatropha, safflower	
G3	Micro- and macroalgae	
G4	Cellulose and starch-containing biomass derived from genetically modified microalgae and Cyanobacteria, as well as some plants	Genetic engineering technologies for the formation of organisms with new metabolic pathways and other useful properties

Figure 2. Scheme of the main steps made during the technological evolution in biofuel (BF) production. G1—first generation technologies [30–32,48–52]; G2—second generation technologies [34,39,49,52]; G3—third generation technologies [36,40–42,53–55]; G4—fourth generation technologies [45,47].

Optimization of the developed technologies has sufficiently accelerated the production processes and also led to an increase in the output of the final product and a decrease in its costs.

For example, a quite high yield of bioethanol has been achieved by the leading countries in conditions of industrial production (75–84% of theoretically possible level and even approximating 95%) [53]. At the same time, the applied technological solutions in the production of BF allowed obtaining an additional some valuable by-products (for instance, glycerin in biodiesel production) which are used in the pharmaceutical, cosmetic and chemical industry. As a result, the by-products began to contribute to the added value generated by the production of certain types of BF [36,56].

One of the particular trends in contemporary bioenergy is the development of fuel pellet production technologies. Wood chips and other timber processing industry wastes are used as raw materials for this [57,58]. The trees growing both in a moderate climate (e.g., *Pinus sylvestris*) and in the tropics may be the source of such raw materials. The relevant products are in a significant demand on the world market due to the environmental component of consumer properties. Currently, the world community is striving to increase the share of REn production and application in the economy, mainly due to the perspective of the environmental safety provision. This striving is determined by the possibility of obtaining additional energy resources and the proven ecological effect from the reduction of CO_2 emission [54,59,60], efficient processing of steadily accumulating biological wastes [61,62] and non-demanded agricultural resources [63–65]. In many countries of the world, there are state plans for development and use of green energy based on REn [66–69] which have confirmed their efficiency. In this vein, it is interesting to note that the growth of electricity capacity with a cumulative result from 2014 to 2018 in global bioenergy, for example, amounted to 30% [11] (Figure 3).

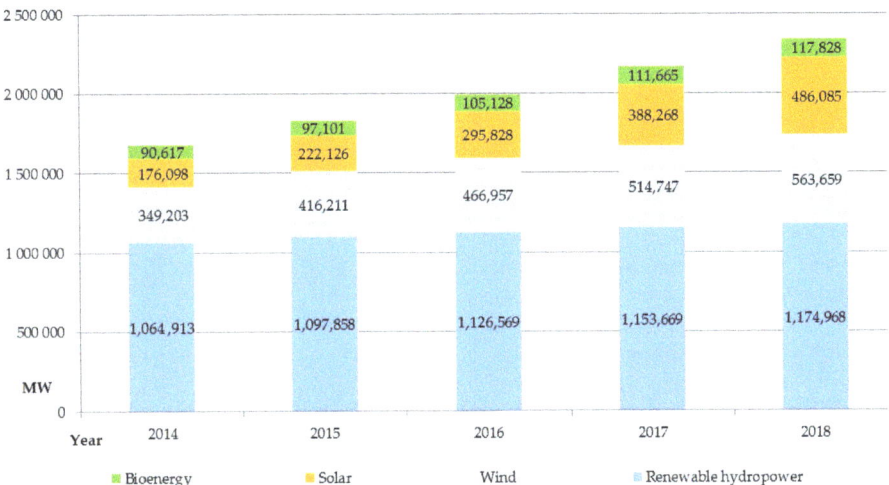

Figure 3. Dynamics of installed capacity of electricity production from renewable sources, megawatt (MW), (2014–2018) [11].

The significant economy indexes already reached in the course of implementation of such plans are confirmed by the materials of the European "Updated Bioeconomy Strategy" [70]. In particular, there are data on the annual growth of added value in the course of production of biochemicals (including BF), which has provided additional €3.5 billion of added value [70]. At the same time, it became known that from 2003 to 2018, world ethanol production increased by 230.5% [69]. Certainly, these results are caused by comprehensive organizational activities by the state and active institutions. Results of ongoing works on regulatory mechanisms improving, standardization of management processes, development, and unification of technological solutions in the area of BF are reflected in the different publications of many scientists from various countries, including Russia [71–77]. It is also worth noting that positive environmental effect from the application of BF is already being registered. For example, according to the available data [78], the CO_2 emissions have reduced by 1.2 million tons since 2015 due to production of electricity using REn facilities. Moreover, a possible reduction of CO_2 emissions may reach 7 million tons per annum due to application of green energy. Currently, there are opening opportunities for a comprehensive economic and environmental analysis from the results of technologies application used in the production of BF [79,80]. The suitable analysis will allow enhancing the evidential base in terms of positive characteristics of the application of BF and diffusion of innovations. As an illustration, the number of publications on economic results, technological features, and environmental consequences of industrial production of BF annually increased. In particular, there are different reports from Spain, Scandinavian countries, Thailand, etc. about the successful experience of BF production from agricultural wastes and timber biomass [81–83].

Due to this, the main direction of REn-based green energy development in Russia may be oriented on the growing opportunities for processing of non-demanded biological resources and various wastes into BF. Thanks to this, it is possible to expect an increase in production efficiency and qualitative change of environmental situation. Besides, several by-products are expected to be obtained [84–86], and these products can contribute to the added value in the production of BF [87].

4. The Results of Analysis of Organisational and Economic Mechanisms Applied for Production of Biofuel in Russia: Status-Quo and Some Prospectives

4.1. Official Regulators and Participants of the Biofuel Market in Russia

The energy sector of Russia based on conventional energy resources is the market barrier on the way of development of renewable energy. The existing production problems of REn facilities connection and high capital expenditures for construction of REn facilities [88] slow down the growth of potential utilization capacity of this type of energy.

The key roles in the economic life of Russia played by the state and other official regulators require special consideration within the analysis of organizational and economic mechanisms applied in BF production. Naturally, the Government and the specialized ministries are located on top of the management pyramid. Also, different programs and projects are developed and officially regulated by appropriate agencies, research, and educational institutions, technological platforms, as well as other organizations accumulating information on key stakeholders of the market. Such organizations may be considered a separate group which provides policy-making, development of significant state and regional programs as well as supervision of their implementation.

The group of the regulator organizations (i) actively cooperates with BF producers which are enterprises with different profiles, sizes, and forms of incorporation. BF producers may be considered as a separate group (ii) of the market participants. Another group of participants (iii) is de facto formed by different consumers. Some of them act as intermediate consumers, converting biofuels, for example, into electricity. Then, this electricity is transferred to end consumers. Finally, it seems appropriate to note various non-profit organizations, profile associations, specialized companies involved in market relations in the course of production and utilization of BF, as a distinct group (iv). The general scheme of relations between the four groups of the BF market participants (organizations) is shown in Figure 4. The analysis of information on these organizations will be given below.

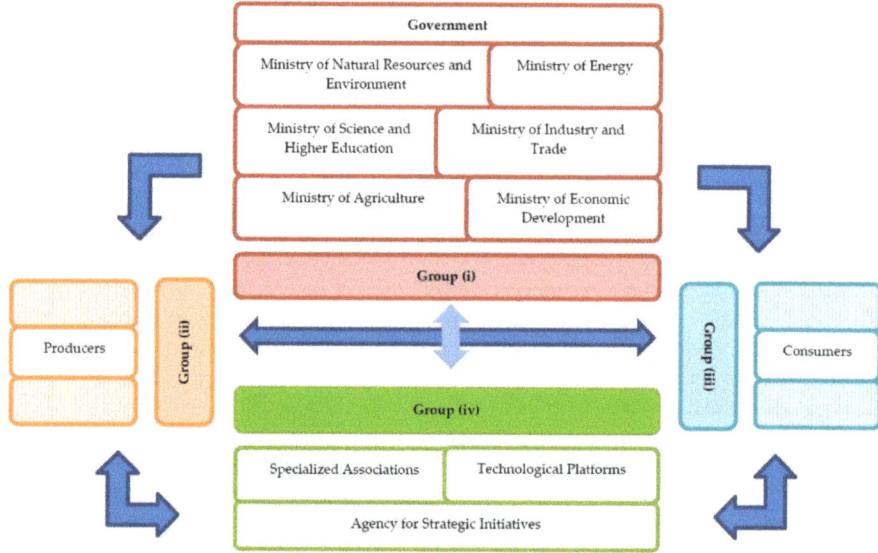

Figure 4. The general scheme of relations between the four groups of the BF market participants group (i) [89–92]; group (ii) [93,94]; group (iii) [94–96]; group (iv) [97–101].

It is worth noting that the Federal Law No. 35-FZ dated on March 26, 2003 "On Electric Energy" already provided different mechanisms of stimulation support of electric energy production by REn-based generating facilities with a capacity not exceeding 25 MW. For example, compensation was provided for the cost of technological connection to electrical networks [102]. The main conditions of the state regulation of the green energy production processes and utilization of renewable energy are applicable since 2009 following the Decree of the Government of the Russian Federation "The Energy Strategy of Development of Russia before 2030" [103].

One of the recent steps for implementation of the green energy development policy in Russia was introduction of amendments to the Federal Law No. 171-FZ dated on November 22, 1995 "On State Regulation of the Production and Distribution of Ethanol, Alcoholic and Alcohol-Containing Products and Restrictions on the Consumption (Drinking) of Alcoholic Products" [104,105]. As a result, some restrictions on bioethanol production were lifted. Also, the procedure of licensing of this type of activity was set out.

Moreover, over the previous 3–4 years, the Government of the Russian Federation and some profile ministries (the Ministry of Energy [89], the Ministry of Economic Development [90], the Ministry of Natural Resources and Environment [91], the Ministry of Industry and Trade [92]) compiled a number of statutory documents aiming at development of BF production in the country. In particular, the Government of the Russian Federation [102,106] has provided for increasing the investment attractiveness of projects for the construction of REn-based generation facility. For this purpose, excessive requirements to the process of design, development, and operation of such generation facilities located in hard-to-reach and isolated areas were excluded. In addition, the Ministry of Energy of Russia prepared some of the measures for the systematization of management in energy and introduced the term "energy management" [107]. Thus, we can talk about the emergence of new opportunities for a system of energy management.

The group of regulator organizations (i) also takes new measures for improvement of the investment environment in the area of BF production. For instance, in 2018, The Ministry of Economic Development of Russia proposed the draft law "On Public Non-Financial Accounting" for public consideration and it currently undergoes expertise [108]. In particular, this document provides the establishment of conditions for the increase of transparency of organizations concerning environmental impact, awareness of the general public of the existing international social responsibility standards, and provision of sustainable development. The list of crucial indexes of the planned accounting will include such essential characteristics as determinants of the weight of pollutants emissions from stationary sources in atmosphere and the estimation of the contribution of energy resources produced employing of REn in the total volume of energy resource production. Moreover, the Government elaborated the plan [109] of reduction of the level of greenhouse gases emission down to 75% of the level of emissions in 1990 by 2020. This plan [109] also provides development of the relevant information base, forecasting the level of greenhouse gases emission for the period before 2020 and before 2030.

The conditions of BF productions may be significantly changed by the initiative of the Government of the Russian Federation for the introduction of amendments to the Federal Law "On Electric Energy" regarding the development of microgeneration [110]. The proposed changes will allow increasing the share of REn-based facilities with an installed capacity not exceeding 15 kW in Russian fuel and energy balance. The following measures are proposed for reaching of this target: (i) to provide an opportunity of sales of excessive produced energy in retail markets for default providers (not to consider such activity as business activity); (ii) to grant authorities for setting of aspects of commercial metering of the produced electric energy, determination of the procedure of payment and the procedure of microgeneration facilities utility connection, etc. to the Government of Russia; (iii) to establish an obligation of execution of an electric energy sale and purchase agreement between the default supplier and the microgeneration facility owner addressed to it and to define the mechanism of price formation for the electric energy purchased by it.

Currently, the use of innovative technologies for the production of BF (I-IV generation) and systematic work on the spread of this type of energy resources have shown that on the scale of individual countries, thus the creating of prerequisites for sustainable economic development [111–114]. Consequently, in Russia, with the development of BF production without the use of food crops, similar results can be achieved. The elements of organizational and economic approaches aiming at reaching the said target presented in Figure 5.

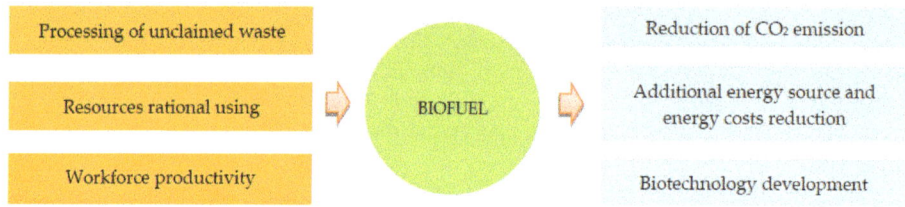

Figure 5. Scheme of organizational and economic approaches to use Biofuel production as a way to sustainable development in Russia regions [4,52,106,108,109].

The significance of waste processing problem in Russia has convincingly demonstrated the data contained in the documents of the Ministry of Natural Resources and Environment of the Russian Federation [91]. For instance, the State Environmental Protection Program states that the volume of waste processing will require to be increased by two times by 2020 and by ten times by 2025 [115]. According to the same Ministry, more than 60 million tons of solid household wastes are generated in Russia per annum. It corresponds to about 400 kg of residues per 1 person per year, and only about 7–8% of the collected wastes of this type are used in economic turnover [116].

According to the available estimations, currently, only 12% of the bioenergy potential of Russia is being used [117]. To combine the efforts of different ministries and agencies for the development of REn, including BF, is considered to appropriate the conduction, a comprehensive assessment of existing biological resources in dynamics. Special attention is paid to prospective of BF production and introduction of the relevant management mechanisms in the global energy and heat management system.

It is believed that the total amount of organic wastes in Russia can reach about 607,000 thousand tons per annum [3]. The available materials regarding the Federal districts of Russia [3,12] allow characterizing the structure of gross potential of organic wastes which may be processed into BF (Figure 6). These and other indexes may be applied in the course of analysis of the potential reduction of CO_2 emissions.

Taking into account the current situation, the Government of the Russian Federation approved the action plan "Development of Biotechnology and Gene Engineering" for 2018–2020 [118]. It specifies the establishment of regulatory, market, and technological conditions for bioenergy development in Russia with emphasis laid on the development of electric energy production.

In general, the efforts of the regulator organizations, lead to optimization of different producers and consumers of BF activity. For instance, the initiative mentioned above of the Government of the Russian Federation on the introduction of amendments to the Federal Law "On Electric Energy" [102] includes a provision that default suppliers are the organizations (including private companies) which conduct power supply. Such organizations shall enter into a power supply agreement and an electric energy (power) sale and purchase agreement with any consumers of electric energy addressing them. Persons acting on their behalf or behalf of another consumer of electrical power and to the benefit of the said consumer of electrical power may act as consumers [102]. As a result, a direct and useful tool for connecting microgeneration facilities to power grids will be.

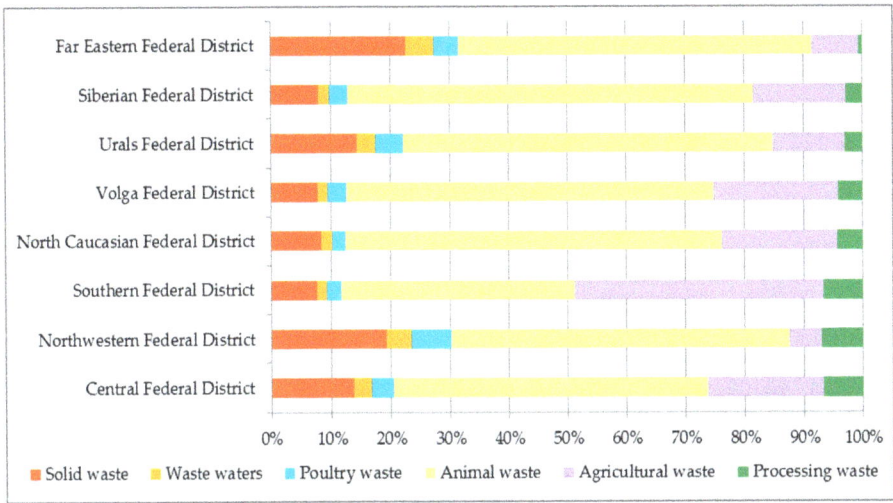

Figure 6. Structure of gross potential of organic wastes in Russian regions, thousand tons per year [3,12].

Some positive initiatives are contained in the draft Federal Law "On Introduction of Amendment to Article 217 of the Tax Code of the Russian Federation" [119]. In particular, it proposes to exempt the income of natural persons received from the sale of energy produced using the microgeneration facilities owned by such person from taxation.

In many cases, the state regulator organizations act as customers of works related to BF production in research and educational institutions. For instance, as part of the state contract, the Geographic Department of Moscow State University [120] and the Joint Institute for High Temperatures of the Russian Academy of Sciences [121] created the special geo-informational portal GIS RENEWABLE ENERGY SOURCES OF RUSSIA [3]. This web resource contains many essential materials on renewable energy in Russia, concerning new technologies and BF production facilities as well as other useful information. This portal will assist in the establishment of an efficient logistics and sales concept of BF utilization in Russia as well as investment planning.

The list of research institutions of the Russian Federation involved in studies of BF creation and utilization includes, in particular, Research Center of Biotechnology RAS [122], National Research Center "Kurchatov Institute" [123], Joint Institute for High Temperatures of the Russian Academy of Sciences [121], etc. Some educational institutions also participate in this work: Moscow State University [120], National University of Oil and Gas "Gubkin University" [124], National Research Nuclear University MEPhI (Moscow Engineering Physics Institute) [125], Moscow Polytechnic University [126]. Also, there are seven specialized departments in educational institutions providing education in the field of REn [78]. However, this is not sufficient for the intentions of the long-term development of this type of production. Nevertheless, the educational standard for bioenergy and biofuel production specialist has already been approved [127].

A functional role is also played by several technology platforms related to production of BF (Bioenergy [97] and Biotech2030 [98]) with their activities also aiming at "utilization of renewable sources of biomass for the purposes of rational and sustainable industrial production and energy provision with lowering of negative environmental impact" etc.

In addition, the Government of Russia established the Agency for Strategic Initiatives [128] to assist in social and professional mobility of young professionals and teams in business and social areas, including by supporting of socially relevant projects and initiatives [128]. To reach this target, the Agency supports the implementation of the national technological initiative [128], which comprises the development of nine markets, including EnergyNet [129]. EnergyNet aims at the creation of smart energy networks, but "development of REn is also included in the wide area of interest" [129]. In the meantime, the roadmap approved by the Government [130] specifies "provision of efficient and reliable energy supplies for isolated and hard-to-reach territories by 2020 using hybrid systems (the optimal combinations of technical solutions: Renewable sources of energy, diesel, gas, local types of fuel, innovative generation, accumulators, electric energy distribution systems, management systems)", including by lifting administrative barriers, among the results planned by June, 2020.

Some non-profit organizations, e.g., Association "Nonprofit Partnership Council for Organizing Efficient System of Trading at Wholesale and Retail Electricity and Capacity Market" (Association "NP Market Council") are known too [131]. This organization certifies the generating facilities which utilize REn and maintains the register of issuance and revoking of certificates confirming the production volume of REn-based electric energy. Over the several years of its activity, the following facilities were included in the list of certified generating facilities utilizing organic raw materials (biomass, wastes): "Bely Ruchei" industrial mini-HES (Heating electrical station) (6 MW, Vologda region), "Luchki" biogas power plant and "Baitsury" biogas station (3.6 MW and 0.526 MW respectively, Belgorod region) and the "Novy Svet-Eko" solid waste landfill active degassing station with a landfill gas (LFG) power plant (2.4 MW, Leningrad region).

In the Russian BF market and the REn market in general, there is an additional but essential group (iv) composing by organizations with different forms of ownership. Their activities are oriented on arrangement and coordination of cooperation between producers and consumers of BF for the provision of rational and sustainable industrial production and energy supplies with lowering of negative environmental impact.

This group (iv) includes Joint-stock company «Trading System Administrator of Wholesale Electricity Market Transactions» (JSC "TSA") [132], a subsidiary of Association "NP Market Council" [131], which is empowered to conduct tenders and ensure settlements between producers and buyers of electricity. Between 2013 and 2018, JSC "TSA" has selected 229 investment projects of construction of REn-based generating facilities with installed capacity exceeding 5 gigawatt (GW) for further support.

Different profile Associations play significant roles as cooperation coordinators. Among them, the Russian Biofuel Association [100], ENBIO Association [101], "Global Energy" Association for Development of International Energy Research and Projects [133] are especially active.

Establishment of Russia Renewable Energy Development Association (RREDA) [78] in 2019 was an essential event for REn development in Russia. Among the Association goals are assistance in the development of REn in Russia and member states of the Eurasian Economic Union, development of international cooperation, popularization of knowledge, increasing of skills in the field of REn, participation in the industry regulations development, etc.

Data about the most active innovative Russian enterprises producing BF and consumers of this type of energy resources will be reviewed in the following subsections with consideration of the regional aspect.

4.2. Liquid Biofuel Producers

The reserves of conventional energy resources available in Russia in conjunction with high capital expenditures in the production of liquid BF do not create conditions for proper development of that industry. Moreover, the legislation is still in transition and requires improvement.

Reprofiling of existing ethanol plants due to carving out of bioethanol from the law "On State Regulation of Production and Turnover of Ethanol" may become an impulse for the development of production of liquid BF [104,105]. For instance, the capacities of 17 ethanol plants in the Republic of North Ossetia-Alania which may be used for the production of bioethanol are unemployed, which is confirmed by the report of the Head of the Republic [134]. At the same time, the international standards, in particular, of fuel bioethanol, mixtures of biodiesel fuel, etc. are applicable in Russia [135–137]. Therefore, there are opportunities for production and using liquid BF in volumes sufficient for utilization as part of 5–10% mixture with gasoline [138–142]. Appearing of corresponding plan indicators may be expected to assist in the development of this type of activity and enhancement of environmental situation.

Development of appropriate biotechnological production may become another institutional breakthrough. For example, the Titan group [143], which is involved in petrochemical and agricultural production, is planning to start producing bioethanol of grain in the coming years [144].

At the same time, according to available data, the raw material supplies base of bioethanol production may be provided by agrarian regions of Russia by the processing of unclaimed agricultural products. On the example of certain regions of South of Russia, it is shown that the economic activity for the development of green energy and utilization of REn has begun as part of the state and regional programs and upon private initiative [52]. However, there are also significant untapped opportunities.

Some active enterprises capable of processing bioethanol into products with the price compared to that of the petrochemical plants production may be an exciting example promising for production and further utilization of liquid BF in Russia [113]. Such activity is already commenced by ETB Catalytic Technologies [145].

According to the basic forecast of IRENA, with target-oriented management, cumulative consumption of liquid BF in Russia may reach 200 petajoule (PJ) by 2030 [12]. Therefore, there is the reason to believe that in the future production of liquid BF will reach a decent level and will assist in the sustainable development of different regions of Russia. The questionnaire survey showed that at least eight Russian enterprises are planning to start producing liquid BF in the near future.

4.3. Producers and Consumers of Solid Biofuel

The conducted analysis showed that current production of solid BF in different regions of Russia is oriented on the following major products: Fuel peat, timber, wood chips, and fuel pellets. It is fuel pellets that are on the leading interest for different producers as they are demanded in the global market and successfully exported. For instance, according to Proskurina et al. [146], the Russian production of fuel pellets will keep its high potential provided this product is sufficiently demanded by EU countries [95,96].

With consideration of the existing achievements, the Strategy of Development of the Russian Timber Complex before 2030 was updated in 2018 [147]. In the course of its implementation, it is planned to achieve significant effects of the measures for stimulation of production, export, and consumption of BF. In particular, the basic scenario implies that the export of pellets will reach 3912 thousand tons by 2030 with a gross potential production volume of 3968 thousand tons. As a comparison, this volume was equal to 1112 thousand tons in 2016, i.e., it is planned to be tripled within 15 years. In case of successful implementation of the updated strategy, the federal and regional budget revenues may reach 1.4 billion rubles per annum in 2030 [147].

The forecast by IRENA suggests that cumulative consumption of solid biofuel in Russia may be equal to 146 PJ by 2030 [12]. Currently, the main part of pellet producers (about 60%) is located in the North-West of Russia, and about 30% is located in the Central region [148]. The leading pellet producing and exporting regions of Russia are Leningrad, Arkhangelsk, Irkutsk regions, Khabarovsk, and Krasnoyarsk krais [149].

In the Leningrad region, the regional state information system (GIS "Energo Efektivnost") is introduced for maintaining of energy-saving and energy efficiency [150]. The leading enterprises of the region are Vyborg Forestry Development Corporation [93], Setnovo (Stora Enso) [151], etc.

According to available data, the annual consumption of BF in the Arkhangelsk region is about 1.2 billion TCE, the share of BF utilization in public energy service reached 42.8% in 2018 whereas the plan indicator is 44% [152]. The region government sets itself the task to expand the use of REn obtained from wood reserves, that will provide up to 30% of the energy supply potential [153]. In the Irkutsk region, it is planned to increase the production of wood pellets by 3000 tons in Biotoplivo-Irkutsk Limited Liability Company (LLC), capital expenditures for production capacities will be equal to

123.9 million rubles [154]. In Krasnoyarsk Krai, BF is utilized in 41 boiler plants, and it is planned to produce 59% of heat energy [155] using this type of fuel.

Production of pellets is intensively developed in many regions of the North-Western Federal District, Siberia and the Far East with the attraction of important investments. For instance, it is reported on plans of construction of a wood torrefaction plant in the Novgorod region, the project volume of investment will be equal to about 14 million dollars [156]. When characterizing producers and consumers of solid BF in some regions of Russia, it is necessary to note a plant functioning in Altai Krai since 2015, which besides sunflower meal and protein concentrate, produces 12,000 tons of fuel pellets [144].

According to the questionnaire survey conducted, 52 Russian enterprises are already producing or plan to start creating solid BF in the nearest future.

4.4. Producers and Consumers of Biogas

The "Luchki" biogas power plant (BPP) owned by AltEnergo (Belgorod region, Luchki settlement) has been efficiently operating in the Russian market for more than seven years [94]. Currently, the "Luchki" BPP annually processes about 95 thousand tons of raw materials using different types of environmental pollutants as raw materials: Solid cattle-breeding wastes (pig and chicken excrements, non-used parts of bodies of livestock animals, etc.), crop production wastes and household wastes. A pig-breeding farm supplying the raw materials using a particular pipeline is located nearby. As a result, according to current estimates, from the time of launch to 2018, the "Luchki" BPP processed about 350 thousand tons of waste [157]. In other words, this BPP achieves a rather significant environmental effect.

The produced biogas is used for generation of electric energy. The capacity of the generating unit reaches 3.6 MW, which allows producing up to 29.3 million kWh per annum. In 2015, LLC "AltEnergo" connected the "Luchki" BPP to the electric network of Interregional Distribution Grid Company of Centre JSC, Belgorodenergo. As part of the stimulation of electrical energy production by REn-based generating facilities, 50% of the capital expenditures for network connection were compensated by the state [102]. Consequently, the "Luchki" BPP currently acts both as a producer of biogas and intermediate consumer, whereas the end-users of the produced electric energy are residents of neighboring settlements.

The output of electric energy produced by the "Luchki" BPP has been growing over the last years. For instance, the "Luchki" BPP produced 1.5 million kWh of electricity more in the first half of 2019 than in the similar period of 2018 [94].

Alongside with electric energy, the "Luchki" BPP created about 27 thousand Gcal of heat energy and obtained 90 thousand tons of biofertilizers as a by-product, which was successfully marketed [94].

For about seven years, Regional Energy Company LLC which produces biogas in the "Baitsury" BPP and processes it into electric energy generating 0.5 MW, has been operating in Belgorod region [158].

It appears that the development of renewable energy in the Belgorod region was provided by taking a set of institutional measures [159]. In particular, some pilot projects were implemented in this field, the regional REn research center was established, and bioenergy specialists were trained in local educational institutions. Last but not least, the relevant production facilities were constructed. A smart grid was established on the basis of AltEnergo and support of certified generating facilities was provided.

Recently, the leadership of the Belgorod region has identified regional target indicators of the development of REn-based green energy before 2020 [160] and specified tax benefits and subsidizing of credit interest rates [161].

It is worth noting that the concept of development of small-scale distributed energy before 2025 is developed and approved in the Belgorod region [162]. According to this plan, the number of generating units capable of processing the agricultural wastes shall rise up to 100 by 2025 (capacity of 223.3 MW). Consequently, the conditions for the efficient disposal of organic wastes will be established, which shall lead to a reduction of human-induced impact on the environment. Moreover, it is planned to widen the energy infrastructure and production energy clusters in rural areas, which will allow increasing the sustainability of power supplies and will limit the growth of prices for heat and electric energy. It is expected that the planned measures will provide sustainable development of the region [162].

To some extent, the positive experience of biogas production in the Belgorod region influences the neighboring regions of the Central Federal District of Russia. For instance, attention to the energy production from renewable raw materials was reflected in the recent Decree of the Head of Tambov Region "On Approval of the Concept and Program of Electric Energy development in Tambov Region in 2019–2023" [163]. This document provides data on the feasibility of deploying BF production from agricultural wastes. For the implementation of this task, it plans to build mini biogas heating and power plants in the existing pig-breeding farms for the production of electric energy and disposal of plant biomass and livestock waste. The designed capacity of the planned facilities for the energy production from renewable raw materials will lead to the production of both heat energy (4 thousand Gcal) and electric energy (8.3 million kW) and by-products in the form of organic fertilizers (14.3 thousand tons). Moreover, JSC "Biokhim" is assumed to be one of the key players [164].

The authorities of the Lipetsk region also plan to use REn-based technologies. According to the analytical data presented in the Decree of the administration of the Lipetsk region [165], investments in solar and wind energy are evaluated as being of little promise for this region. It is preferred to develop renewable energy based on the processing of agricultural waste and construct small hydropower plants. It is believed that the gross bioenergy potential of crop waste in the Lipetsk region is 1153.5 thousand TCE and livestock: 104.4 thousand TCE. This raw material can ensure the production of 529.29 MW of energy and even lead to energy independence of the agricultural sector in this region.

According to the questionnaire survey conducted, fourteen Russian enterprises are already producing, or plan to start producing biogas in the near future.

4.5. Prospects of Utilization of Algae and Wastewaters as Raw Material for Biofuel Production in Russia

Reportedly, over the first two decades of the 21st century, the tracked climatic changes and growing impact of human-induced factors have led to, in particular, an increase of biomass of cyanobacterial algal. This biomass is considered as promising raw material for the production of BF [40,166]. Cyanobacteria is a large and rather heterogeneous group of prokaryotic organisms, some species of which are toxic. In some regions of the world, including in Russia, these limnetic, and marine microorganisms have begun to affect the environment significantly and even potentially threaten human health.

It is known that Cyanobacteria and some eukaryotic microalgae have complex photosynthetic systems. These systems allow using solar energy for the accumulation of various organic compounds which may be used for the production of both BF and a number of valuable by-products [40,167,168]. There are several reports on the creation of the third generation of BF production technologies which allow processing algae, including specifically cultivated ones, into BF [169–173]. At the same time, some authors pose the question of whether it is possible to use contemporary technologies of algae processing effectively for environmentally clean production of BF and cleaning of wastewater [46,174]. It is also noted that currently there are different problems which make it complicated to give a positive answer to this question without any reasonable doubt.

For instance, according to some authors, the industrial utilization of cyanobacterial algal is not economically sustainable due to the comparatively high cost of production of required biomass [175]. Although there is a broad spectrum of methods for manufacturing and accumulation of these microorganisms, including separation of cells from surrounding liquid, apparently, there is still no universal technology. According to Singh & Patidar [175], the efficient technique for this purpose may be a combination of different methods matching the following six criteria: Provision of biomass quality, economic cost, the quantity of biomass, appropriate processing time, maintenance of species specificity and nontoxicity. Singh & Patidar [175] pay special attention to the methods based on different methods of filtering and ultra filtering stressing that they demonstrate high efficiency and economic cost, do not require any chemical actions and allow using filtered water (water recycles).

The significance of emerging opportunities of algae and wastewaters processing may be presented using the example of the Volga river basin, the largest river in the European part of Russia. In particular, the long-term research of Cyanobacteria in the river Volga and many other water reservoirs of the Volga basin showed that the content of different species of these organisms reached rather high values in summer. For example, it was shown that average quantity and biomass of Picocyanobacteria in Gorky and Cheboksary basins of the river Volga might range between 34–322 × 10^3 cells per mL and 38–455 mg/mL^3 respectively in summer [176]. In addition, molecular genetics methods and immunoassay, the populations of toxic Cyanobacteria synthesizing hepatotoxin (microcystin), were discovered in Upper Volga reservoirs [177]. It was also noted that non-indigenous species of planktonic algae previously inhabiting other regions of Eurasia or even North America appear in Volga water [178].

One might think that the conditions of the Volga reservoirs with a slow flow of water and relatively shallow water contribute to the intensive growth of planktonic Cyanobacteria. As a result, there is a very negative phenomenon called "water bloom" [176,179]. These processes are also significantly affected by the so-called diffusive pollution of the Volga basin water objects. The anthropogenic sources of it shall be studied, and the effect of them shall be minimized to enhance the environmental situation in this vast region [180]. Cyanobacteria accumulate in wastewaters and water treatment plants, like excessive planktonic microorganisms, shall be disposed of and they may be utilized, in particular, for production of BF and some by-products.

It is believed that currently the environmental situation is rather tense in the Volga basin due to the anthropogenic pollution [180]. Due to excessive discharges of polluted water, in particular, the active development of Cyanobacteria occurs [181]. For solving of the existing problems, the Presidium of Presidential Council of the Russian Federation elaborated and approved the federal project "Preservation and Prevention of Pollution of the River Volga" in 2017 [180,182]. By this project, a set of measures shall be taken for reduction of the volume of discharge of polluted water in the river Volga basin by 80% before 2025. In the course of implementation of the project, it is planned to construct several engineering, sanitary-hygienic facilities. Also defined is a list of treatment facilities that must be built in 16 regions of the Russian Federation involved in the project.

The initiatives, which are comprised by this project, may facilitate the development of unique, economic opportunities for production of BF directly of algae, wastewater and organic raw materials, that may be produced during wastewater treatment by the algae-bacterial consortium [183].

By the reference data [183,184], there are consortiums of eukaryotic and prokaryotic algae, which are capable to provide a high level of synthetic wastewaters treatment. There is the opinion that these consortiums be able to eliminate up to 96% of chemical oxygen demand (COD) [183,184].

The use of technology with the participation of such consortia of microorganisms seems to be quite promising given the volume of polluted wastewaters discharged by the constituent entities of the Russian Federation in the river Volga. It is estimated this amount could reach up to 3175 thousand km^3 per annum [184].

Some propositions for determination of growing points for the production of BF in the Volga river basin are justified by the conducted studies and technical solutions related to the processing of Cyanobacteria into BF [184–187].

Thus, the orientation on the utilization of Cyanobacteria and eukaryotic algae in conjunction with the introduction of the third-generation BF production technologies is rather promising at least for some regions of Russia.

In conclusion, the analysis of the Status-Quo of BF production in Russia show that the creation of BF is performed with a relatively low level of subsidizing and at certain market limitations. Given this, the efforts aiming at the processing of different wastes and non-demanded raw materials seem to be promising. As a result, a significant economic effect which will allow for making a big step towards "establishment of the renewable energy sector" [129] is possible.

Apart from distinct environmental, social and economic advantages, according to some authors, production of BF may open a door for the transition of economy from the linear model to the closed cycle model or "circular economy" [188–192]. It is believed that transition to the circular economy is the most efficient method to provide sustainable development.

Hence, ongoing with the taken measures, it is necessary to assess the reached effects and possible risks in dynamics. Solving of such problems requires the formation of assessment, measurement instruments [189], and the determination of application conditions.

4.6. Six Indexes for Assessment of the Biofuel Production Development

The conducted analysis of organizational and economic mechanisms applied for production of BF in Russia and the obtained data on several producers allowed for proposing six individual indexes to the BF production development assessment. The proposed set of assessment tools is intended for use in the BF production management system in Russia.

First, at least for Russia, the level of raw material reserves may be one of the important indexes. Index assessment is proposed to be carried using the k(M) ratio:

$$k(M) = M_{BF}/M_n, \qquad (1)$$

where M_{BF} is the amount of available raw materials for BF production, M_n is the number of raw materials required for the continuous production of BF.

It is expected that calculation of the index (1) will allow systematizing the information about available raw materials, their types, territorial belonging and major technical and economic characteristics as well as on opportunities to use the existing reserves.

Secondly, the availability of BF for end consumers may be of significant importance. This characteristic may be assessed employing the index (2) represented by BF consumption ratio k(CON). For the corresponding calculation, it is necessary to define the share of energy consumption, produced by BF, in the total energy consumption:

$$k(CON) = CON_{BF}/CON_t, \qquad (2)$$

where CON_{BF} is a volume of consumed electricity (in MW) or consumed heat energy (in gigajoule (GJ)) produced from BF, CON_t is the total consumption of electric or heat energy (in MW or GJ, respectively).

One might propose that determination of the index (2) in dynamics will allow for combining the data on BF sales, the territorial potential of use, technological connection to electric networks, and opportunities of interregional distribution.

It is suggested that, based on results of the index (2) analysis, opportunities for taking different measures aiming at the formation of the sustainable target market will appear [193].

Another one means of assessment may be monitoring of tariffs on energy resources produced from BF. For this purpose, it is proposed to compare the cost of a unit of energy produced from BF with conventional analog by calculating the cost ratio k(V):

$$k(V) = V_{BF}/V, \qquad (3)$$

where V_{BF} is the cost of a unit of energy produced from BF, V is the cost of a unit of energy produced by conventional sources.

It is evident that for the development of efficient production of BF in Russia, it is necessary to give preference to competitive technologies of raw material processing, as well as to lower the number of subsidies and other financial incentives.

The following proposed index (4) is intended for assessment of technical support of BF production. This problem may be solved using the technical support ratio k(F):

$$k(F) = F_{BF}/F_t, \qquad (4)$$

where F_{BF} is the number of machinery and equipment for the production of BF, F_t is the total number of energy machinery and equipment. The index (4) will allow characterizing the increase of the number of machinery and equipment for the production of BF in dynamic.

At the stage of mass production and application of BF, it will be important to assess how it may be influenced by changes in the levels of state subsidizing. Hence, the special index (5) using the subsidizing ratio k(G) for analysis is introduced:

$$k(G) = G_{BF}/G_t \qquad (5)$$

where G_{BF} is the number of subsidies for the development of BF, G_t is the amount of state subsidies for innovative activities.

Lastly, with consideration of the fact that the role of the workforce capacity of BF production will rise, it is necessary to conduct monitoring of specialists involved in this production. The corresponding index (6) may be obtained using the ratio: k(H), which reflects the number of employees with specialized education in the area of BF production.

$$k(H) = H_{BF}/H_t, \qquad (6)$$

where H_{BF} is the number of employees with specialized education in BF production, H_t is the total number of employees.

Currently, the estimated number of employments in the BF production sector is 1.9 million, and Brazil is justly considered the leader in terms of the number of employees [13]. In terms of this data, Russia significantly lags behind the leading countries. It is evident that an increase in employees' qualifications in BF production will be associated with long-term expenditures in the sphere of education for Russia. Management of this index (6) will allow providing human resources for BF production over the long term.

The proposed indexes are intended for detection of different necessary measures, which will be planned for the long-term development of BF production. Using the proposed indexes, various measures aiming at the development of BF production possible to plan and to take. The appropriate management may be implemented by analyzing the dynamics of these indexes and the application of targeted transmission mechanisms.

In particular, the index (1) may be considered convenient for determination of the balance between the amounts of available raw materials and the amounts required for maintaining of continuous production of BF. In this case, if it turns out that k(M) < 1, one should expect the suspension of production, with k(M) > 1, overstocking is possible. Therefore, the significant (indicator) value of the index (1) is 1.0. For correction of the situation and maintenance of continuous BF production, measures for optimization of generating facilities deployment and raw material base may be taken. Moreover, it is possible to change the methods of transfer or products transportation with the consideration of the location of the end-user.

The results of BF accessibility for end consumers determination using the index (2) may be useful for arrangement of measures aiming at the increase of the consumption ratio k(CON). The proposed significant (indicator) value of the index (2) is 0.05. This value is based on an estimated share of BF consumption in the total amount of energy resources in Russia [4,12,78]. Among the measures which may influence this index, it is recommended to use the mixture of gasoline with addition of 10% of liquid BF, to enhance the indirect BF costs control mechanism ("green tariff"), to keep tax benefits due to lack of mineral extraction tax, to consider possibility of quota arrangement of BF consumption. Also, the positive effect can provide the specialized information base containing data on the availability of raw materials for BF production, location of producers, licensors of technological solutions and possible consumers of BF in the territory of Russia.

Moreover, the measures aiming at achieving this goal may probably include different educational events for organizations producing and processing the crop products, timber industry enterprises as well as for concerned parties representing the private business. Participation of executive bodies and investors in such events is believed to be an essential condition for the achievement of expected effects.

The index (3) may be useful for taking measures aiming at the reduction of the cost of a unit of BF energy produced by BF. The possible ideal indicative value of this index is 1.0, i.e., equalizing of the cost of energy produced by renewable and conventional sources of energy. In this case, it is considered necessary to estimate the economic effect of CO_2 emission prevention which is reached by using BF as compared to conventional hydrocarbon sources of energy.

The index (4) is intended for monitoring of increase by the number of machinery and equipment involved in the production by BF. The possible significant (indicator) value of this index is 0.05 as it reflects the share of equipment required in BF production in the total number of energy equipment in Russia with specific assumptions.

Development of this index may be stimulated by the provision of maximum equivalence of connection and introduction of environmental requirements for processing of waste and unnecessary agricultural products in BF. At the same time, it is also necessary to define the enterprise's fixed assets threshold deterioration level and efficient ways of possible interaction with the centralized distribution network.

The use of this index may assist in the popularization of knowledge on available BF generation technologies and equipment with different capacity. For example, the relevant information on existing mini-plants will be useful for organizing of the BF production in hard-to-reach or rural areas.

The results of the application of the index (5), using the subsidizing coefficient, apparently will be useful for efficient cooperation between BF producers and the regulator organizations. In case of positive economic effect and increase of the number of power supply agreements, it is proposed to balance the number of state subsidies and to consider the base trend of this type of investment the downtrend with use of the subsidizing ratio k(G). It appears that assessment of the significant (indicator) value of the index (5) is not possible so far.

The index (6) applying the k(H) ratio is intended for monitoring of specialists involved in BF production and may significantly influence the processes of training of such specialists. For instance, the obtained data on necessity of additional high-skilled specialists involvement in production of BF or on their retraining with consideration of new technologies and equipment will stimulate a set of measures, in particular, those aiming at updating of existing educational standards, modernization of educational programs and establishment of other conditions for additional BF-related education. The opportunity of scientific and educational support employing association of industrial enterprises with regional scientific and educational centers or specialized consulting facilities may play an essential role. It will become possible to define the significant (indicator) value of the index (6) as far as similar experience is accumulated.

Naturally, alongside the target measures causing influence on the process of creation and application of BF, it is essential to assess possible risks which may cause a negative impact on the development of BF production.

4.7. Application of the Proposed Indexes for Analysis of Potential Risks of Biofuel Production and Utilisation in Russia

The assessment of risks faced by companies producing BF in different countries of the world is an essential component for rational decision-making [24–26]. Respectively, it appeared to be necessary to analyze the above-listed indexes for a description of potential risks which may arise in this sphere of activity in Russia. The risks were categorized and grouped according to the assumption that the state currently provides public assistance to this sector of the economy, and the political risks are minimal.

To define attitudes towards specific types of risk, the value of possible losses (P_q) and possibility of arising (I) were estimated by reflecting the risk rank (I_r) as the product of these values: $I_r = P_q \times I$. Risks ranging by degree and level of impact was undertaken by the E. Kulikova's method of determining the probability of risks [27]. The achieved results with the specification of opportunities to manage these risks are presented in Table 1.

Table 1. Assessment of potential risks of biofuel production in Russia.

№	Risks	P_q [1]	I [2]	I_r [3]	Addiction	
					Influence Degree	Influence Level
	operational, resource					
1	Variable raw material supplies	2	4	8	Minor	Reasonable
2	Narrow range of raw material application using existing equipment	4	3	12	Significant	Intolerable
	distributive					
3	Market stagnation	3	3	9	Moderate	Reasonable
4	Lack of relevant target market, insufficient supply, low volumes of sales and distribution	2	5	10	Moderate	Reasonable
	financial					
5	High expenditures for popularization events	2	1	2	Ignorable	Acceptable
6	Growth of the energy unit cost	4	3	12	Significant	Intolerable
7	Low efficiency of state subsidies	2	5	10	Moderate	Reasonable
8	Low efficiency of investments	2	5	10	Moderate	Reasonable
	human resource					
9	Insufficient human resources and low level of trained human resources	5	3	15	Significant	Intolerable
10	Low quality of educational programs of main and additional education	3	3	9	Moderate	Reasonable

[1] P_q—possible losses; [2] I—possibility of arising; [3] I_r—risk rank.

The several application of the presented indexes for analysis of corresponding risks is presented below.

For maintenance of positive dynamics of the index (1), it is necessary to determine the following risks: (i) variability of raw material suppliers; (ii) narrow range of raw material application by a specific enterprise using existing equipment. For index (2) with ratio k(CON), the risks may be caused by market stagnation and high expenditures for popularisation events.

The growth of the cost of an energy unit may be identified as the risk for index (3). The risks for the index (4) are the following: Lack of sales and distribution corresponding to production level, insufficient supply, low sales, and distribution.

Despite the fact that the significant (indicator) value of the index (5) and (6) currently appear to be impossible to be determined, studying of their changes over time and in conjunction with different other data on the state of BF production may become the basis for additional classification of some risks. For instance, based on the results of such studies of the index (5), it appears to be possible to note the risks caused by the low efficiency of state subsidies and low efficiency of investments. In turn, accumulation of information on human resources and the existing qualification level of available human resources which is reflected by the index (6) in dynamics, may become a basis for determination of relevant risks. With consideration of these reasons, some potential risks associated with the application of indexes (5) and (6) are also included in Table 1.

The materials accumulated in Table 1 allowed for proposing some methods for potential risk management of BF production in Russia (Table 2). Representation and grouping of these methods correspond to the order in Table 1.

Table 2. Methods of risk management of biofuel production in Russia.

№	Methods of Risk Management of Biofuel Production in Russia
1	Allocation of production in accordance with the accessibility of the raw material base Obtainment of authentic information on suppliers Entering into supply agreements with several suppliers Planning of the required amount of raw materials for the provision of the supply system (when raw materials are supplied using own production) Insurance (of contractual obligations or own production of raw materials) Selection of reserve suppliers and opportunities for supplies of alternative raw materials
2	Creation of production associations Enhancement of production capabilities for the utilized types of raw materials
3	Stimulation of work performance by state mechanisms Support of innovative, science, technical potential
4	Monitoring of cost characteristics of products
5	State control of market barriers lifting Establishment of a uniform informational source with free access for end consumers
6	Reduction of costs of energy production State regulation of price-formation
7	Reduction of expenditures per unit of product
8	Increase of production output
9	Formation of human resource reserve by creating training programs for specialists with higher and secondary education for the provision of all production systems Optimization of personnel work depending on the level of employee efficiency Supervision of the structure and amount of salaries Compliance of the salary level of BF production enterprises active in a region with at least average regional salary Determination of incentive payments in the payroll fund for timely staff motivation
10	Involvement of representatives of technological platforms acting in the area of energy, scientific organizations conducting fundamental and applied studies, producers of BF and BF-production equipment, educational institutions and other concerned parties in formation and implementation of educational programs State approval of the educational programs and their constant improvement depending on the necessity of practical implementation of technological solutions Increase of qualification of enterprises' personnel using their own capital and the federal budget

5. Discussion

The gathered materials on the state and perspectives of development of renewable energy in Russia confirm that positive growth of production of power using different types of BF may be expected in the near future. Figure 7 summarizes the results of the conducted analysis and forecasting of the electric energy production volume by certified REn-based facilities with consideration of the discovered risks.

The expected growing trends of production using biogas, biomass, including solid wastes, and landfill gas utilization are shown by dashed lines which are built based on the calculations performed using the tendency function and the least square method [131,194].

The presented calculations and some noted-above data show that significant growth of production may be provided, in particular, by fuel pellets demanded in the global market. Biogas production and landfill gas processing also demonstrate positive dynamics provided targeted management is performed.

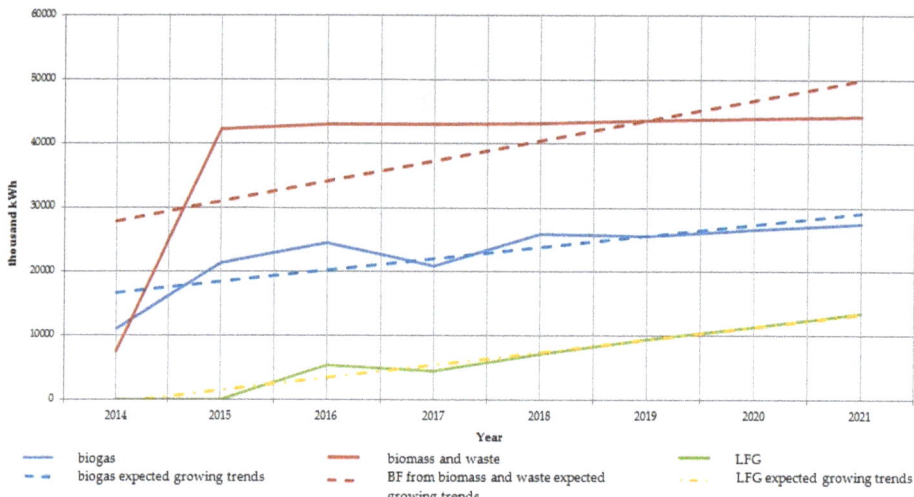

Figure 7. Forecasting of certificates confirmed electricity production by qualified renewable energy facilities (retail and wholesale markets), 2014–2021, thousand kWh [131,194].

Based on the existing data, it is possible to assume that institutional changes are required for the activation of liquid BF production. In addition, it appears to be necessary to grant the respective power to related ministries and agencies for the lifting of bureaucratic barriers. Moreover, it is essential to systematize the studies and available data on raw material reserves and technologies which may be applied in Russia. The positive impact may be caused by the resuming of existing but frozen ethanol plants operation. Conducting of full-scale works for fuel mixtures utilization may be of help too.

Generally, it seems that monitoring of the proposed indexes and other indicators influencing the production of BF may be useful for an increase in work efficiency. A significant role may be played by the creation of a specialized information base, which will provide all market participants with authentic information on dynamic changes in this area. Formulation of multiplicative (aggregate) coefficient which may be useful for reflection of relations between the discovered indexes with the assignment of weighting coefficients may be another organizational and economic mechanism contributing in the development of BF production in Russia [195,196].

6. Conclusions

The progress in development and application of innovative technologies in BF production has become a significant factor contributing in the establishment of conditions for sustainable development of economy among the leading countries. Relevant environmental results have already been achieved: Effective recycling of various industrial, agricultural, and household wastes is carried out, greenhouse gas emissions in the environment are reduced, etc.

With this in mind, although in Russia the fuel and energy sector of the economy continues to be based mainly on conventional sources of energy, the production of biofuels is also beginning to deploy. In a relatively short time, the necessary legislative and regulatory system was largely created; dozens of enterprises that produce various types of biofuels appeared. Some types of produced BF (pellets) became exported goods.

The conducted analysis of organizational and economic mechanisms applied for BF production in Russia, and the obtained data on some producing organizations allowed proposing six distinct indexes for the assessment of such production performance. These indexes were used to describe potential risks, which may be faced in this sphere of activity in Russia.

Therefore, the gathered materials allow us to suggest that BF production in Russia will develop and contribute to the sustainable development of a number of the country regions in the nearest future.

Moreover, taking into account the existing world experience, it can be assumed that in the coming years, the development aimed at improving biofuel production technologies will continue actively. In parallel, progress can be expected in managing the production and use of biofuels to achieve better economic results.

Acknowledgments: This research did not receive any specific grant from funding agencies in the public, commercial, or not for profit sectors.

Conflicts of Interest: The authors declare no conflict of interest.

Abbreviation

BF	Biofuel
REn	Renewable energy
TCE	Tons of coal equivalent
MW	Megawatt
HES	Heating electrical station
LFG	Landfill gas
JSC	Joint Stock Company
GW	Gigawatt
PJ	Petajoule
LLC	Limited Liability Company
BPP	Biogas power plant
GJ	Gigajoule
COD	Chemical oxygen demand

References

1. United Nations. Available online: https://www.un.org/sustainabledevelopment/ru/sustainable-development-goals/ (accessed on 20 June 2019).
2. Ars Technica. Available online: https://arstechnica.com/science/2019/01/renewables-led-by-wind-provided-more-power-than-coal-in-germany-in-2018/ (accessed on 20 June 2019).
3. GIS Renewable Energy Sources of Russia (GIS RESR). Available online: http://gisre.ru/tables#bio (accessed on 9 August 2019).
4. Proskuryakova, L.N.; Ermolenko, G.V. The future of Russia's renewable energy sector: Trends, scenarios and policies. *Renew. Energy* **2019**, *143*, 1670–1686. [CrossRef]
5. Nuno, A.; John, F.A.S. How to ask sensitive questions in conservation: A review of specialized questioning techniques. *Biol. Conserv.* **2015**, *189*, 5–15. [CrossRef]
6. Balagué, C.; de Valck, K. Using Blogs to Solicit Consumer Feedback: The Role of Directive Questioning Versus No Questioning. *J. Interact. Mark.* **2013**, *27*, 62–73. [CrossRef]
7. Scoboria, A.; Memon, A.; Trang, H.; Frey, M. Improving responding to questioning using a brief retrieval training. *J. Appl. Res. Mem. Cogn.* **2013**, *2*, 210–215. [CrossRef]
8. Bondarchuk, M.M. Questioning as a method of identifying requirements for qualifications of workers in the textile industry. *Educ. Sci. Russ. Abroad* **2019**, *51*, 98–101.
9. Dolzhenko, Y.; Pozdnyakova, A. Online questionnaire survey as a modern and effective way of research. *Transp. Bus. Russ.* **2015**, *1*, 109–110.
10. International Renewable Energy Agency (IRENA). Available online: http://resourceirena.irena.org/gateway/dashboard/?topic=4&subTopic=18 (accessed on 19 July 2019).
11. International Renewable Energy Agency (IRENA). Available online: http://resourceirena.irena.org/gateway/dashboard/?topic=4&subTopic=16 (accessed on 19 July 2019).
12. International Renewable Energy Agency (IRENA). REMAP 2030. Available online: https://www.irena.org/-/media/Files/IRENA/Agency/Publication/2017/Apr/IRENA_REmap_Russia_paper_2017.pdf (accessed on 19 July 2019).

13. Renewable Energy Policy Network for the 21st Century (REN21). Renewables 2018. Global Status Report. Available online: http://www.ren21.net/gsr-2018/chapters/chapter_01/chapter_01/#sub_4 (accessed on 17 June 2019).
14. Il'enkova, S.D.; Gohberg, L.M.; Kuznecov, V.I.; Yagudin, S.Y. *Innovation Management*; Moscow International Institute of Econometrics, Informatics, Finance and Law: Moscow, Russia, 2003.
15. Dutta, S.; Reynoso, R.E.; Bernard, A.L. The Global Innovation Index 2015: Effective Innovation Policies for Development. Available online: https://www.wipo.int/edocs/pubdocs/en/wipo_pub_gii_2015-chapter1.pdf (accessed on 23 July 2019).
16. Global Innovation Index 2018. Annex 2—Adjustments to the Global Innovation Index Framework and Year-on-Year Comparability of Results. Available online: https://www.wipo.int/edocs/pubdocs/en/wipo_pub_gii_2018-annex2.pdf (accessed on 23 July 2019).
17. Crespo, N.F.; Crespo, C.F. Global innovation index: Moving beyond the absolute value of ranking with a fuzzy-set analysis. *J. Bus. Res.* **2016**, *69*, 5265–5271. [CrossRef]
18. National Innovation Report in Russia. Available online: https://www.rvc.ru/upload/iblock/c64/RVK_innovation_2017.pdf (accessed on 23 July 2019).
19. Aleskerov, F.; Egorova, L.; Gokhberg, L.; Myachin, A.; Sagieva, G. Pattern Analysis in the Study of Science, Education and Innovative Activity in Russian Regions. *Procedia Comput. Sci.* **2013**, *17*, 687–694.
20. Miremadi, I.; Saboohi, Y.; Jacobsson, S. Assessing the performance of energy innovation systems: Towards an established set of indicators. *Energy Res. Soc. Sci.* **2018**, *40*, 159–176. [CrossRef]
21. Binz, C.; Truffer, B. Global Innovation Systems—A conceptual framework for innovation dynamics in transnational contexts. *Res. Policy* **2017**, *46*, 1284–1298. [CrossRef]
22. Janger, J.; Schubert, T.; Andries, P.; Rammer, C.; Hoskens, M. The EU 2020 innovation indicator: A step forward in measuring innovation outputs and outcomes? *Res. Policy* **2017**, *46*, 30–42. [CrossRef]
23. Dziallas, M.; Blind, K. Innovation indicators throughout the innovation process: An extensive literature analysis. *Technovation* **2019**, *80–81*, 3–29. [CrossRef]
24. Dong, J.; Xue, G.; Feng, T.; Liu, D. System Dynamics Modelling of Renewable Power Generation Investment Decisions under Risk. *Int. J. Simul. Syst. Sci. Technol.* **2016**, *17*, 1–10.
25. Pries, F.; Talebi, A.; Schillo, R.S.; Lemay, M.A. Risks affecting the biofuels industry: A US and Canadian company perspective. *Energy Policy* **2016**, *97*, 93–101. [CrossRef]
26. Abdullah, B.; Muhammad, S.A.F.S.; Shokravi, Z.; Ismail, S.; Kassim, K.A.; Mahmood, A.N.; Aziz, M.A. Fourth generation biofuel: A review on risks and mitigation strategies. *Renew. Sustain. Energy Rev.* **2019**, *107*, 37–50. [CrossRef]
27. Kulikova, E.E. *Risk Management. Innovation Aspect*; Berator-Publishing: Moscow, Russia, 2008.
28. Gokhberg, L.; Sokolov, A. Technology foresight in Russia in historical evolutionary perspective. *Technol. Forecast. Soc.* **2017**, *119*, 256–267. [CrossRef]
29. Hu, R.; Skea, J.; Hannon, M.J. Measuring the energy innovation process: An indicator framework and a case study of wind energy in China. *Technol. Forecast. Soc.* **2018**, *127*, 227–244. [CrossRef]
30. Guo, M.; Song, W.; Buhain, J. Bioenergy and biofuels: History, status, and perspective. *Renew. Sustain. Energy Rev.* **2015**, *42*, 712–725. [CrossRef]
31. Lopes, M.L.; Paulillo, S.C.; Godoy, A.; Cherubin, R.A.; Lorenzi, M.S.; Giometti, F.H.; Bernardino, C.D.; Amorim Neto, H.B.; Amorim, H.V. Ethanol production in Brazil: A bridge between science and industry. *Braz. J. Microbiol.* **2016**, *47*, 64–76. [CrossRef]
32. Saracevic, E.; Frühauf, S.; Miltner, A.; Karnpakdee, K.; Munk, B.; Lebuhn, M.; Wlcek, B.; Leber, J.; Lizasoain, J.; Friedl, A.; et al. Utilization of Food and Agricultural Residues for a Flexible Biogas Production: Process Stability and Effects on Needed Biogas Storage Capacities. *Energies* **2019**, *12*, 2678. [CrossRef]
33. Robak, K.; Balcerek, M. Review of Second Generation Bioethanol Production from Residual Biomass. *Food Technol. Biotechnol.* **2018**, *56*, 174–187. [CrossRef] [PubMed]
34. Castro, R.C.A.; Ferreira, I.S.; Roberto, I.C.; Mussatto, S.I. Isolation and physicochemical characterization of different lignin streams generated during the second-generation ethanol production process. *Int. J. Biol. Macromol.* **2019**, *129*, 497–510. [CrossRef] [PubMed]
35. Ruiz Olivares, A.; Carrillo-González, R.; del Carmen A. González-Chávez, M.; Soto Hernandez, R.M. Potential of castor bean (*Ricinus communis* L.) for phytoremediation of mine tailings and oil production. *J. Environ. Manag.* **2013**, *114*, 316–323. [CrossRef] [PubMed]

36. Moncada, J.; Cardona, C.A.; Rincon, L.E. Design and analysis of a second and third generation biorefinery: The case of castorbean and microalgae. *Bioresour. Technol.* **2015**, *198*, 836–843. [CrossRef]
37. Titova, E.; Bondarchuk, N.; Romanova, E. Economic aspects of plants cultivation used as raw materials for biofuel production. *Int. Agric. J.* **2017**, *1*, 54–61.
38. Miskolczi, N.; Buyong, F.; Angyal, A.; Williams, P.T.; Bartha, L. Two stages catalytic pyrolysis of refuse derived fuel: Production of biofuel via syncrude. *Bioresour. Technol.* **2010**, *101*, 8881–8890. [CrossRef]
39. Islam, Z.U.; Yu, Z.; Hassan, E.B.; Chang, D.; Zhang, H. Microbial conversion of pyrolytic products to biofuels: A novel and sustainable approach toward second-generation biofuels. *J. Ind. Microbiol. Biotechnol.* **2015**, *42*, 1557–1579. [CrossRef]
40. Sarsekeyeva, F.; Zayadan, B.K.; Usserbaeva, A.; Bedbenov, V.S.; Sinetova, M.A.; Los, D.A. Cyanofuels: Biofuels from Cyanobacteria. Reality and perspectives. *Photosynth. Res.* **2015**, *125*, 329–340. [CrossRef]
41. Chen, W.H.; Lin, B.J.; Huang, M.Y.; Chang, J.S. Thermochemical conversion of microalgal biomass into biofuels: A review. *Bioresour. Technol.* **2015**, *184*, 314–327. [CrossRef]
42. Wang, Y.; Ho, S.H.; Yen, H.W.; Nagarajan, D.; Ren, N.Q.; Li, S.; Hu, Z.; Lee, D.J.; Kondo, A.; Chang, J.S. Current advances on fermentative biobutanol production using third generation feedstock. *Biotechnol. Adv.* **2017**, *35*, 1049–1059. [CrossRef]
43. Lakatos, G.E.; Ranglova, K.; Manoel, J.C.; Grivalsky, T.; Kopecky, J.; Masojidek, J. Bioethanol production from microalgae polysaccharides. *Folia Microbiol. (Praha)* **2019**, 1–18. [CrossRef] [PubMed]
44. Abo, B.O.; Odey, E.A.; Bakayoko, M.; Kalakodio, L. Microalgae to biofuels production: A review on cultivation, application and renewable energy. *Rev. Environ. Health* **2019**, *34*, 91–99. [CrossRef] [PubMed]
45. Farrokh, P.; Sheikhpour, M.; Kasaeian, A.; Asadi, H.; Bavandi, R. Cyanobacteria as an eco-friendly resource for biofuel production: A critical review. *Biotechnol. Prog.* **2019**, *35*, e2835. [CrossRef] [PubMed]
46. Vo Hoang Nhat, P.; Ngo, H.H.; Guo, W.S.; Chang, S.W.; Nguyen, D.D.; Nguyen, P.D.; Bui, X.T.; Zhang, X.B.; Guo, J.B. Can algae-based technologies be an affordable green process for biofuel production and wastewater remediation? *Bioresour. Technol.* **2018**, *256*, 491–501. [CrossRef]
47. McLeod, C.; Nerlich, B.; Mohr, A. Working with bacteria and putting bacteria to work: The biopolitics of synthetic biology for energy in the United Kingdom. *Energy Res. Soc. Sci.* **2017**, *30*, 35–42. [CrossRef]
48. Schlör, H.; Venghaus, S.; Märker, C.; Hake, J.F. Managing the resilience space of the German energy system—A vector analysis. *J. Environ. Manag.* **2018**, *218*, 527–539. [CrossRef]
49. Golberg, A.; Sack, M.; Teissie, J.; Pataro, G.; Pliquett, U.; Saulis, G.; Stefan, T.; Miklavcic, D.; Vorobiev, E.; Frey, W. Energy-efficient biomass processing with pulsed electric fields for bioeconomy and sustainable development. *Biotechnol. Biofuels* **2016**, *9*, 94. [CrossRef]
50. Leksin, V.N.; Porfiryev, B.N. Socio-Economic Priorities of Sustainable Development of Russian Arctic Macro-Region. *Econ. Reg.* **2017**, *13*, 985–1004. [CrossRef]
51. Kuzmenkova, V.D. Sustainable development of Russian regions. *Proc. Vor. State Univ. Eng. Technol.* **2016**, *2*, 257–261. [CrossRef]
52. Bondarchuk, N.V.; Titova, E.S. Renewable energy prospects as one of sustainable development direction in some South Russian regions. *South Russ. Ecol. Dev.* **2017**, *4*, 12–31. [CrossRef]
53. Mulko, L.; Rivarola, C.R.; Barbero, C.A.; Acevedo, D.F. Bioethanol production by reusable *Saccharomyces cerevisiae* immobilized in a macroporous monolithic hydrogel matrices. *J. Biotechnol.* **2016**, *233*, 56–65. [CrossRef] [PubMed]
54. Alalwan, H.A.; Alminshid, A.H.; Aljaafari, H.A.S. Promising evolution of biofuel generations. Subject review. *Renew. Energy Focus* **2019**, *28*, 127–139. [CrossRef]
55. Sikarwar, V.S.; Zhao, M.; Fennell, P.S.; Shah, N.; Anthony, E.J. Progress in biofuel production from gasification. *Prog. Energy Combust. Sci.* **2017**, *61*, 189–248. [CrossRef]
56. Laker, F.; Agaba, A.; Akatukunda, A.; Gazet, R.; Barasa, J.; Nanyonga, S.; Wendiro, D.; Wacoo, A.P. Utilization of Solid Waste as a Substrate for Production of Oil from Oleaginous Microorganisms. *J. Lipids* **2018**, *2018*, 1578720. [CrossRef] [PubMed]
57. Lu, X.; Withers, M.R.; Seifkar, N.; Field, R.P.; Barrett, S.R.; Herzog, H.J. Biomass logistics analysis for large scale biofuel production: Case study of loblolly pine and switchgrass. *Bioresour. Technol.* **2015**, *183*, 1–9. [CrossRef] [PubMed]
58. Purohit, P.; Chaturvedi, V. Biomass pellets for power generation in India: A techno-economic evaluation. *Environ. Sci. Pollut. Res. Int.* **2018**, *25*, 29614–29632. [CrossRef]

59. Rafaj, P.; Kiesewetter, G.; Gul, T.; Schopp, W.; Cofala, J.; Klimont, Z.; Purohit, P.; Heyesa, C.; Amann, M.; Borken-Kleefeld, J.; et al. Outlook for clean air in the context of sustainable development goals. *Glob. Environ. Chang.* **2018**, *53*, 1–11. [CrossRef]
60. Lyng, K.-A.; Brekke, A. Environmental Life Cycle Assessment of Biogas as a Fuel for Transport Compared with Alternative Fuels. *Energies* **2019**, *12*, 532. [CrossRef]
61. Teigiserova, D.A.; Hamelin, L.; Thomsen, M. Review of high-value food waste and food residues biorefineries with focus on unavoidable wastes from processing. *Resour. Conserv. Recycl.* **2019**, *149*, 413–426. [CrossRef]
62. Benato, A.; Macor, A. Italian Biogas Plants: Trend, Subsidies, Cost, Biogas Composition and Engine Emissions. *Energies* **2019**, *12*, 979. [CrossRef]
63. He, K.; Zhang, J.; Zeng, Y. Knowledge domain and emerging trends of agricultural waste management in the field of social science: A scientometric review. *Sci. Total Environ.* **2019**, *670*, 236–244. [CrossRef] [PubMed]
64. Dai, Y.; Sun, Q.; Wang, W.; Lu, L.; Liu, M.; Li, J.; Yang, S.; Sun, Y.; Zhang, K.; Xu, J.; et al. Utilizations of agricultural waste as adsorbent for the removal of contaminants: A review. *Chemosphere* **2018**, *211*, 235–253. [CrossRef] [PubMed]
65. Enes, T.; Aranha, J.; Fonseca, T.; Lopes, D.; Alves, A.; Lousada, J. Thermal Properties of Residual Agroforestry Biomass of Northern Portugal. *Energies* **2019**, *12*, 1418. [CrossRef]
66. Yang, X.J.; Hu, H.; Tan, T.; Li, J. China's renewable energy goals by 2050. *Environ. Dev.* **2016**, *20*, 83–90. [CrossRef]
67. Liu, J. China's renewable energy law and policy: A critical review. *Renew. Sustain. Energy Rev.* **2019**, *99*, 212–219. [CrossRef]
68. Rueda-Bayona, J.G.; Guzmán, A.; Cabello Eras, J.J.; Silva-Casarín, R.; Bastidas-Arteaga, E.; Horrillo-Caraballo, J. Renewables energies in Colombia and the opportunity for the offshore wind technology. *J. Clean. Prod.* **2019**, *220*, 529–543. [CrossRef]
69. Bórawski, P.; Bełdycka-Bórawska, A.; Szymańska, E.J.; Jankowski, K.J.; Dubis, B.; Dunn, J.W. Development of renewable energy sources market and biofuels in The European Union. *J. Clean. Prod.* **2019**, *228*, 467–484. [CrossRef]
70. European Commission. A Sustainable Bioeconomy for Europe: Strengthening the Connection between Economy, Society and the Environment: Updated Bioeconomy Strategy. Available online: https://ec.europa.eu/research/bioeconomy/pdf/ec_bioeconomy_strategy_2018.pdf (accessed on 23 February 2019).
71. Baral, N. What socio-demographic characteristics predict knowledge of biofuels. *Energy Policy* **2018**, *122*, 369–376. [CrossRef]
72. Liu, B.; Shumway, C.R.; Yoder, J.K. Lifecycle economic analysis of biofuels: Accounting for economic substitution in policy assessment. *Energy Econ.* **2017**, *67*, 146–158. [CrossRef]
73. Nugroho, Y.K.; Zhu, L. Platforms planning and process optimization for biofuels supply chain. *Renew. Energy* **2019**, *140*, 563–579. [CrossRef]
74. Ratner, S.V.; Ratner, M.D. Developing Regional Environmental Management Models in Compliance with the Innovative Format of ISO 14001:2015. *Dig. Financ.* **2017**, *22*, 163–174. [CrossRef]
75. Chen, R.; Qin, Z.; Han, J.; Wang, M.; Taheripour, F.; Tyner, W.; O'Connor, D.; Duffield, J. Life cycle energy and greenhouse gas emission effects of biodiesel in the United States with induced land, use change impacts. *Bioresour. Technol.* **2018**, *251*, 249–258. [CrossRef] [PubMed]
76. Rybicka, J.; Tiwari, A.; Leeke, G.A. Technology readiness level assessment of composites recycling technologies. *J. Clean. Prod.* **2016**, *112*, 1001–1012. [CrossRef]
77. Esteves, E.M.M.; Herrera, A.M.N.; Peçanha Esteves, V.P.; Rosário Vaz Morgado, C. Life cycle assessment of manure biogas production: A review. *J. Clean. Prod.* **2019**, *219*, 411–423. [CrossRef]
78. Russia Renewable Energy Development Association. Available online: https://rreda.ru/bulletin (accessed on 8 August 2019).
79. Budzianowski, W.M.; Budzianowska, D.A. Economic analysis of biomethane and bioelectricity generation from biogas using different support schemes and plant configurations. *Energy* **2015**, *88*, 658–666. [CrossRef]
80. Budzianowski, W.M. A review of potential innovations for production, conditioning and utilization of biogas with multiple-criteria assessment. *Renew. Sustain. Energy Rev.* **2016**, *54*, 1148–1171. [CrossRef]
81. Ruiz, D.; San Miguel, G.; Corona, B.; Gaitero, A.; Domínguez, A. Environmental and economic analysis of power generation in a thermophilic biogas plant. *Sci. Total Environ.* **2018**, *633*, 1418–1428. [CrossRef]

82. Mustapha, W.F.; Trømborg, E.; Bolkesjø, T.F. Forest-based biofuel production in the Nordic countries: Modelling of optimal allocation. *For. Policy Econ.* **2019**, *103*, 45–54. [CrossRef]
83. Nutongkaew, P.; Waewsak, J.; Riansut, W.; Kongruang, C.; Gagnon, Y. The potential of palm oil production as a pathway to energy security in Thailand. *Sustain. Energy Technol.* **2019**, *35*, 189–203. [CrossRef]
84. de Jong, S.; Hoefnagels, R.; Wetterlund, E.; Pettersson, K.; Faaij, A.; Junginger, M. Cost optimization of biofuel production—The impact of scale, integration, transport and supply chain configurations. *Appl. Energy* **2017**, *195*, 1055–1070. [CrossRef]
85. Brosowski, A.; Krause, T.; Mantau, U.; Mahro, B.; Noke, A.; Richter, F.; Raussen, T.; Bischof, R.; Hering, T.; Blanke, C.; et al. How to measure the impact of biogenic residues, wastes and by-products: Development of a national resource monitoring based on the example of Germany. *Biomass Bioenergy* **2019**, *127*, 105275. [CrossRef]
86. Tanzer, S.E.; Posada, J.; Geraedts, S.; Ramírez, A. Lignocellulosic marine biofuel: Technoeconomic and environmental assessment for production in Brazil and Sweden. *J. Clean. Prod.* **2019**, *239*, 117845. [CrossRef]
87. Matraeva, L.; Solodukha, P.; Erokhin, S.; Babenko, M. Improvement of Russian energy efficiency strategy within the framework of "green economy" concept (based on the analysis of experience of foreign countries). *Energy Policy* **2019**, *125*, 478–486. [CrossRef]
88. Popel, O.S. Renewable energy sources: The role and place in modern and promising energy. *Russ. Chem. J.* **2008**, *6*, 95–106.
89. Ministry of Energy of Russian Federation. Available online: https://minenergo.gov.ru/node/987 (accessed on 8 August 2019).
90. Ministry of Economic Development of the Russian Federation. Available online: http://economy.gov.ru/minec/activity/sections (accessed on 8 August 2019).
91. Ministry of Natural Resources and Environment of the Russian Federation. Available online: http://www.mnr.gov.ru/docs/ (accessed on 8 August 2019).
92. Ministry of Industry and Trade of the Russian Federation. Available online: http://minpromtorg.gov.ru/projects/general/ (accessed on 8 August 2019).
93. Vyborg Forestry Development Corporation. Available online: http://vfdc.ru/ru (accessed on 17 August 2019).
94. LLC "AltEnergo". Biogas Station. Available online: http://altenergo.su/biogas/ (accessed on 21 June 2019).
95. Sun, L.; Niquidet, K. Elasticity of import demand for wood pellets by the European Union. *For. Policy Econ.* **2017**, *81*, 83–87. [CrossRef]
96. Pristupa, A.O.; Mol, A.P.J. Renewable energy in Russia: The take off in solid bioenergy? *Renew. Sustain. Energy Rev.* **2015**, *50*, 315–324. [CrossRef]
97. Technological Platform "Bioenergy". Available online: http://www.tp-bioenergy.ru/about/management/ (accessed on 25 July 2019).
98. Technological Platform "Bioindustry and Bioresources" (BioTech2030). Available online: http://biotech2030.ru/deyatelnost/ (accessed on 25 July 2019).
99. Agency for Strategic Initiatives. Available online: https://asi.ru/about_agency/ (accessed on 31 July 2019).
100. Russian Biofuel Association (RBA). Available online: http://www.biotoplivo.ru/o-nas/ (accessed on 25 July 2019).
101. ENBIO. Association of Biofuel Market Participants. Available online: http://enbio.ru/rubric/novosti (accessed on 25 July 2019).
102. Federal Law of Russian Federation No. 35-FZ of March 26, 2003 "On Electric Energy". Available online: http://www.consultant.ru/document/cons_doc_LAW_41502/ (accessed on 16 August 2019).
103. Decree of the Russian Federation Government of November 13, 2009 No. 1715-r "On the Energy Strategy of Development of Russia before 2030". Available online: https://www.garant.ru/products/ipo/prime/doc/96681/ (accessed on 11 June 2019).
104. Federal Law of Russian Federation No. 171-FZ of November 22, 1995 "On Amendments in the State Regulation of Production and Turnover of Ethyl Alcohol, Alcoholic and Alcohol-Containing Products and Restriction of Alcoholic Products Consumption". Available online: http://www.consultant.ru/document/cons_doc_LAW_8368/ (accessed on 16 August 2019).
105. Federal Law of Russian Federation No. 448-FZ of November 28, 2018 "On State Regulation of Production and Turnover of Ethyl Alcohol, Alcoholic and Alcohol-containing Products and Restriction of Alcoholic Products Consumption". Available online: http://www.consultant.ru/document/cons_doc_LAW_312102/ (accessed on 16 August 2019).

106. Decree of the Russian Federation Government No. 1145 of September 27, 2018 "On Amending of Certain Acts of the Government of the Russian Federation Regarding the Promotion of the Use of Renewable Energy Sources". Available online: http://www.consultant.ru/document/cons_doc_LAW_307870/ (accessed on 1 June 2019).
107. Ministry of Energy of Russian Federation. News Report "Alexander Novak Report on the Development of Energy Efficiency and Energy Saving". Available online: https://minenergo.gov.ru/node/9591 (accessed on 8 August 2019).
108. Draft of the Decree of the Russian Federation Government "On Approval of the List of Key (Basic) Indicators of Public Non-Financial Accounting". Available online: http://www.consultant.ru/cons/cgi/online.cgi?req=doc;base=PNPA;n=46279#08990458648385347 (accessed on 11 June 2019).
109. Decree of the Russian Federation Government of April 2, 2014 No. 504-r "On Approval of an Action Plan to Ensure by 2020 the Reduction of Greenhouse Gas Emissions to No More Than 75 Percent of the Volume of These Emissions in 1990". Available online: http://www.consultant.ru/document/cons_doc_LAW_161475/ (accessed on 11 June 2019).
110. Order of the Russian Federation Government of November 3, 2018 No. 2381-r "On Amendments to the Federal Law "On Electric Power" Regarding the Development of Micro-Generation". Available online: http://government.ru/activities/selection/301/34594/ (accessed on 11 June 2019).
111. Acheampong, M.; Ertem, F.C.; Kappler, B.; Neubauer, P. In pursuit of Sustainable Development Goal (SDG) number 7: Will biofuels be reliable? *Renew. Sustain. Energy Rev.* **2017**, *75*, 927–937. [CrossRef]
112. Tsita, K.G.; Kiartzis, S.J.; Ntavos, N.K.; Pilavachi, P.A. Next generation biofuels derived from thermal and chemical conversion of the Greek transport sector. *Therm. Sci. Eng. Prog.* **2019**, 100387. [CrossRef]
113. Boboescu, I.Z.; Chemarin, F.; Beigbeder, J.-B.; de Vasconcelos, B.R.; Munirathinam, R.; Ghislain, T.; Lavoie, J.-M. Making next-generation biofuels and biocommodities a feasible reality. *Curr. Opin. Green Sustain. Chem.* **2019**, *20*, 25–32. [CrossRef]
114. Sun, C.H.; Fu, Q.; Liao, Q.; Xia, A.; Huang, Y.; Zhu, X.; Reungsang, A.; Chang, H.-X. Life-cycle assessment of biofuel production from microalgae via various bioenergy conversion systems. *Energy* **2019**, *171*, 1033–1045. [CrossRef]
115. Resolution of the Government of Russian Federation No. 326 of April 15, 2014 "On Approval of the State Environmental Protection Program for 2012–2020". Available online: http://www.consultant.ru/document/cons_doc_LAW_162183/ (accessed on 21 May 2019).
116. Order of the Ministry of Natural Resources and Ecology of the Russian Federation of August 14, 2013 No. 298 "On Approval of a Comprehensive Strategy for the Treatment of Solid Municipal (Household) Waste in the Russian Federation". Available online: https://www.garant.ru/products/ipo/prime/doc/70345114/ (accessed on 11 June 2019).
117. Namsaraev, Z.B.; Gotovtsev, P.M.; Komova, A.V.; Vasilov, R.G. Current status and potential of bioenergy in the Russian Federation. *Renew. Sustain. Energy Rev.* **2018**, *81*, 625–634. [CrossRef]
118. Decree of the Russian Federation Government No. 337-r of February 28, 2018 "On Approval of the Action Plan ("Road Map") "Development of Biotechnologies and Gene Engineering" for 2018–2020". Available online: http://www.garant.ru/products/ipo/prime/doc/71792682/ (accessed on 11 June 2019).
119. Federal Law Draft "On Amending Article 217 of the Tax Code of the Russian Federation". Available online: http://government.ru/activities/selection/302/37660/ (accessed on 21 July 2019).
120. Moscow State University. Available online: https://www.msu.ru/science/sci-dir-1.html (accessed on 8 August 2019).
121. Joint Institute for High Temperatures of the Russian Academy of Sciences. Available online: https://jiht.ru/science/ (accessed on 8 August 2019).
122. The Federal Research Centre "Fundamentals of Biotechnology" of the Russian Academy of Sciences (Research Center of Biotechnology RAS). Available online: https://www.fbras.ru/napravleniya-nauchnyx-issledovanij/nauchnaya-deyatelnost (accessed on 25 July 2019).
123. National Research Center "Kurchatov Institute". Available online: http://www.nrcki.ru/catalog/index.shtml?g_show=7732&path=3977,7732 (accessed on 8 August 2019).
124. National University of Oil and Gas "Gubkin University". Available online: https://gubkin.ru/faculty/chemical_and_environmental/chairs_and_departments/general_and_inorganic_chemistry/ (accessed on 8 August 2019).

125. University MEPhI (Moscow Engineering Physics Institute). Available online: https://mephi.ru/science/osnovnye_nauchnye_napravleniya.php (accessed on 8 August 2019).
126. Moscow Polytechnic University. Available online: https://mospolytech.ru/index.php?id=7 (accessed on 8 August 2019).
127. Decree of Ministry of Labor and Social Protection of Russian Federation No. 1047n of December 21, 2015 "On Approval of the Professional Standard "Specialist in the Organization of Bioenergy and Biofuel Production". Available online: http://base.garant.ru/71312996/ (accessed on 14 June 2019).
128. Order of the Russian Federation Government of August 11, 2011 No. 1393-r "On the Establishment of an Autonomous Non-Profit Organization "Agency of Strategic Initiatives for Promoting New Projects". Available online: https://asi.ru/upload/medialibrary/3e7/1393-upd2.pdf (accessed on 15 June 2019).
129. EnergyNet. Available online: http://www.nti2035.ru/markets/energynet (accessed on 31 July 2019).
130. Decree of the Russian Federation Government No. 830-r of April 28, 2018 "On Approval of the Plan "Roadmap" to Legislation Improving and Administrative Barriers Removing to Ensure the Implementation of the National Technology Initiative "EnergyNet". Available online: http://government.ru/docs/32548/ (accessed on 31 May 2019).
131. Association "Nonprofit Partnership Council for Organizing Efficient System of Trading at Wholesale and Retail Electricity and Capacity Market" (Association "NP Market Council"). Available online: https://www.np-sr.ru/ru/market/vie/index.htm (accessed on 14 June 2019).
132. Trading System Administrator of Wholesale Electricity Market Transactions (TSA). Available online: http://www.atsenergo.ru/vie (accessed on 14 June 2019).
133. Global Energy. Association for the Development of International Research and Projects in the Field of Energy. Available online: https://globalenergyprize.org/ru/about-us/about-us (accessed on 25 July 2019).
134. Ministry of Economic Development of the Republic of North Ossetia-Alania. Available online: http://economy.alania.gov.ru/news/278 (accessed on 14 June 2019).
135. Interstate Standard (GOST 33872-2016). *Fuel Denatured Bioethanol*; Standartinform: Moscow, Russia, 2017; Available online: http://docs.cntd.ru/document/1200145331 (accessed on 11 June 2019).
136. Interstate Standard (GOST 33131-2014). *Mixtures of Biodiesel (B6–B20)*; Standartinform: Moscow, Russia, 2016; Available online: http://docs.cntd.ru/document/1200121053 (accessed on 11 June 2019).
137. Interstate Standard (GOST 33113-2014). *B100 Basic Biodiesel Fuel and Biodiesel Mixtures*; Standartinform: Moscow, Russia, 2016; Available online: http://docs.cntd.ru/document/1200121050 (accessed on 11 June 2019).
138. Ershov, M.A.; Grigoreva, E.V.; Habibullin, I.F.; Emelyanov, V.E.; Strekalina, D.M. Prospects of bioethanol fuels E30 and E85 application in Russia and technical requirements for their quality. *Renew. Sustain. Energy Rev.* **2016**, *66*, 228–232. [CrossRef]
139. Joseph, O.O.; Loto, C.A.; Joseph, O.O.; Dirisu, J.O. Comparative Assessment of the Degradation Behaviour of API 5l X65 And Micro-Alloyed Steels in E20 Simulated Fuel Ethanol Environment. *Energy Procedia* **2019**, *157*, 1320–1327. [CrossRef]
140. Johansen, L.C.R.; Hemdal, S.; Denbratt, I. Comparison of E10 and E85 spark ignited stratified combustion and soot formation. *Fuel* **2017**, *205*, 11–23. [CrossRef]
141. Tao, J.; Yu, S.; Wu, T. Review of China's bioethanol development and a case study of fuel supply, demand and distribution of bioethanol expansion by national application of E10. *Biomass Bioenergy* **2011**, *35*, 3810–3829. [CrossRef]
142. Yusri, I.M.; Mamat, R.; Najafi, G.; Razman, A.; Awad, I.O.; Azmi, W.H.; Ishak, W.F.W.; Shaiful, A.I.M. Alcohol based automotive fuels from first four alcohol family in compression and spark ignition engine: A review on engine performance and exhaust emissions. *Renew. Sustain. Energy Rev.* **2017**, *77*, 169–181. [CrossRef]
143. Group of Companies "Titan". Available online: http://www.titan-group.ru/about/ (accessed on 21 August 2019).
144. Osmakova, A.; Kirpichnikov, M.; Popov, V. Recent biotechnology developments and trends in the Russian Federation. *New Biotechnol.* **2018**, *40*, 76–81. [CrossRef]
145. ETB "Catalytic Technologies". Available online: http://www.etbcat.com/technology/ (accessed on 17 August 2019).
146. Proskurina, S.; Heinimö, J.; Mikkilä, M.; Vakkilainen, E. The wood pellet business in Russia with the role of North-West Russian regions: Present trends and future challenges. *Renew. Sustain. Energy Rev.* **2015**, *51*, 730–740. [CrossRef]

147. Resolution of the Government of Russia No. 1989-p of September 20, 2018 "On Approval of the Timber Complex of Russia Development Strategy until 2030". Available online: http://government.ru/docs/34064/ (accessed on 15 March 2019).
148. Rakitova, O.; Kholodkov, V. *Final Report of the Pellet Market and Wood Resources in the North-West of Russian*; Baltic 21 Lighthouse Project, the Baltic Sea Bioenergy Promotion Project; Biocenter: Saint-Petersburg, Russia, 2009.
149. Pirus, M. Working on the future: Congress and exhibition "Biomass: Fuel and energy". *Lesprominform* **2018**, *4*, 134–137.
150. Committee on the Fuel and Energy Complex of the Leningrad Region. Available online: http://power.lenobl.ru/o-komitete/informatsionnye-sistemy/modul-gis-energoeffektivnost/ (accessed on 21 July 2019).
151. The Renewable Materials Company "STORAENSO". Available online: https://www.storaenso.com/ (accessed on 20 July 2019).
152. INFOBIO. News Reports "Annually in the Arkhangelsk Region Introduced by Several Biofuel Boiler Houses". Available online: http://www.infobio.ru/news/4577.html (accessed on 19 July 2019).
153. Resolution of the Arkhangelsk Region Government of October 15, 2013 No. 487-pp "On Approval of the State Program of the Arkhangelsk Region "Development of Energy and Housing and Communal Services of the Arkhangelsk Region (2014–2024)". Available online: http://docs.cntd.ru/document/462608457 (accessed on 14 June 2019).
154. INFOBIO. News Reports "One More Pellet Production to Appear in Irkutsk Region". Available online: http://www.infobio.ru/news/4586.html (accessed on 4 August 2019).
155. INFOBIO. News Reports "In the Territory of the Krasnoyarsk Region Biofuels Apply to 41 Boiler". Available online: http://www.infobio.ru/news/4589.html (accessed on 4 August 2019).
156. Perederyi, S. And again about torrefaction. Borreal BioEnergy sawmill acquires for the production of black pellets. *Lesprominform* **2019**, *4*, 148–149.
157. ROSTEPLO. Biogas Station "Luchki" Showed Historic Maximum Power Generation. Available online: http://www.rosteplo.ru/news/2018/02/20/1519067728-biogazovaya-stanciya-luchki-pokazala-istoricheskij (accessed on 15 June 2019).
158. Resolution of the Belgorod Region Governor No. 52 of April 25, 2018 "On Approval of the Scheme and Program for the Development of the Electric Power Industry of the Belgorod Region for 2019–2023". Available online: http://publication.pravo.gov.ru/Document/View/3100201804280001 (accessed on 14 June 2019).
159. Akimova, V.V. Institutional factor of renewable energy development in the Belgorod Region. *Bull. Mosc. Univ. (Ser. 5 Geogr.)* **2017**, *6*, 18–24.
160. Decree of the Belgorod Region Government No. 427-pp of October 29, 2012 "On Approving of the Long-Term Target Program "Development of Renewable Energy Sources for 2013–2015 and for the Period until 2020". Available online: http://base.garant.ru/26350607/ (accessed on 14 June 2019).
161. Decree of the Belgorod Region Government No. 475-pp of November 25, 2013 "On Amendments to the Belgorod Region Government No. 427-pp of October 29, 2012". Available online: http://docs.cntd.ru/document/428669040#loginform (accessed on 14 June 2019).
162. Order of the Belgorod Region Government No. 574-rp of December 8, 2014 "On Approving the Concept for the Development of Small Distributed Energy in the Belgorod Region until 2025". Available online: http://altenergo-nii.ru/docs/574-rp.pdf (accessed on 14 June 2019).
163. Decree of the Head of the Administration of the Tambov Region of April 25, 2018 No. 107 "On Approval of the Concept and Program of Electric Energy development in Tambov Region in 2019–2023". Available online: http://gkh.tmbreg.ru/DOC/Pravo/2018/PGAO_107.pdf (accessed on 14 June 2019).
164. Joint Stock Company "Biokhim" (JSC "Biokhim"). Available online: http://biohim68rsk.ucoz.ru (accessed on 14 May 2019).
165. Decree of the Lipetsk Region Administration No. 319 of April 23, 2018 "On the Approval of Schemes and Programs for the Development of the Electric Power Industry of the Lipetsk region for 2019–2023". Available online: http://publication.pravo.gov.ru/Document/View/4800201804250001 (accessed on 14 June 2019).
166. Lang-Yona, N.; Kunert, A.T.; Vogel, L.; Kampf, C.J.; Bellinghausen, I.; Saloga, J.; Schink, A.; Ziegler, K.; Lucas, K.; Schuppan, D.; et al. Fresh water, marine and terrestrial Cyanobacteria display distinct allergen characteristics. *Sci. Total Environ.* **2018**, *612*, 767–774. [CrossRef] [PubMed]

167. Singh, R.; Parihar, P.; Singh, M.; Bajguz, A.; Kumar, J.; Singh, S.; Singh, V.P.; Prasad, S.M. Uncovering Potential Applications of Cyanobacteria and Algal Metabolites in Biology, Agriculture and Medicine: Current Status and Future Prospects. *Front. Microbiol.* **2017**, *8*, 515. [CrossRef] [PubMed]
168. Luan, G.; Lu, X. Tailoring cyanobacterial cell factory for improved industrial properties. *Biotechnol. Adv.* **2018**, *36*, 430–442. [CrossRef]
169. Taparia, T.; Mvss, M.; Mehrotra, R.; Shukla, P.; Mehrotra, S. Developments and challenges in biodiesel production from microalgae: A review. *Biotechnol. Appl. Biochem.* **2016**, *63*, 715–726. [CrossRef]
170. Chena, J.; Li, J.; Dong, W.; Zhang, X.; Tyagi, R.D.; Drogui, P.; Surampalli, R.Y. The potential of microalgae in biodiesel production. *Renew. Sustain. Energy Rev.* **2018**, *90*, 336–346. [CrossRef]
171. Mathimani, T.; Baldinelli, A.; Rajendran, K.; Prabakar, D.; Matheswaran, M.; Van Leeuwen, R.P.; Pugazhendhi, A. Review on cultivation and thermochemical conversion of microalgae to fuels and chemicals: Process evaluation and knowledge gaps. *J. Clean. Prod.* **2019**, *208*, 1053–1064. [CrossRef]
172. Collotta, M.; Champagne, P.; Mabee, W.; Tomasoni, G. Wastewater and waste CO_2 for sustainable biofuels from microalgae. *Algal Res.* **2018**, *29*, 12–21. [CrossRef]
173. Antonov, I.A.; Kotelev, M.S.; Afonin, D.S.; Gushin, P.A.; Ivanov, E.V. Isoprenoids oil of microalgae hydrocracking with production of winter and arctic diesel fuels. *Bashkir Chem. J.* **2012**, *19*, 170–172.
174. Knoot, C.J.; Ungerer, J.; Wangikar, P.P.; Pakrasi, H.B. Cyanobacteria: Promising biocatalysts for sustainable chemical production. *J. Biol. Chem.* **2018**, *293*, 5044–5052. [CrossRef]
175. Singh, G.; Patidar, S.K. Microalgae harvesting techniques: A review. *J. Environ. Manag.* **2018**, *217*, 499–508. [CrossRef] [PubMed]
176. Kopylov, A.; Romanenko, A.V.; Zabotkina, E.A.; Mineeva, N.M.; Krylova, I.N.; Maslennikova, T.S. Picocyanobacteria in eutrophic reservoirs of the Middle Volga: Abundance, production, viral infection. *Zhurnal Obs. Biol.* **2014**, *75*, 234–244.
177. Sidelev, S.I.; Fomichev, A.A.; Babanazarova, O.V.; Zubishina, A.A. The detection of microcystin-producing Cyanobacteria in the Upper Volga watersheds. *Mikrobiologiia* **2013**, *82*, 370–371. [PubMed]
178. Korneva, L.G. Invasions of alien species of planktonic algae into Holarctic freshwaters (review). *Russ. J. Biol. Invasions* **2014**, *1*, 5–33.
179. Yoshida, M.; Yoshida, T.; Kashima, A.; Takashima, Y.; Hosoda, N.; Nagasaki, K.; Hiroishi, S. Ecological dynamics of the toxic bloom-forming cyanobacterium Microcystis aeruginosa and its cyanophages in freshwater. *Appl. Environ. Microbiol.* **2008**, *74*, 3269–3273. [CrossRef] [PubMed]
180. IZVESTIYA. Dirty Story. How the Volga became the Dirtiest River in Russia? Available online: https://iz.ru/660518/izdaleka-dolgo (accessed on 10 June 2019).
181. Zaytseva, N.V. The Problem of Development of Blue-Green Algae in the Votkinsky and Izhevsk Reservoirs. *Mod. Sci. Res. Innov.* **2014**, *6*. Available online: http://web.snauka.ru/issues/2014/06/36048 (accessed on 16 May 2019).
182. Passport of Federal Project "Improving the Volga". 2018. Available online: http://www.minstroyrf.ru/docs/17662/ (accessed on 17 July 2019).
183. Ho, S.-H.; Chen, Y.-D.; Qu, W.-Y.; Liu, F.-Y.; Wang, Y. Chapter 8—Algal culture and biofuel production using wastewater. In *Biomass, Biofuels, Biochemicals, Biofuels from Algae*, 2nd ed.; Pandey, A., Chang, J.-S., Soccol, C.R., Lee, D.-J., Chisti, Y., Eds.; Elsevier: Amsterdam, The Netherlands, 2019; pp. 167–198.
184. Guieysse, B.; Borde, X.; Munoz, R.; Hatti-Kaul, R.; Nugier-Chauvin, C.; Patin, H.; Mattiasson, B. Influence of the initial composition of algal-bacterial microcosms on the degradation of salicylate in a fed-batch culture. *Biotechnol. Lett.* **2002**, *24*, 531–538. [CrossRef]
185. Milyutkin, V.A.; Borodulin, I.V.; Agarkov, E.A.; Rozenberg, G.S.; Kudinova, G.E. Technical Solution for Processing Blue-Green Algae in Biofuel. In Proceedings of the 35th Anniversary of the Volga Basin Institute of Ecology of the Russian Academy of Sciences and the 65th Anniversary of the Kuibyshev Biological Station "Ecological Problems of Large River Basins", Togliatti, Russia, 15–19 October 2018; Rozenberg, G.S., Saksonov, S.V., Eds.; Anna: Togliatti, Russia, 2018.
186. Milyutkin, V.A.; Topelkin, S.A.; Borodulin, I.V.; Agarkov, E.A. Renewable Energy Sources (REn)—Biofuels from the Blue-Green Algae Biomass—Cyanobacteria. In Proceedings of the II All-Russian (National) Scientific and Practical Conference "Priority Areas of Energy Development in the Agro-Industrial Complex", Kurgan, Russia, 22 February 2018; Sukhanova, S.F., Maltsev, T.S., Eds.; Kurgan State Agricultural Academy: Kurgan, Russia, 2018.

187. ECOVOLGA. Development of Unified Cyanobacteria Database in Saratov Storage Reservoir. Available online: http://ecovolga.com/ (accessed on 10 June 2019).
188. Olabi, A.G. Circular economy and renewable energy. *Energy* **2019**, *181*, 450–454. [CrossRef]
189. Moraga, G.; Huysveld, S.; Mathieux, F.; Blengini, G.A.; Alaerts, L.; Van Acker, K.; De Meester, S.; Dewulf, J. Circular economy indicators: What do they measure? *Resour. Conserv. Recycl.* **2019**, *146*, 452–461. [CrossRef]
190. Svensson, N.; Funck, E.K. Management control in circular economy. Exploring and theorizing the adaptation of management control to circular business models. *J. Clean. Prod.* **2019**, *233*, 390–398. [CrossRef]
191. Laso, J.; García-Herrero, I.; Margallo, M.; Bala, A.; Fullana-i-Palmer, P.; Irabien, A.; Aldaco, R. LCA-Based Comparison of Two Organic Fraction Municipal Solid Waste Collection Systems in Historical Centres in Spain. *Energies* **2019**, *12*, 1407. [CrossRef]
192. Ratner, S.V. Circular economy: Theoretical bases and practical applications in the regional economy and management. *Innovations* **2018**, *9*, 29–37.
193. Trabelsi, A.; Elouedi, Z.; Lefevre, E. Decision tree classifiers for evidential attribute values and class labels. *Fuzzy Set Syst.* **2019**, *366*, 46–62. [CrossRef]
194. Dunjic, S.; Pezzutto, S.; Zubaryeva, A. Renewable energy development trends in the Western Balkans. *Renew. Sustain. Energy Rev.* **2016**, *65*, 1026–1032. [CrossRef]
195. Parchomenko, A.; Nelen, D.; Gillabel, J.; Rechberger, H. Measuring the circular economy—A Multiple Correspondence Analysis of 63 metrics. *J. Clean. Prod.* **2019**, *210*, 200–216. [CrossRef]
196. Avdiushchenko, A.; Zajac, P. Circular Economy Indicators as a Supporting Tool for European Regional Development Policies. *Sustainability* **2019**, *11*, 3025. [CrossRef]

© 2019 by the author. Licensee MDPI, Basel, Switzerland. This article is an open access article distributed under the terms and conditions of the Creative Commons Attribution (CC BY) license (http://creativecommons.org/licenses/by/4.0/).

Article

Incorporation of a Non-Constant Thrust Force Coefficient to Assess Tidal-Stream Energy

Lilia Flores Mateos [1,2,3,*,†] **and Michael Hartnett** [1,2,3,†]

1. College of Engineering and Informatics, NUI Galway, H91TK33 Galway, Ireland; michael.hartnett@nuigalway.ie
2. Ryan Institute, NUI Galway, H91TK33 Galway, Ireland
3. Centre for Marine and Renewable Energy Ireland (MaREI), P43C573 Cork, Ireland
* Correspondence: l.floresmateos1@nuigalway.ie
† These authors contributed equally to this work.

Received: 16 September 2019; Accepted: 25 October 2019; Published: 31 October 2019

Abstract: A novel method for modelling tidal-stream energy capture at the regional scale is used to evaluate the performance of two marine turbine arrays configured as a fence and a partial fence. These configurations were used to study bounded and unbounded flow scenarios, respectively. The method implemented uses turbine operating conditions (TOC) and the parametrisation of changes produced by power extraction within the turbine near-field to compute a non-constant thrust coefficient, and it is referred to as a momentum sink TOC. Additionally, the effects of using a shock-capture capability to evaluate the resource are studied by comparing the performance of a gradually varying flow (GVF) and a rapidly varying flow (RVF) solver. Tidal-stream energy assessment of bounded flow scenarios through a full fence configuration is better performed using a GVF solver, because the head drop is more accurately simulated; however, the solver underestimates velocity reductions due to power extraction. On the other hand, assessment of unbounded flow scenarios through a partial fence was better performed by the RVF solver. This scheme approximated the head drop and velocity reduction more accurately, thus suggesting that resource assessment with realistic turbine configurations requires the correct solution of the discontinuities produced in the tidal-stream by power extraction.

Keywords: tidal-stream energy; thrust force coefficient; momentum sink; unbounded flow; open channel flows; shock-capturing capability

1. Introduction

Resource characterisation is the initial step of any tidal energy project. However, conventional methodologies to assess tidal-stream power are based on infinite extent flow and the representation of a turbine as actuator disc [1–6]. This approach excludes the influence of boundary conditions such as free-surface and seabed in the power extraction analysis, which leads to the Lanchester-Betz–Joukowsky limit and indicates that in optimal conditions turbines can convert up to 59% (or 16/27) of kinetic energy into mechanical energy [7]. To better represent the aquatic medium of tidal turbines analytical approaches introduced features of a finite flow; firstly by introducing a rigid lid surface and lately by considering a deformable surface and a turbine downstream region where turbine wake mixing occurs. This analytical model is known as the linear momentum actuator disc in open channel flow (LMAD-OCH) [8]. Analysis of an actuator disc within a finite flow unveiled the importance of turbine bypass flow, blockage ratio (B), and the turbine downstream mixing region; where B indicates the fraction of a channel cross-sectional area occupied by the turbine.

A large number of potential regions for tidal power extraction are located in coastal areas; in these regions the use of finite flow is more appropriated as tidal streams are strongly influenced by the seabed and free-surface conditions. The consideration of these boundaries is relevant because they can significantly increase the amount of power extracted from the flow [9–11]. This is because in shallow waters the seabed and free surface constraint effect in the turbine is more significant and produces a stronger blockage effect, which in turn increases the maximum power extractable by the turbine. On the other hand, a recent methodology, which undertakes natural boundaries in the power extraction analysis, uses a rather expensive computational technique and constrains the head drop produced by power extraction. This methodology uses a rapidly varying flow solver and shock fitting techniques to implement a line sink of momentum [11,12]. The line sink of momentum is based on the LMAD-OCH analytical model, which enable to determine a relation between upstream and downstream depth-average flow velocities and depths as a function of the turbine operating conditions. This relation is given by the relative change of head drop across the turbine. Line sink of momentum approach uses an RVF solver to compute the rapid changes produced by the tidal turbine power extraction in the water depth and flow velocities. The authors of [11,12] used a Godunov-type scheme, which implements a shock-fitting technique. The momentum extracted by turbines is simulated by computing the modification of mass and energy fluxes across the line sink of momentum. The region occupied by a turbine array acts as an interface between upstream and downstream conditions of the flow. This interface represents the elevation and velocity discontinuities produced by power extraction. Computation of the discontinuity which initially separates upstream and downstream conditions is equivalent to solving a Riemann problem [13]; consequently, the Riemann solution indicates the flux through the discontinuity and therefore the shock propagation. To complete the line sink of momentum numerical representation, a condition on the head drop across the array is required [14]. This condition is given by the relative change of the head drop provided by the LMAD-OCH. This method was used to assess the tidal-stream potential of turbines configured as a fence in an idealised channel [11]. Such a configuration was used in the Pentland Firth to estimate maximum power extracted, 4.2 GW, and the extractable power at the sub-channels [15]. It was found that power availability varies according to the device operating conditions within the fences. A further refinement of the Pentland Firth estimation is given by [16]; they reported a more conservative upper limit of 1.9 GW based on a large, but a viable blockage ratio ($B = 0.4$). Contrary to the line sink of momentum approach, in this research the momentum extracted by the turbines was simulated considering the turbine operating conditions and implementing a method that does not constrain the water elevation change. For this task the momentum sink TOC was used [17], the method is based in LMADT-OCH and computes the axial component of the thrust force exerted by turbines to the flow by solving a sink term in the momentum equations. Consequently, momentum sink TOC incorporates the natural constraints of the coastal tidal-stream, the operating conditions of the turbine, and does not condition the head drop produced by power extraction. In this research, the momentum sink TOC was incorporated in a (1) conventional and (2) an innovative numerical solver to assess the resource. The conventional scheme solves gradually varying flows, which do not experience strong spatial gradients and are characterised by small Froude numbers. On the other hand, the innovative scheme solves rapidly varying flows that enable the simulation of flows, which experience discontinuities such as hydraulic jumps, flood waves, and shock waves. Therefore, a GVF solver simulates a flow that experiences smoother and slower changes than RVF. The rapidly varying flows solver implemented in this research is a shock-capturing model that consists of the algebraic combination of first-order and second-order upwind schemes. In this way, the model uses a low-order scheme to simulate the regions where strong gradients exist and spurious numerical solutions are likely to appear. This solver constitutes an efficient approach for simulating energy capture in rapidly varying flows because the scheme is not required to solve the Riemann problem at each grid, where an array of turbines are defined, to compute the discontinuities produced in the flow due to power removal. Solving the Riemann problem represents an expensive procedure in computational terms [18]

Two turbine configurations were selected to assess the resource at a regional scale: A fence and a partial fence. Partial-fence configuration covers a selected part of a channel cross-section producing flow diversion. Part of the stream passes through the array, experiencing velocity reduction due to the momentum loss, and the rest bypass the array, presenting velocity intensification when circumventing the partial fence. This scenario is referred to as unbounded flow and two analytical theories describe power extraction with this type of flow: (i) The LMAD-OCH and (ii) two scales models, their main characteristics are described below. LMAD-OCH studies a partial fence as a long row [11], and in this way, upstream conditions can be assumed to be uniform, and marine turbines within the array can be represented by actuator discs. This approach estimates the momentum extracted by a partial fence and provides information on energy lost only within the turbine's near field region (L_v) [11], i.e., just downstream of the turbine. LMAD-OCH is also referred to as a single-scale model [19] because the quasi-inviscid flow assumption used in the theory allows important turbulent mixing to occur at a far downstream region (L_h) where the pressure equilibrates across the channel cross-section. The L_v region is smaller than the far downstream region ($L_v < L_h$). Within the L_v region the quasi-inviscid assumption of LMAD-OCH enables the estimation of turbine-scale wake mixing, which occurs just downstream of the turbine and it is not significant in comparison to the array-scale wake mixing. This quasi-inviscid model is consistent with three-dimensional actuator disc computations of mixing just downstream of the turbines [20]. Therefore, the LMAD-OCH theory enables the parametrisation of turbine-wake mixing within the near-field region, in this region it is assumed that elevation and velocity perturbations due to power extraction occur. This assumption captures vertical flow variations produced by horizontal mixing effects at the turbine scale [21]. In contrast to the LMAD-OCH theory, the two scale model studies energy extraction via a partial fence considering two scales of flow: (a) A turbine scale flow, which describes the flow around a single turbine and the wake of the turbine, studied by [8] and (b) a array scale, which is the large-scale flow around the turbine array and the array wake [22]. The two scales' separation implies that the flow bypassing the turbine occurs faster than the horizontal expansion of the flow bypassing the entire array, and analyses power extraction as two quasi-inviscid open channel flow problems. The two scales are linked via the upstream boundary condition of the turbine-scale flow because power extraction produces a flow diversion around the array and reduces the mass flux through the array and this mass flux reduction provides the upstream boundary to the turbine scale [23]. Initially, a single long turbine row and a rigid lid surface approximation which did not account for the effect of changes in water depth were considered [22]. Later, a better accounting for the interaction of the device- and array-scale flow events enabled the analytical and numerical study of short rows [20]. Recently, [24] included a deformable free-surface in the two-scale analysis by incorporating finite Froude (F_r) numbers in the analysis and consequently accounting for a deformable surface.

Analytical models provide an understanding of the physics involved in tidal-stream power extraction; however, numerical simulations are required to provide a more complete analysis of the effects of tidal energy extraction. In this context, the two-dimensional approach, the line sink of momentum was developed to assess tidal-stream resource with a long partial fence, through the solution of shallow water equations. The method was used to numerically estimate the force applied by a porous disc in scale experiments [25], and to identify the power available at an idealised coastal headland [26] and at Anglesey Skerries, off the Welsh coast [12]. On the other hand, attempts to numerically couple the turbine scale with the array scale are limited to three-dimensional simulations of small turbine arrays [27], which require high-fidelity turbine scale simulations; this methodology has not been applied to regional scale [21].

In this research, tidal-stream energy resource at regional scale was evaluated in two scenarios: bounded and unbounded flows, which were represented with a fence and a partial-fence turbine array configuration, respectively. The energy capture by the arrays was simulated with the momentum sink TOC method, which numerically implements the LMAD-OCH theory. To investigate the role of numerical solution schemes for simulating tidal energy capture, the momentum sink TOC was implemented in two hydrodynamic models that use a conventional (GVF) and up-to-date (RVF) solution scheme. The novelty of this research lays in the numerical representation of energy capture by arrays of turbines considering (i) a flow more representative of coastal regions, (ii) not conditioning of head drop across a turbine array, (iii) use of hydrodynamic models which are able to represent realistic coastal scenarios, and (iv) implementation of an efficient RVF solver that uses a shock-capturing scheme. The methodology proposed will enable the identification of a less computationally expensive numerical tool that provides a reliable evaluation of the resource in realistic scenarios.

2. Methodology

The assessment of tidal-stream resource at regional scale requires the specification of three aspects: A numerical model approach, the tidal energy extraction representation, and domain size [21]. These specifications follow.

2.1. Modelling Approach

Momentum sink TOC was implemented in two numerical hydrodynamic models based on the depth integrated velocity and solute transport (DIVAST) model [28]. DIVAST was developed to simulate hydrodynamic solute and sediment transport processes in rivers, estuaries, and coastal waters, and it incorporates a flooding and drying capability [29–31].

2.1.1. Gradually Varying Flows

The first model uses an alternating direction implicit (ADI) methodology, solves two-dimensional shallow water equations (2D-SWEs), and simulates gradually varying flows, which are characterised by small Froude numbers. Hereafter ADI, represents the conventional strategy used to assess tidal-stream energy. The 2D-SWEs were used to describe the evolution of a tidal-stream and power extraction, the depth integrated continuity (Equation (1)) and x-component momentum (Equation (2), a similar expression was used for the y-component) equations to solve an inviscid flow, neglect Coriolis force, and to omit wind forcing. The inviscid character of the governing equations allows the neglection of vortical structures associated with the turbine's blade root and tip. The viscous terms omission is justified if bottom friction prevails over viscosity effects [32]. Moreover, a scaling analysis indicates that viscous terms can be disregarded if the horizontal scale of the domain is large and tidal currents change smoothly over the length of the domain [14]. Furthermore, viscous terms neglection has been considered in the numerical simulation of regional scale tidal-stream energy extraction [11,16].

$$\frac{\partial \zeta}{\partial t} + \frac{\partial q_x}{\partial x} + \frac{\partial q_y}{\partial y} = 0 \qquad (1)$$

$$\frac{\partial q_x}{\partial t} + \frac{\partial \left(\frac{\beta q_x^2}{H}\right)}{\partial x} + \frac{\partial \left(\frac{\beta q_x q_y}{H}\right)}{\partial y} = -gH\frac{\partial \zeta}{\partial x} - \frac{g q_x \sqrt{q_x^2 + q_y^2}}{H^2 C_e^2} - F_{Tx} \qquad (2)$$

where $q_x = UH$ and $q_y = UH$ indicate depth-integrated velocity flux component in the x- and y-direction; t stands for time and β is the momentum correction factor for non-uniform vertical velocity profile. The surface elevation change with respect to mean water depth h is represented by ζ; therefore, total water depth is defined as $H = h + \zeta$. The depth-integrated turbulence model considers only the effects of the bed shear stress, which is a function of Chezy roughness coefficient (C_e) and gravity (g).

2.1.2. Rapidly Varying Flows

The second model combines a MacCormack and a symmetric five point total variation diminishing (TVD) schemes to solve strong spatial gradients in the flow using an efficient shock-capturing method [30,33]. Henceforth, TVD allows the computation of discontinuities likely to appear in rapidly varying flows simulation without treating the discontinuity as a Riemann problem. To simulate flows which experience discontinuities TVD solves the conservative form of 2D-SWEs, [17] describe this form of the governing equations in detail. Some examples of this kind of flows are hydraulic jumps, storm surges, flood waves, and shock waves.

2.2. Numerical Methods

ADI and TVD use a finite difference spatial discretisation to approximate a solution of the governing equations. The models discretise the equations onto a square structured grid, where the edges of the grid cells are oriented parallel to the Cartesian coordinates [34,35].

The GVF solver uses an ADI's semi-implicit finite difference scheme which splits a single time-step solution into two time steps [36]. The final finite difference equations for each half time step (HFDT) are solved using the method of Gaussian elimination and back substitution [34]. Meanwhile, the RVF solver uses an explicit scheme to approximate the solution of the governing equations at the current time and convergence to the solution is conditioned to a maximum time step. RVF solver uses an algebraic combination of the first-order and second-order upwind schemes, where a kind of artificial viscosity is included to smooth the solution close to the shock [30,37]. The proportion of the contribution of each scheme is adjusted depending on the nature of the flow; if the solution is sub-critical (smooth), the second-order scheme is implemented otherwise, if the solution is trans- or super-critical, the lower-order scheme is implemented [30].

The stability of the numerical models solution was assured by satisfying Courant-Fredrichs-Lewy condition. To identify the most efficient spatial and temporal resolution a range of time steps and grid sizes were tested. The elected spatial discretisation was $\Delta X = 150$ m in both models. The semi-implicit character of ADI allowed a relatively large time step ($\Delta t = 12$ s), but the explicit scheme of TVD did not tolerate time steps larger than 1.50 s.

2.3. Turbine Representation

The method used to simulate energy capture by marine turbines is momentum sink TOC and it implements numerically the LMADT-OCH theory using a sink approach. The mathematical representation of energy capture was introduced as a sink term $\vec{F_T}$ in the momentum equations (see x-component, Equation (2)). This term represents the turbine thrust force (T) responsible for tidal stream momentum loss (Equation (3)). The thrust force is a function of the turbine cross-sectional area A, thrust coefficient C_T, and depth-average velocity $\vec{U}$.

$$T = \frac{1}{2}\rho \vec{U}^2 A C_T \tag{3}$$

Figure 1a illustrates the cross section of the turbine area A, indicated by a dotted-red line, and the thrust force exerted on the flow. The components of the thrust force are given by:

$$F_{Tx} = T|\sin\theta| \tag{4}$$
$$F_{Ty} = T|\cos\theta| \tag{5}$$

where it is assumed that the thrust force makes an angle θ with the y-axis. Numerically, the sink term is the thrust force per unit-grid and per unit-mass. The components of the thrust force introduced as sink terms in the momentum equations and solved by the models were:

$$F_{Tx} = \frac{T}{\Delta x \Delta y}|\sin\theta| = \frac{1}{\Delta x \Delta y}\frac{1}{2}\rho A_x C_T U^2 \tag{6}$$
$$F_{Ty} = \frac{T}{\Delta x \Delta y}|\cos\theta| = \frac{1}{\Delta x \Delta y}\frac{1}{2}\rho A_y C_T V^2$$

Furthermore, the thrust force computed with momentum sink TOC is a function of the thrust force coefficient (Equation (7)), which depends on the specification of wake-induction (α_4) factor and the bypass flow coefficient (β_4).

$$C_T = \beta_4^2 - \alpha_4^2 \tag{7}$$

LMADT-OCH theory allows to relate turbine operating conditions to momentum captured by the turbine and to link changes produced by power extraction to thrust forces within the turbine's near field length L_v [14]. LMADT-OCH analytical model assumes that the upstream flow passes through the fence, mixes, and returns to a vertical profile similar to that upstream over a length L_v. A sub-critical tidal-stream is assumed and energy capture produces a tidal flow division into core-flow and bypass-flow components, a head drop across the array (Δh), and energy loss by turbine wake mixing. The core-flow refers to the stream that passes through the turbine, which experiences a velocity reduction due to energy extraction, represented by the turbine velocity coefficient α_2. Core-flow presents a further velocity reduction downstream of the turbine due to turbine wake mixing dissipation, denoted by the wake induction factor α_4. On the other hand, the bypass-flow circumvents the turbine and presents a velocity magnitude intensification denoted by the bypass induction factor β_4. Based on LMADT-OCH energy capture within L_v region is parameterised by: bypass induction factor (β_4), wake induction factor, turbine velocity coefficient (α_2), and water depth drop across the array (Δh).

Calculation of thrust force required: (i) election of turbine operating conditions, (ii) identification of upstream conditions of the flow, (iii) and parametisation of changes produced by power extraction within the turbine near-field region. These requirements are sketched in Figure 1a. Turbine operating conditions were defined by the blockage ratio B and wake induction factor α_4. Factor α_4 is an indicator of velocity rate reduction due to wake mixing at the turbine scale [38]. In ADI and TVD models, the blockage ratio implemented is the ratio of the cumulative turbine area per cell-grid (A) over the grid-cell cross-section area ($H \Delta X$) [17]. Parameterisation of flows' depth and velocity changes within L_v required the specification of α_4, B, and upstream Froude number ($F_r = U/\sqrt{gH}$). These parameters enabled the calculation of β_4, which is the physical admissible root of a quartic polynomial [10]. An eigenvalue method was used to find the roots of the polynomial [17]. Finally, is possible to calculate $C_T(\alpha_4, \beta_4)$ and thrust force (Figure 1b).

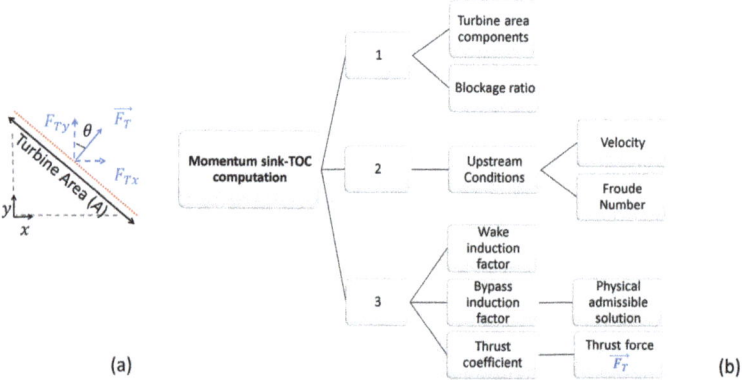

Figure 1. Two-dimensional representation of the thrust force ($\vec{F_T}$) exerted by the turbine to the incident tidal-stream (**a**) and the momentum sink turbine operating conditions (TOC) computation procedure (**b**).

2.4. Tidal Channel and Turbine Array

A tidal fence and a partial-fence configuration were implemented in a large domain, which corresponded to a large channel connecting two large basins (Figure 2). The channel is relatively narrow and long with 12 km length (L) and 3 km wide (W) and it is characterised by an aspect ratio $W/L = 0.25$. The basin extension is 4.5 times the length of the channel. The domain configuration was selected based on previous studies of tidal-stream energy assessment [11]. The channel was forced with a standing wave with semi-diurnal frequency M_2 on the western boundary (Equation (8)). The amplitude of the forcing was introduced as a ramped-up wave over two M_2 tidal periods (Equation (9)); the gradual introduction of the forcing removed the initial numerical noise in the simulation and helped to reach a time varying steady state faster. This state was reached when tidal circulation became regular.

$$\zeta = A(t) \cos(\omega t) \cos(kx) \tag{8}$$

$$A(t) = A\left(-2\left(\frac{t}{2T_{M_2}}\right)^3 + 3\left(\frac{t}{2T_{M_2}}\right)^2\right), t < 2T_{M_2} \tag{9}$$

At the eastern boundary, water elevation was set to constant and velocity was set to zero. At the walls, the boundaries satisfied a no-slip condition. The water depth is relatively shallow, it was set to 40 m = 2.5 d, where $d = 16$ m is the turbine diameter. A small and constant bottom friction, defined as a function of a non-dimensional drag coefficient, was implemented ($C_d = 0.0025$). This coefficient has been shown to best reproduce field measurements of velocity and elevation of M_2 tidal currents in the vicinity and far-field of Rathlin Sound [39].

Simulations started from quiescent initial conditions and a steady periodic flow was reached after the second tidal period. The simulation time was 50 h, equivalent to four M_2 tidal periods ($4\ T_{M_2}$). It is worth mentioning that evaluations of 8 T_{M_2} were consistent with conditions reported at the fourth tidal cycle. The consistent boundary conditions and forcing implementation between the models were assured by obtaining consistent hydrodynamic conditions at the natural state. This state refers to marine turbines omission and nil energy extraction.

Turbine configurations were deployed in the middle of the channel (Figure 2). The fence corresponds to an array of turbines distributed as a single row that fully extends across the channel cross-section. On the other hand, the partial fence covers 40% of channel cross-section; the length of the array is 1200 m defined over eight grid-cells, where each grid-cell contains a cluster of turbines.

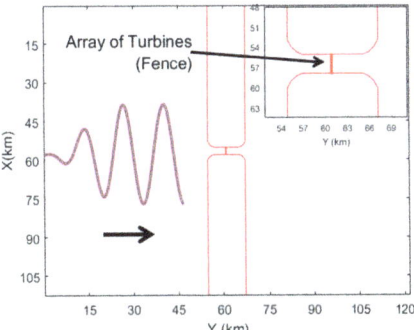

Figure 2. Tidal channel model domain. A channel zoom out (upper-right location) shows the array of turbines deployed. Forcing implemented is represented by the ramping up elevation in magenta.

3. Resource Assessment

Though a significant amount of power could be extracted if high blockage ratios were used, this procedure presents adverse effects such as (i) significant flow rate reduction [40], (ii) tidal hydrodynamics affectation, and (iii) mixing and transport processes impact at turbine scale and regional scale [6,21,41,42]. As a result of this evidence, evaluation of tidal-stream resource is performed within realistic blockage ratios ($0 < B \leq 0.4$) as a post-processing stage.

Indicators used to assess the energy resource were the following metrics: Total power extracted, power dissipated by turbine wake mixing, and power available for electrical generation. The procedure used to calculate them follows. The initial step was the election of turbine operating conditions and calculation of turbine velocity coefficient (Equation (10)). Subsequently, thrust (Equation (7)) and power (Equation (12)) coefficients associated to the turbine were estimated. The next step was identification of head drops across the array and turbine efficiency (Equation (13)). These two parameters enable estimation of total power extracted (Equation (14)), power dissipated by turbine wake mixing (Equation (15)), and power available for electrical generation (Equations (11) and (16)).

$$\alpha_2 = \frac{2(\beta_4 + \alpha_4) - (\beta_4 - 1)^3(B\beta_4^2 - B\beta_4\alpha_4)^{-1}}{4 + (\beta_4^2 - 1)(\alpha_4\beta_4)^{-1}}. \quad (10)$$

$$P = \frac{1}{2}\rho U^3 A C_P; \quad (11)$$

$$C_P = \alpha_2(\beta_4^2 - \alpha_4^2) \quad (12)$$

$$\eta \approx \alpha_2 \left(1 - \frac{1}{2}\frac{\Delta h}{h}\right) \quad (13)$$

$$P_T = \rho g U \frac{A}{B} \Delta h \left(1 - F_r^2 \frac{1 - \Delta h/2h}{(1 - \Delta h/h)^2}\right) \quad (14)$$

$$P_W = P_T(1 - \eta) \quad (15)$$

$$P_* = \eta P_T \quad (16)$$

The operating conditions of the turbine and the turbine configurations evaluated with ADI and TVD models are specified in the four scenarios reported in Table 1.

Table 1. Scenarios simulated and initial parameters specification.

Model	Configuration	Scenario	B	α_4	Δt (sec)
ADI	Fence	1	$0 \leq B \leq 0.4$	$\alpha_4 = 1/3$	12
	Partial-Fence	2	$0 \leq B \leq 0.4$	$\alpha_4 = 1/3$	12
TVD	Fence	3	$0 \leq B \leq 0.4$	$\alpha_4 = 1/3$	1.5
	Partial-Fence	4	$0 \leq B \leq 0.4$	$\alpha_4 = 1/3$	1.5

3.1. Thrust and Power Coefficients Time Series

Calculation of C_P and C_T required specification of the wake induction factor, which depends on the location of the turbine within the array and the local tidal dynamics [26] as both affect turbine wake mixing. However, to maximise power available for electrical generation the downstream core flow ($\alpha_4 U$) was adjusted by setting the wake induction factor to $\alpha_4 = 1/3$ [43,44]. A constant α_4 was used during the simulation; however, a slightly better power available is obtained if a variable wake induction factor is used [12].

Thrust and power coefficients are functions of the upstream conditions through the Froude number used in β_4 calculation; therefore, C_T and C_P are modulated by the tidal cycle. Thrust and power coefficients time series were obtained from both models using a fence configuration constituted by turbines with a relatively large blockage ratio ($B = 0.3$). The time series from an M_2 tidal cycle and taken from the middle of the channel, where the array was deployed, are shown in Figure 3a,b. C_T and C_P time series present two maximum values that correspond to maximum flood tide and maximum ebb tide velocities. Comparison between the solutions obtained from the models indicate that ADI produces slightly higher magnitudes of C_P ($1.21 < C_P < 1.238$) and C_T ($2.38 < C_T < 2.43$) than TVD's $1.21 < C_P < 1.235$ and $2.36 < C_T < 2.40$. Time-average of C_T and C_P obtained from ADI and TVD are consistent with LMADT-OCH analytical model and they indicate upper limits for these coefficients.

Additionally, time series obtained with ADI present a phase delay evident in the maximum values. ADI's phase delay was evident at stream-wise velocity component of M_2 tidal current at natural state. This behaviour was consistent along the domain, i.e., it was not restricted to the channel, suggesting that the delay may be related to the different solution scheme employed by the models. However, further understanding of M_2 phase simulated by both models would require examination of field observations of free surface elevation and depth-averaged flow velocity. It is worth noticing that the parameters to be presented in the following sections refer to values averaged over a tidal period; therefore, the phase delay reported by ADI time series does not influence the results presented.

The larger magnitude of C_T reported by ADI is associated with the smaller affectation of the momentum extraction on the bypass velocity reduction (represented by β_4). ADI underestimation of velocity reduction due to power extraction was reported by [45]. In the case of C_P, the turbine velocity coefficient reduces the difference between the magnitudes obtained from the models.

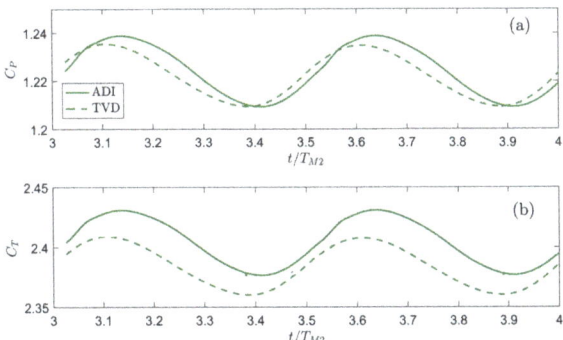

Figure 3. Time-series of C_P (**a**) and C_T (**b**) coefficients obtained from alternating direction implicit (ADI) (continuous-line) and total variation diminishing (TVD) (dash-line) schemes for $B = 0.3$.

3.2. Upstream Velocities

Contrary to a tidal fence configuration, where energy extraction produce uniform conditions along the array and a middle-fence cell capture this dynamic, the implementation of a partial fence produces a scenario where energy extraction occurs along a limited section of the cross-channel, indicating the existence of array-bypass flows. To identify the effects of a partial fence on the upstream velocities, the time-averaged stream-wise component of velocity was calculated at each grid cell that constitutes the partial fence. Upstream velocities normalised by natural state velocities and obtained for increasing values of blockage ratio computed with ADI and TVD are presented in Figure 4. Upstream velocities along the fence tend to be uniform for small blockage ratios ($B \leq 0.15$), but further increase of B influence differently the grid-cells located at the middle and at the edges of the partial fence. Large values of B produce higher velocity decreases in the middle of the array, meanwhile the edges exhibit smaller velocity reduction. A comparison of results obtained from both models shows that upstream velocities from ADI present smaller reductions with increasing B than TVD. In addition, for small B, velocities at the edges of the partial fence simulated with ADI show a slight magnitude enhance. This suggests that the faster array-bypass flow computed with ADI model is influencing the velocities at the partial-fence edges. The smaller velocity reduction presented in ADI's velocity solutions is consistent with [45]. They indicated that a gradual varying flow scheme under-estimate the velocity decrease produced by momentum extraction; meanwhile, a rapidly varying flow scheme approximates more accurately this reduction due to its capacity to solve the velocity spatial gradients generated during energy extraction.

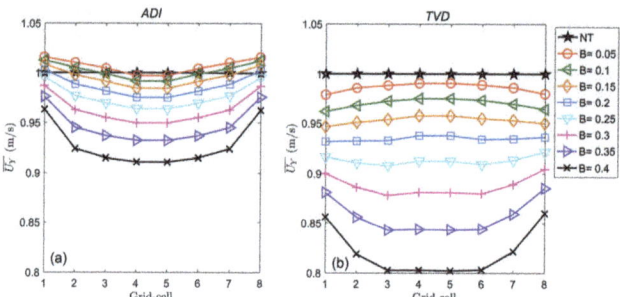

Figure 4. Blockage ratio effect on normalised upstream velocities along a partial fence, ADI (**a**) and TVD (**b**). Natural state flow condition is indicated by NT.

3.3. Head Drop and Turbine Efficiency

Estimation of turbine efficiency required calculation of head drop across the array. Δh was calculated using an analytical and numerical approach. Analytical solutions were obtained by solving a cubic polynomial, which represents a condition on the head drop across the array obtained from LMADT-OCH [8]. Coefficients of the polynomial are a function of blockage ratio, thrust coefficient, and Froude number. C_T and F_r were obtained from TVD simulations, as this scheme approximates the velocity reduction more accurately for a fence scenario [45]. Meanwhile, numerical solutions correspond to head drops obtained from depth differences across the array at upstream and downstream locations. These locations correspond to the neighbouring upstream and downstream cell. In the case of a 16 m diameter device, the mean axial upstream velocities are recovered to about 80% [46,47] within 160 m; therefore, the grid size used in this research (ΔX = 150 m) is of the order of the turbine wake length suggesting that the neighbouring cell approximates the upstream conditions of the flow. Maximum head drops within a tidal cycle were calculated for increasing B, Δh_{max} obtained analytically and numerically using a fence (Figure 5a) and a partial fence (Figure 5b) were compared. For bounded-flows, analytical and numerical solutions are similar; however, a GVF scheme (Figure 5a.1) produces a solution more consistent with the analytical solution (Figure 5a). In the case of unbounded flows, the analytical solution of LMAD-OCH indicates a homogeneous water drop along the partial fence for given blockage ratio (Figure 5b). This solution is consistent with the 1D assumption of the theoretical model [9,43]. Conversely, numerical solutions indicate non-uniform head drops along the array for increasing B values. Within the partial fence, maximum values of head drop are found in the middle cells, while smaller depth change are exhibited towards the edge of the array. Comparison of Δh_{max} obtained from TVD and ADI models for a partial fence indicate that TVD simulates larger head drops (Figure 5b.2). Additionally, for small B, Δh_{max} at the middle-cells of the array obtained from TVD are similar to analytical solutions; however, for $B \geq 0.30$, the head drop magnitudes become smaller than the analytical solution. These results indicate that rapidly varying flow solver provides a reasonable insight into the effect of small blockage ratios on the head drop across the partial fence, where Δh is influenced by the bypass flow producing a head drop reduction at the edges of the partial fence.

Turbine efficiencies were estimated for increasing blockage ratios. Analytical and numerical solutions of η for flows characterised by small Froude numbers were calculated, as the solutions were found to be consistent only numerical solutions are reported. These solutions are functions of the head drops obtained with ADI and TVD. Time-averaged and array-averaged turbine efficiencies calculated for both configurations: fence and partial fence are presented in Figure 6a,b, respectively. Turbine efficiencies were plotted against the time-averaged, array-averaged thrust coefficients. Increasing B values indicate a gradual efficiency reduction but thrust coefficient augmentation. Maximum turbine efficiency found ($\bar{\eta} = 0.63$) correspond to the smaller blockage ratio tested ($B = 0.05$) and this value is associated to $\overline{C_T} = 1.0$; meanwhile, blockage ratio increases to $B = 0.4$ produce a turbine-efficiency reduction ($\overline{\eta} = 0.47$) and thrust coefficient increase ($\overline{C_T} = 3.5$). Efficiency decrease with B augmentation is explained by the larger energy dissipated due to turbine-wake mixing [48]. On the other hand, the analytical study of further blockage ratio increase ($B > 0.4$) indicates a slow turbine-efficiency decrease rate [14]. This trend indicates that for a very large B, subsequent dimension augmentation produces a small efficiency reduction; such a scenario is attractive; however, large blockage ratios are not a practical option.

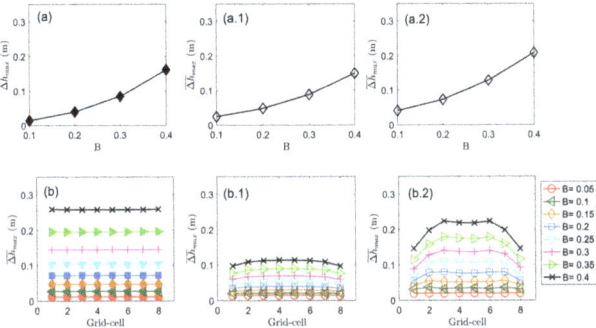

Figure 5. Maximum head drops across a fence (**a**) and partial fence (**b**); analytical (filled markers) and numerical solutions (un-filled markers) obtained from ADI (*.1) and TVD (*.2) models.

Comparison of ADI and TVD solutions from a partial fence shows similar results; however, slightly higher efficiency values are reported with TVD. As the efficiency is a function of Δh, higher η's are explained by the larger head drops obtained with RVF scheme. Furthermore, comparison of the solutions obtained from both configurations indicate a high degree of similarity. This consistency suggest that the partial fence is long enough to resemble conditions similar to a fence at the turbine's near-field region. It is worth mentioning that turbine efficiencies calculated for a partial fence do not capture the power dissipated by the array-wake mixing, because LMAD-OCH theory only describes the energy extraction dynamic within the turbine's near-field region [20,23]. In this region, wake mixing produced by individual turbines occurs and flow's uniform conditions are recovered; therefore, wake mixing generated by the array is not included in the analysis.

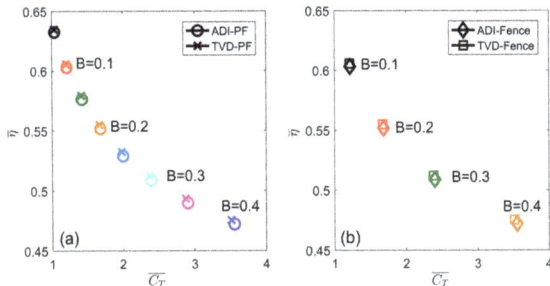

Figure 6. Turbine efficiency and thrust coefficient for partial fence (**a**) and fence (**b**) configurations obtained from ADI and TVD.

3.4. Power Metrics

To compare the performance of a partial fence with a fence, time-averaged and array-averaged power metrics were obtained from both configurations. Total power extracted estimated with ADI and TVD are presented in Figure 7(a.1,a.2), respectively. P_T is a function of water drop and the increase of blockage ratio produces a head drop augmentation which is responsible of larger values of total power extracted. For bounded flows (F), [45] indicated that P_T is accurately approximated by an ADI model because a GVF solver approximates better the head drops across a fence. For unbounded flows scenario (PF), P_T magnitudes obtained from ADI and TVD are smaller than the fence. This result is explained by the head drop reported by the models for the partial fence; smaller Δh magnitudes were obtained for this configuration due to the reduced head drops at the edges of the array. The comparison of P_T obtained by the models for the partial fence indicate that TVD scheme produces larger magnitudes of total power extracted. These solutions are more reliable as TVD approximates better the head drop across a partial fence. The power dissipated by turbine-wake mixing estimated with both ADI and TVD models are shown in Figure 7(b.1,b.2), respectively. P_W depends on the turbine efficiency and the total power extracted, which in turn is strongly influenced by the head drop across the turbine array. This dependency indicates that P_W solutions are also explained by head drop values. For bounded flows, P_W values obtained with ADI are reliable solutions [45]. In the case of unbounded flows scenario, P_W magnitudes obtained from the models are slightly smaller than the fence. This result is expected as smaller Δh_{max} were obtained for the partial fence. The comparison of P_W obtained from the models for the partial fence, indicate that TVD produces larger magnitudes of power lost. TVD solutions are more accurate due to the better approximation of the head drop for unbounded flows provided by the model.

Regarding the power removed by the turbine, two metrics were calculated. First, in function of turbine efficiency P_* and second, in terms of cubic velocity P. These metrics are expected to be consistent as they describe the amount of power available for electricity generation.

Power available in terms of turbine efficiency estimated with ADI and TVD are presented in Figure 8(a.1,a.2), respectively. P_* is a function of total power extracted and consequently depend on the head drop. For bounded flows, P_* is better approximated by an ADI scheme; however, for unbounded flows scenario, this metric is better simulated by an TVD scheme.

On the other hand, power available in terms of cubic velocity calculated with ADI and TVD are presented in Figure 8(b.1,b.2), respectively. P magnitude depends on the upstream velocity and for bounded flows, [45] reported the incapacity of ADI model to satisfactorily approximate the velocity reduction due to power extraction. This underestimation of velocity decrease for increasing values of blockage ratio explains the large values of P reported by ADI for both configurations. Meanwhile, P obtained with TVD for a fence and a partial-fence configuration present magnitudes similar to P_* obtained from both ADI and TVD models. This result indicates that TVD scheme better approximates the consistency between P and P_* for both configurations. ADI fails to represent this consistency as ADI underestimate the velocity reduction due to power extraction and consequently report higher values of Power than TVD. On the other hand, TVD simulates better this velocity reduction, so powers in function of velocity are better solved by TVD and these values are more consistent with the power in function of turbine efficiency.

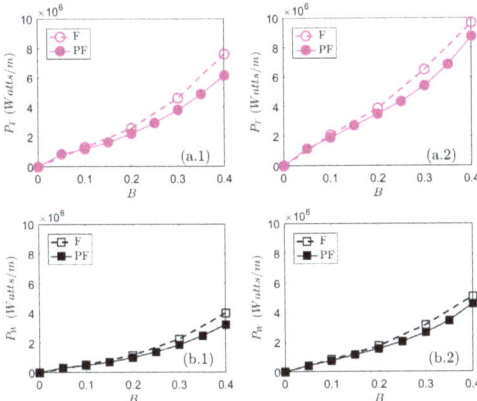

Figure 7. Effect of blockage ratio on P_T (**a**) and P_W (**b**) for a fence (dash-line) and partial-fence (continuous-line) configuration. Solutions obtained from ADI (*.1) and TVD (*.2).

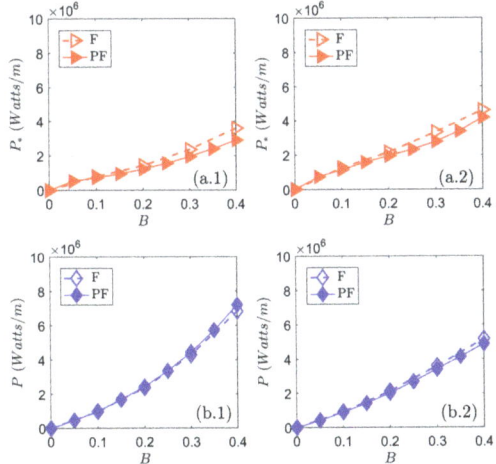

Figure 8. Effect of blockage ratio on P_* (**a**) and P (**b**) for a fence (dash-line) and partial-fence (continuous-line) configuration. Solutions obtained from ADI (*.1) and TVD (*.2).

4. Discussion and Conclusions

This research implemented momentum sink TOC method in two hydrodynamic models that can solve realistic coastal scenarios to simulate marine turbine configurations and to perform tidal-stream resource assessment. Contrary to the conventional consideration of constant thrust coefficient [4,6,49], the use of turbine operating conditions to calculate the thrust force, imparted by the turbine on the stream, enabled the calculation of non-constant thrust and power coefficients. These coefficients undertake the tidal regime specific to a potential site and allow the use of tailored values of C_T and C_P to assess the resource, therefore providing a more realistic numerical representation of the turbine.

In contrast to [11], the head drop across the array used to assess the resource was obtained from the depth difference between the array upstream and downstream location. The use of a no-fixed head drop in a partial fence enabled the identification of non-uniform conditions along the array. This configuration represented the simulation of an unbounded flow, which favours the existence of two regions: array-bypass flow and array-wake flow. These regions describe the velocity intensification when bypassing the array, and the significant velocity diminution downstream of the array due to array-wake mixing. Flow conditions along the partial fence are not uniform as array-bypass flows influence the upstream velocity and water depth at the edges of the array. These features illustrates the effect of a partial fence on the free flow as reported by [23,50] and indicate the advantage of using numerical tools to evaluate the tidal-stream resource and complement analytic models.

In terms of turbine efficiency, both turbine configurations showed consistent efficiencies. The similarity of the results could be explained by the use of the same wake-induction factor ($\alpha_4 = 1/3$), which is used for the turbine velocity coefficient (α_2) calculation [45]. The turbine efficiency used in this research considers the turbine-wake mixing within the near-field region. Over the near-field length scale, it is assumed that flow passing through the fence and partial fence mixes and leads to a smooth vertical profile similar to upstream profiles [43]. This assumption captures vertical flow variations produced by horizontal mixing effects at the turbine scale [21]. The models compute the turbulence generated by turbine-wake mixing correctly by implementing (i) turbulence induced by the bottom friction and (ii) inviscid flow assumption [51].

This research assesses upper limits for tidal-stream resource assessment in a semi-narrow tidal channel using the M_2 tidal constituent. Additionally, evaluation of two turbine configurations allowed the identification of the numerical scheme that performs more accurate tidal-stream energy resource evaluation. Opposite to conventional methodologies to assess the resource, where tidal energy extracted by marine turbines was approximated with a bottom roughness [40,52,53] and a momentum sink [6,47,49,54], the numerical implementation of LMADT-OCH enabled the calculation of the turbine efficiency. This parameter provided a distinction between total power extracted, power available for electrical energy generation, and power dissipated by turbine-wake mixing. Evaluation of the conventional scheme to assess tidal-stream (ADI model which solves GVF) and a scheme with shock-capturing capability (TVD model which solves RVF), indicate that unbounded flow scenarios are better assessed with a TVD scheme. The better approximation of the head drops and velocities reduction by a TVD model for a partial-fence scenario implies that a rapidly varying flow solver is required to calculate an upper bound for tidal-energy extraction when a realistic configurations of turbines is used. The conclusions of this work are as follows:

- Depth perturbations due to power extraction were solved by both models ADI and TVD. A satisfactory head drop was approximated with TVD scheme for both turbine configurations, indicating that water depth gradients are better solved by the shock-capturing capability.
- The total power extracted (P_T), available power in terms of the turbine efficiency (P_*), and power dissipated by turbine-wake mixing (P_W), were correctly calculated by the models. For a tidal fence, power metrics were better computed by the scheme that solves slow and smooth flows, because the head drop across the array simulated by this scheme was more accurate. In the case of a partial fence, the power metrics were better approximated by solving a rapidly varying flow solution.
- Simulation of unbounded flows scenario with sink TOC method and RVF solver indicates non-uniform conditions along the partial fence, contrary to one-dimensional LMAD-OCH.
- In terms of power removed by turbines as a function of the upstream velocity (P), this metric is better solved by an RVF solver as this model correctly represents the flow velocity reduction produced by power extraction.

Author Contributions: Conceptualization, L.F.M. and M.H.; methodology, L.F.M.; validation, L.F.M.; formal analysis, L.F.M.; resources, M.H.; writing–original draft preparation, L.F.M.; writing–review and editing, M.H.; visualization, L.F.M.; funding acquisition, M.H.

Funding: This material is based upon works supported by the Science Foundation Ireland under Grant no. 12/RC/2302 through MaREI, the National Centre for Marine Renewable Energy Ireland.

Acknowledgments: The authors thank the Irish Centre for High-End Computing for the provision of computational facilities and support.

Conflicts of Interest: The authors declare no conflict of interest.

References

1. Fraenkel, P. Power from marine currents. *Proc. Inst. Mech. Eng. Part J. Power Energy* **2002**, *216*, 1–14. [CrossRef]
2. Blunden, L.S.; Bahaj, A.S. Tidal energy resource assessment for tidal stream generators. *Proc. Inst. Mech. Eng. Part J. Power Energy* **2006**, *221*, 137–146. [CrossRef]
3. Sun, X.; Chick, J.P.; Bryden, I.G. Laboratory-scale simulation of energy extraction from tidal currents. *Renew. Energy* **2008**, *33*, 1267–1274. [CrossRef]
4. Ahmadian, R.; Falconer, R.; Bockelmann-Evans, B. Far-field modelling of the hydro-environmental impact of tidal stream turbines. *Renew. Energy* **2012**, *38*, 107–116. [CrossRef]
5. Neill, S.; Jordan, J.R.; Couch, S.J. Impact of tidal energy converter (TEC) arrays on the dynamics of headland sand banks. *Renew. Energy* **2012**, *37*, 387–397. [CrossRef]
6. Fallon, D.; Hartnett, M.; Olbert, A.; Nash, S. The effects of array configuration on the hydro-environmental impacts of tidal turbines. *Renew. Energy* **2014**, *64*, 10–25. [CrossRef]
7. Van Kuik, G.A.M. The Lanchester–Betz–Joukowsky limit. *Wind Energy* **2007**, *10*, 289–291. [CrossRef]
8. Houlsby, G.; Vogel, C. The power available to tidal turbines in an open channel flow. *Proc. Inst. Civ. Eng. Energy* **2017**, *170*, 12–21. [CrossRef]
9. Garrett, C.; Cummins, P. The power potential of tidal currents in channels. *Proc. R. Soc. Math. Phys. Eng. Sci.* **2005**, *461*, 2563–2572. [CrossRef]
10. Whelan, J.I.; Graham, J.M.R.; Peiró, J. A free-surface and blockage correction for tidal turbines. *J. Fluid Mech.* **2009**, *624*, 281. [CrossRef]
11. Draper, S.; Houlsby, G.T.; Oldfield, M.L.G.; Borthwick, A.G.L. Modelling tidal energy extraction in a depth-averaged coastal domain. *IET Renew. Power Gener.* **2010**, *4*, 545. [CrossRef]
12. Serhadlıoğlu, S.; Adcock, T.A.A.; Houlsby, G.T.; Draper, S.; Borthwick, A.G.L. Tidal stream energy resource assessment of the Anglesey Skerries. *Int. J. Mar. Energy* **2013**, *3–4*, e98–e111. [CrossRef]
13. Khan, F.A. Two-Dimensional Shock Capturing Numerical Simulation of Shallow Water Flow Applied to Dam Break Analysis. Ph.D. Thesis, Loughborough University, London, UK, 2010.
14. Draper, S. Tidal Stream Energy Extraction in Coastal Basins. Ph.D. Thesis, St Catherine's College, University of Oxford, Oxford, UK, 2011.
15. Draper, S.; Nishino, T.; Adcock, T.A.; Houslby, G.T. Wind and tidal turbines in uniform flow. In *3rd Oxford Tidal Energy Workshop*; University of Oxford: Oxford, UK, 2014.
16. Adcock, T.A.A.; Draper, S.; Houlsby, G.T.; Borthwick, A.G.L.; Serhadlioglu, S. The available power from tidal stream turbines in the Pentland Firth. *Proc. R. Soc. Math. Phys. Eng. Sci.* **2013**, *469*, 20130072. [CrossRef]
17. Flores Mateos, L.; Hartnett, M. Depth average tidal resource assessment considering an open channel flow. In *Advances in Renewable Energies Offshore, Proceedings of the 3rd International Conference on Renewable Energies Offshore (RENEW)*; Soares Guedes, C., Ed.; Taylor & Francis Group: Abingdon, UK, 2018; p. 10.
18. Shyue, K.M. An Efficient Shock-Capturing Algorithm for Compressible Multicomponent Problems. *J. Comput. Phys.* **1998**, *142*, 208–242. [CrossRef]
19. Takafumi, N.; Willden, R.H. The Efficiency of Tidal Fences: A Brief Review and Further Discussion on the Effect of Wake Mixing. In Proceedings of the International Conference on Ocean, Offshore and Arctic Engineering, Nantes, France, 9–14 June 2013; p. 10.
20. Nishino, T.; Willden, R.H.J. Two-scale dynamics of flow past a partial cross-stream array of tidal turbines. *J. Fluid Mech.* **2013**, *730*, 220–244. [CrossRef]
21. Adcock, T.A.; Draper, S.; Nishino, T. Tidal power generation–A review of hydrodynamic modelling. *Proc. Inst. Mech. Eng. Part J. Power Energy* **2015**, *229*, 755–771. [CrossRef]
22. Nishino, T.; Willden, R.H.J. The efficiency of an array of tidal turbines partially blocking a wide channel. *J. Fluid Mech.* **2012**, *708*, 596–606. [CrossRef]
23. Vogel, C.R.; Willden, R.H.; Houslby, G.T. A Correction for Depth-Averaged Simulations of Tidal Turbine Arrays. In Proceedings of the 10th European Wave and Tidal Energy Conference (EWTEC), Aalborg, Denmark, 2–5 September 2013.
24. Vogel, C.R.; Houlsby, G.T.; Willden, R.H.J. Effect of free surface deformation on the extractable power of a finite width turbine array. *Renew. Energy* **2016**, *88*, 317–324. [CrossRef]

25. Draper, S.; Stallard, T.; Stansby, P.; Way, S.; Adcock, T. Laboratory scale experiments and preliminary modelling to investigate basin scale tidal stream energy extraction. In Proceedings of the 10th European Wave and Tidal Energy Conference (EWTEC), Aalborg, Denmark, 2–5 September 2013; p. 12.
26. Draper, S.; Borthwick, A.G.; Houlsby, G.T. Energy potential of a tidal fence deployed near a coastal headland. *Philos. Trans. Math. Phys. Eng. Sci.* **2013**, *371*, 20120176. [CrossRef]
27. Nishino, T.; Willden, R.H.J. Energetics of marine turbine arrays - extraction, dissipation and diminution. *arXiv* **2013**, arXiv:1308.0940.
28. Falconer, R. Temperature distributions in tidal flow field. *J. Environ. Eng.* **1984**, *110*, 1099–1116. [CrossRef]
29. Falconer, R.A.; Asce, M. Water quality simulation study of a natural harbor. *J. Waterw. Port, Coast. Ocean. Eng.* **1986**, *112*, 15–34. [CrossRef]
30. Liang, D.; Falconer, R.A.; Lin, B. Comparison between TVD-MacCormack and ADI-type solvers of the shallow water equations. *Adv. Water Resour.* **2006**, *29*, 1833–1845. [CrossRef]
31. Kvočka, D.; Falconer, R.A.; Bray, M. Appropriate model use for predicting elevations and inundation extent for extreme flood events. *Nat. Hazards* **2015**, *79*, 1791–1808. [CrossRef]
32. Kvočka, D. Modelling Elevations, Inundation Extent and Hazard Risk for Extreme Flood Events. Ph.D. Thesis, Cardiff University, Cardiff, UK, 2017.
33. Liang, D.; Lin, B.; Falconer, R.A. Simulation of rapidly varying flow using an efficient TVD–MacCormack scheme. *Int. J. Numer. Methods Fluids* **2007**, *53*, 811–826. [CrossRef]
34. Falconer, R.; Lin, B. *Divast Model Reference Manual*; Cardiff University: Cardiff, UK, 2001.
35. Kvocka, D. *DIVAST-TVD User Manual*; Technical Report; Hydro-environmental Research Centre School of Engineering Cardiff University: Cardiff, UK, 2015.
36. Chung, T.J. *Computational Fluid Dynamics*, 2nd ed.; Cambridge University Press: Cambridge, UK, 2010.
37. Gustafsson, B. *Fundamentals of Scientific Computing*; Texts in Computational Science and Engineering; Springer: Uppsala, Sweden, 2011; Volume 8, p. 337.
38. Johnson, B.; Francis, J.; Howe, J.; Whitty, J. Computational Actuator Disc Models for Wind and Tidal Applications. *J. Renew. Energy* **2014**, *2014*, 1–10. [CrossRef]
39. Perez-Ortiz, A.; Borthwick, A.; McNaughton, J.; Avdis, A. Characterization of the Tidal Resource in Rathlin Sound. *Renew. Energy* **2017**. [CrossRef]
40. Sutherland, G.; Foreman, M.; Garrett, C. Tidal current energy assessment for Johnstone Strait, Vancouver Island. *Proc. Inst. Mech. Eng. Part J. Power Energy* **2007**, *221*, 147–157. [CrossRef]
41. Bonar, P.A.J.; Bryden, I.G.; Borthwick, A.G.L. Social and ecological impacts of marine energy development. *Renew. Sustain. Energy Rev.* **2015**, *47*, 486–495. [CrossRef]
42. Nash, S.; Phoenix, A. A review of the current understanding of the hydro-environmental impacts of energy removal by tidal turbines. *Renew. Sustain. Energy Rev.* **2017**, *80*, 648–662. [CrossRef]
43. Houlsby, G.T.; Draper, S.; Oldfield, M.L.G. *Application of Linear Momentum Actuator Disc Theory to Open Channel Flow*; Technical Report OUEL2296/08; Oxford University Engineering Laboratory UK: Oxford, UK, 2008.
44. Vennell, R. Tuning tidal turbines in-concert to maximise farm efficiency. *J. Fluid Mech.* **2011**, *671*, 587–604. [CrossRef]
45. Flores Mateos, L.; Hartnett, M. Tidal-stream power assessment–A novel modelling approach. In Proceedings of the 6th International Conference on Energy and Environment Research, ICEER 2019, Aveiro, Portugal, 22–25 July 2019; Energy Reports, In Press.
46. Stallard, T.; Collings, R.; Feng, T.; Whelan, J. Interactions between tidal turbine wakes: Experimental study of a group of three-bladed rotors. *Philos. Trans. Math. Phys. Eng. Sci.* **2013**, *371*, 20120159. [CrossRef] [PubMed]
47. Nash, S.; Olbert, A.; Hartnett, M. Towards a Low-Cost Modelling System for Optimising the Layout of Tidal Turbine Arrays. *Energies* **2015**, *8*, 13521–13539. [CrossRef]
48. Serhadlioglu, S. Tidal Stream Resource Assessment of the Anglesey Skerries and the Bristol Channel. Ph.D. Thesis, University of Oxford, Oxford, UK, 2014.
49. Ahmadian, R.; Falconer, R.A. Assessment of array shape of tidal stream turbines on hydro-environmental impacts and power output. *Renew. Energy* **2012**, *44*, 318–327. [CrossRef]
50. Perez-Campos, E.; Nishino, T. Numerical Validation of the Two-Scale Actuator Disc Theory for Marine Turbine Arrays. In Proceedings of the 11th European Wave and Tidal Energy Conference, Nantes, France, 6–11 September 2015; p. 8.

51. Nishino, T.; Willden, R.H.J. Effects of 3-D channel blockage and turbulent wake mixing on the limit of power extraction by tidal turbines. *Int. J. Heat Fluid Flow* **2012**, *37*, 123–135. [CrossRef]
52. Karsten, R.H.; McMillan, J.M.; Lickley, M.J.; Haynes, R.D. Assessment of tidal current energy in the Minas Passage, Bay of Fundy. *Proc. Inst. Mech. Eng. Part J. Power Energy* **2008**, *222*, 493–507. [CrossRef]
53. Pérez-Ortiz, A.; Borthwick, A.; McNaughton, J.; Smith, H.C.; Xiao, Q. Resource characterization of sites in the vicinity of an island near a landmass. *Renew. Energy* **2017**, *103*, 265–276. [CrossRef]
54. Nash, S.; O'Brien, N.; Olbert, A.; Hartnett, M. Modelling the far field hydro-environmental impacts of tidal farms–A focus on tidal regime, inter-tidal zones and flushing. *Comput. Geosci.* **2014**, *71*, 20–27. [CrossRef]

© 2019 by the authors. Licensee MDPI, Basel, Switzerland. This article is an open access article distributed under the terms and conditions of the Creative Commons Attribution (CC BY) license (http://creativecommons.org/licenses/by/4.0/).

Article

Assessment and Day-Ahead Forecasting of Hourly Solar Radiation in Medellín, Colombia

Julián Urrego-Ortiz [1,*], J. Alejandro Martínez [1], Paola A. Arias [1] and Álvaro Jaramillo-Duque [2]

[1] Grupo de Ingeniería y Gestión Ambiental (GIGA), Escuela Ambiental, Facultad de Ingeniería, Universidad de Antioquia, Calle 67 No. 53-108, Medellín 050010, Colombia; john.martinez@udea.edu.co (J.A.M.); paola.arias@udea.edu.co (P.A.A.)
[2] Research Group in Efficient Energy Management (GIMEL), Departamento de Ingeniería Eléctrica, Facultad de Ingeniería, Universidad de Antioquia, Calle 67 No. 53–108, Medellín 050010, Colombia; alvaro.jaramillod@udea.edu.co
* Correspondence: julian.urrego@udea.edu.co

Received: 29 July 2019; Accepted: 11 November 2019; Published: 19 November 2019

Abstract: The description and forecasting of hourly solar resource is fundamental for the operation of solar energy systems in the electric grid. In this work, we provide insights regarding the hourly variation of the global horizontal irradiance in Medellín, Colombia, a large urban area within the tropical Andes. We propose a model based on Markov chains for forecasting the hourly solar irradiance for one day ahead. The Markov model was compared against estimates produced by different configurations of the weather research forecasting model (WRF). Our assessment showed that for the period considered, the average availability of the solar resource was of 5 PSH (peak sun hours), corresponding to an average daily radiation of ~5 kWh/m^2. This shows that Medellín, Colombia, has a substantial availability of the solar resource that can be a complementary source of energy during the dry season periods. In the case of the Markov model, the estimates exhibited typical root mean squared errors between ~80 W/m^2 and ~170 W/m^2 (~50%–~110%) under overcast conditions, and ~57 W/m^2 to ~171 W/m^2 (~16%–~38%) for clear sky conditions. In general, the proposed model had a performance comparable with the WRF model, while presenting a computationally inexpensive alternative to forecast hourly solar radiation one day in advance. The Markov model is presented as an alternative to estimate time series that can be used in energy markets by agents and power-system operators to deal with the uncertainty of solar power plants.

Keywords: global horizontal irradiance (GHI); forecasting; clearness coefficient; Markov chains; weather research and forecasting model; solar resource

1. Introduction

The performance of solar power plants depends essentially on an adequate characterization of the variations of the incoming solar radiation over land surface [1]. This variation is mainly associated with the interaction between clouds and the incoming solar radiation, which leads to attenuation values that, in some cases, can reach 80% or higher [2]. Solar resource variations cause subsequent changes of the output at solar power plants that could not only affect the electric infrastructure but also the revenue models that govern energy supply [3].

In the case of the day-ahead energy market in Colombia, plants bid to offer energy blocks to the energy market national operator, XM (www.xm.com.co), one day before obtaining the market clearing results [4]. The biddings for a certain day must be offered before 8 a.m. of that day. Based on these biddings, the energy-market operator determines which plants will supply the demand at each hour of the next day. This shows that the short-term operation of solar power plants depends on the correct forecasting of the incoming solar radiation and respective electricity generation. According to the

UMPE (the Colombia governmental entity in charge of designing energy expansion plans in Colombia), in the 2021–2029 planning horizon, there are scenarios where it is expected to have 329 MW of new solar plants distributed, that is about 2% of the actual power capacity in Colombia [5].

In addition, due to setbacks during the start up of the Hidroituango hydraulic plant [6], in the short-term it is expected to have new solar plants and distributed generation based on solar photovoltaic energy in Colombia.

Information about the hourly evolution of global horizontal irradiance (GHI) can be obtained through the characterization of solar radiation in the site of study [4,5,7]. Such characterization is also useful to provide benchmarking information that can be used to assess the performance of the GHI estimates obtained with different models (e.g., dynamical models [1,8] statistical models [9]). When using these models, characterization of solar resource is also necessary for model calibration as well as the quantification of their related uncertainties.

In general, the GHI can be estimated using two types of models: dynamical models and statistical models (including machine-learning techniques). The dynamical models are physically driven models that estimate the GHI from the physical relationships that exist between solar radiation and other atmospheric variables. Dynamical models like the weather research and forecasting (WRF) model [10] estimate the state of the atmosphere by numerically solving the atmosphere Equations for large horizontal domains (i.e., synoptic- and meso-scales) that are discretized in elements that usually comprehend several kilometers, even with convection-permitting resolution. However, the use of dynamical models is still a challenge, not only in terms of the hardware, computational time and knowledge required; but also because, in spite of producing physically-consistent results, their estimates may exhibit large biases [11,12].

As a counterpart, there exist the statistical models for forecasting GHI. These models include elements from time-series analysis and assume that the future series are statistically similar to the past series. This means that the estimates of a statistical model will mostly reflect the common features of the measured series used to train it. Therefore, statistical models depend on the size and quality of the available measurements to produce realistic estimates of the GHI. Different statistical and machine-learning models can be used to estimate GHI and the decision of which model should be used greatly depends on the features of the estimates (i.e., the temporal resolution of the estimates and the lead times at which they are needed). Among the simplest statistical models are the regression-based models. These models show an adequate performance for forecasts that require estimates with very coarse temporal resolutions, ranging from weeks to months. It is also a common practice to use these models not as forecasting models but rather diagnostic models that predict the daily or monthly solar radiation based on other available atmospheric variables, like daily or monthly temperature, which could have been previously measured or forecasted with a different model [13–15].

Currently, the use of machine-learning models like artificial neural networks (ANN) have gained a greater importance in the forecasting of solar radiation. For instance, reference [16] uses a multi-layer perceptron for estimating the global daily radiation for one day ahead. Reference [17] used a combination of an Auto Regressive Moving Average model (ARMA) and a time-delay neural network (TDNN) to forecast GHI for a lead time of 1 h, with a time resolution of 10 min. Reference [18] used an ARMA-based model and an ANN model to forecast GHI, using two types of forecasts: a first forecast with a temporal resolution of 10 min and different lead times, starting from 10 min to 60 min, and a second forecast with a temporal resolution of 1 h and lead times from 1 h to 6 h. As noted by [19] and the previously mentioned studies, the majority of these approaches exhibit their highest performance at estimating the GHI when the temporal resolutions of the estimates are of the same order than the lead time of the forecast.

Other types of models that have been used previously for estimating the GHI are the Markov based models, which mainly focus on estimating the clearness coefficient and from it, calculate the corresponding GHI values. These models have been widely used for generating synthetic GHI series that exhibit the same statistical characteristics as the GHI records in the site of study at different time resolutions [9,20–22]. Therefore, this type of model has become very useful for modeling solar energy systems in addition to being computationally inexpensive, which allows them to produce several realizations in a very short time. One example of this is the work of [9], where minutely radiation series were used to train a Markov chains models to produce synthetic series of GHI at high temporal resolution. In that work, even when the error values were high, the intention was to produce series that would behave, statistically, in the same way as the measured series, which poses as an advantage for photovoltaic (PV) system simulations considering the high temporal resolution of the realizations. In more recent works, Markov based approaches for modeling GHI are now being used to forecast the GHI values and not to only generate synthetic GHI series. One example is the case of [23], who proposed a hybrid model based on Markov chains and the Myecielski approach to forecast 1-h ahead GHI and found a satisfactory performance, in some cases even exceeding ANN models. Reference [24] proposes a Markov switching model that can be used to produce day-ahead forecasting of GHI. The model proves to have an adequate behavior at estimating the hourly data. However, it is based on a persistence approach for estimating the initial state of the next day to be estimated, which is an approach that does not work satisfactorily for sites with large variability in cloudiness. Reference [25] proposed a combined model based on k-means and Markov chains to statistically model the transition of the daily solar irradiance and characterize the transition probabilities among different states. In [26], a Markov-chain mixture model was proposed. The model was formulated to perform very high temporal resolution forecast of the clear-sky index (minutely resolutions).

In general, Markov-chain based approaches are not frequently used in the mid-term forecast (day-ahead forecasting) of the hourly series of GHI, which is a type of forecasting that is fundamental in the operation of generation systems based on solar radiation. They are rather used for generating synthetic GHI series for solar system simulations, or to perform forecasts with time horizons near the time resolution of the model (i.e., few time steps into the future). However, through correct implementation, a simple Markov based model could be used to perform mid-term forecasts of GHI hourly data with satisfactory performance and still represent a simple and computational inexpensive solution.

Therefore, as an alternative to obtain mid-term forecasts of the hourly GHI, we implemented a two-part model based on discrete Markov processes to estimate GHI in Medellín, Colombia, a highly populated city in the tropical Andes. This model is a modified version of the formulations found in the works of [9,21,22,27] and is capable of forecasting diurnal series of hourly GHI for one day-ahead, by taking into consideration the current seasonal effects over the behavior of the variable. Although the Markov transition matrices (MTM) calculated in this work reflect the particular features of the clearness coefficients in the region of study, the criteria followed for its construction and training can be applied to other locations. In order to provide a point of reference regarding the performance of the proposed Markov model, we used two additional models as benchmarking for the GHI estimates. One is the numerical weather prediction (NWP) model called the weather research forecasting model (WRF), and the second one is a modified Markov model based on persistence.

This work also provides insights on the intra-day behavior of the solar resource in Medellín, Colombia, which is not currently available for this region, using GHI measurements from a pyranometer operated by the Early Alert System of the Aburrá Valley (SIATA; https://siata.gov.co) during the period March 2016 to February 2017. This information is helpful since tropical regions exhibit a relative uniform income of radiation throughout the year compared to higher latitudes but can exhibit relatively high variability throughout the day. Also, knowledge about the intra-daily variability of the GHI is fundamental when formulating and calibrating models to characterize and forecast the incoming solar radiation.

Additionally, even when only one year of data was initially available and no inferences regarding annual and inter-annual variations of the GHI can be made, a single year of hourly records contains enough samples of hourly GHI series under different sky conditions, thus providing an important source for studying the intra-day behavior of the GHI in the region of interest. This variability can be related to the presence of mountains since they are related to localized formation of clouds due to orographic lifting and elevated heat sources, which adds complexity to the simulation and forecast of GHI [28].

Thus, we developed a benchmarking study for the further assessment of the accuracy of solar radiation estimates obtained from different types of models in Medellín, Colombia, as well as for providing information about the behavior of the intra-day GHI. The key contributions of our work are the following:

- We performed an initial assessment of the intra-day and daily variability of the solar resource in Medellín, Colombia, which is a piece of information that was not available in the site of study and that is necessary for determining the intra-day generation potentials.
- We proposed a statistical model based on Markov chains for forecasting the hourly GHI series for one day-ahead lead time, with low computational costs. We also used an NWP model (WRF) and a persistence-based Markov model as a benchmarking for the proposed Markov model.
- We evaluated the performance of the Markov model at estimating the hourly GHI and daily clearness coefficient considering different cloud covers in a tropical climate region.
- The performance of the Markov model was also evaluated under local atypical and synoptic atypical cloudy conditions.
- The method used for the formulation and training of the Markov model can be extended to other locations with different climatological conditions.
- The magnitude of the errors obtained with the Markov model are comparable to the errors obtained with other models identified in the literature.

2. Data and Methods

2.1. Pyranometer Data

We used in situ measurements provided by SIATA (https://siata.gov.co). The instruments used to retrieve these measurements correspond to Kipp and Zonen SMP11 pyranometers, which record GHI values with a precision of 1 W/m^2. The pyranometer considered (hereafter SIATA station) is located in Medellín, Colombia, at 6.2593° latitude and −75.5887° longitude (Figure 1). The SIATA pyranometers are inspected monthly and are calibrated according to the ISO 9847:1992. For the hourly values, the expected uncertainty is of 3% of the true value of radiation. Measurements are provided at a one-minute time resolution.

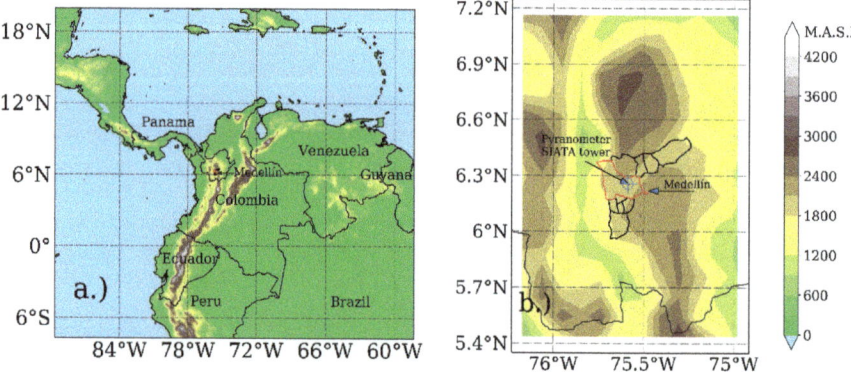

Figure 1. Location of the site of study and the pyranometer used. (a) Location of Colombia in South America. (b) Location of Medellín in Colombia. Color shades represents topography.

We performed a statistical analysis of the data provided by the SIATA station. The period of analysis ranges from 1 March 2016, to 28 February 2017. Although a single year is not enough for characterizing the seasonal variability of GHI in Medellín, it contains enough samples of hourly GHI series useful for studying the intra-day variability of GHI under different sky conditions. Consequently, the statistical analysis of GHI developed in this work is focused on the intra-day variations of the hourly GHI and not on the monthly or annual variations of the solar radiation. However, given the seasonal changes of solar radiation in the region, we discriminated our statistical analysis for each month of the year. In central Colombia, where our region of study is located, precipitation shows two wet seasons, one during March–April–May and a second one during September–October–November. By contrast, two dry seasons are observed during June–July–August and during December–January–February. This behavior is mainly associated with the latitudinal migration of the Intertropical Convergence Zone [29–32].

Since the characterization presented here is focused on the hourly values of GHI, the 1-min records provided by the SIATA pyranometer are averaged around each of the day-time hours. In this case, if 10% or more of the 1-min records within an hour were missing values, the corresponding hourly average value was set as a missing value. Less than 8% of the total hourly data corresponds to missing values, which was considered a small fraction. However, when a missing value was identified, usually the remaining values of the corresponding diurnal cycle were also missing values. This means that if data imputation is attempted, records corresponding to entire days would have to be produced. This is undesired since it could result in introducing high biases in the statistical analysis and, therefore, in the performance of the stochastic model.

2.2. Clearness Coefficient Estimation

The clearness index or clearness coefficient, k_t, is a dimensionless index that can be used to indirectly describe the behavior of solar radiation and is calculated as shown in Equation (1):

$$k_t = \frac{GHI}{I_{ext}} \tag{1}$$

where I_{ext} is the extraterrestrial radiation calculated as,

$$I_{ext} = I_{sc}\left(1 + 0.033 \times \cos\left(\frac{360 \times day\ of\ the\ year}{365}\right)\right) \times \cos\theta_z \tag{2}$$

In Equation (2), θ_z is the zenith angle, which depends on the declination angle δ, the latitude φ and the hour angle ω. I_{sc} is the solar constant and is equal to 1367 W/m². The zenith angle can be estimated at the same resolution of the GHI measurements.

The clearness coefficient represents the fraction of the extraterrestrial radiation that reaches the land surface after traversing the atmosphere. According to [33], the GHI can be assumed to consist of two components: a deterministic component, which is represented by the extraterrestrial radiation and the effects of the air mass coefficient (A.M) on the GHI, and a stochastic component which is primarily associated to the effects of clouds over the GHI. The effects of the A.M (a deterministic component) and the stochastic component are both inherited by the clearness coefficient when calculated as shown in Equation (1).

The deterministic component of k_t is, in general, better understood than the stochastic component and is mostly associated to changes in the path length that solar radiation must traverse before reaching the land's surface. This length changes during the day and is a function of the zenith angle, θ_z. The corresponding geometrical change in the solar radiation path is what is represented by A.M, which according to [34,35], is calculated as shown in Equation (3).

$$A.M = \begin{cases} \frac{1}{\cos\theta_z} & \text{For } 0° \leq \theta_z < 70° \\ \frac{\exp(-0.000118 \times Altitude(sea\ level))}{\cos\theta_z + 0.5057(96.080 - \theta_z)^{-1.634}} & \text{For } 70° \leq \theta_z < 90° \end{cases} \tag{3}$$

On the other hand, the stochastic component is still not represented by simple Equations. Different studies have identified that the stochastic contribution over the GHI is mainly associated to the interactions between clouds and radiation [33,36–38]. An estimate of the stochastic component of k_t can be obtained through the normalization expression proposed by [39], who outlined the dependency between the hourly k_t and the A.M. According to [39], the k_s is obtained by removing the dependency of the k_s to the A.M as:

$$k_s = \frac{k_t}{1.031 \exp\left(\frac{-1.4}{0.9 + \frac{9.4}{A.M}}\right) + 0.1} \tag{4}$$

The normalized clearness coefficient, k_s, represents solely the effect of clouds over the GHI and behaves as a stochastic variable. In this work, we used a Markov model to simulate hourly values of k_S.

In addition to the calculation of k_S, another clearness coefficient is obtained at daily temporal resolutions and is referred to as the daily clearness coefficient, k_d. This value is also calculated from Equation (1) using the daily values of GHI and I_{ext}, and is then used to represent the average sky conditions of a given day.

2.3. Discrete Markov Chain Model

Markov chains describe the transition process of a random variable, where the probability distribution of the following state of the variable depends on its previous states. The degree of the process indicates the number of previous observations on which the next state of the variable statistically depends. Given that the random variable, X, has a set S with a finite number of possible states m, so that $S = \{s_1, s_2, s_3, \ldots, s_m\}$, the random variable has a state i at a time step t when $X_t = i$. The transition probabilities between all possible states are stored in a MTM, P. In a first-degree Markov process, the next state of the variable X depends only on its current state and so the transition probability of the variable from state s_i to state s_j, P_{ij} in a single time-step can be written as:

$$P_{ij} = P(X_t = j | X_{t-1} = i, X_{t-2} = i_{t-2}, \ldots, X_0 = i_0) = P(X_t = j | X_{t-1} = i) \tag{5}$$

The MTM of such a process, P, is a squared matrix with dimensions $m \times m$, where the row values indicate the current state of the variable and the column values correspond to the next possible states of the variable. Therefore, there are m elements in each row of P e.g., the first row should be $p_{11}, p_{12}, p_{13}, p_{14}, \ldots, p_{1m}$ as shown in (6).

$$P = \begin{bmatrix} p_{11} & p_{12} & \cdots & p_{1m} \\ p_{21} & p_{22} & \cdots & p_{2m} \\ \vdots & \vdots & \ddots & \vdots \\ p_{m1} & p_{m2} & \cdots & p_{mm} \end{bmatrix} \tag{6}$$

Since the Markov chain transitions between discrete states, the clearness coefficient must be transformed from a continuous variable to a discrete variable. This is done by dividing the range of the clearness coefficient (i.e., 0–1) into equally-spaced intervals or bins. Each interval is enumerated in ascending order, with each number corresponding to a discrete state of the clearness coefficient. A state of the clearness coefficient refers to the interval in which its continuous value is contained.

The number of states is seen to greatly improve the performance of the model up to a certain number of states. Initially, ref [40] demonstrated that a set of discrete states $m = 20$, corresponded to the smallest interval subdivision that would still result in a regular behavior of the MTM. Furthermore, in a more recent study, ref [41] found through a cross-validation process, that the performance of a Markov based model did not improve the performance of the model significantly for a number of states larger than 20. In other words, it was evidenced that the model did not improve when the bin width of the discretization process went below 0.05. These findings are consistent with the discretization widths used in previous works, regarding the modeling of the clearness coefficient using Markov based models [9,20,22,36,42,43]. For these reasons, we discretized both the normalized clearness coefficient (k_s) and the daily clearness coefficient (k_d), into 20 discrete states of width 0.05. The resulting discrete states are shown in Table 1.

Table 1. Discretization of the daily and hourly clearness coefficients.

Discrete State	Continuous Interval	Discrete State	Continuous Interval	Discrete State	Continuous Interval	Discrete State	Continuous Interval
State 1	0–0.05	State 6	0.25–0.3	State 11	0.5–0.55	State 16	0.75–0.8
State 2	0.05–0.1	State 7	0.3–0.35	State 12	0.55–0.6	State 17	0.8–0.85
State 3	0.1–0.15	State 8	0.35–0.4	State 13	0.6–0.65	State 18	0.85–0.9
State 4	0.15–0.2	State 9	0.4–0.45	State 14	0.65–0.7	State 19	0.9–0.95
State 5	0.2–0.25	State 10	0.45–0.5	State 15	0.7–0.75	State 20	0.95–1

Additionally, the works of [44,45] demonstrated that the statistical behavior of a set of k_s depends, predominantly, on the overall sky conditions of the day they belong to (i.e., on the state of the corresponding k_d). This indicates that the modeling of the k_s must be discriminated by the states of the k_d. This procedure has been used before in other studies using discrete states to estimate GHI. Initially, reference [43] introduced the idea of a library of MTMs, where they calculated an MTM for the daily clearness coefficient, for each state of the monthly clearness coefficient. This choice was made based on the similarities of the shapes of the probability functions alone. Furthermore, reference [20] showed that the probability density function (PDF) of sets of k_s for days that have k_d values grouped in intervals of 0.05 displayed the same statistical behavior and therefore could be modeled through the same function. Following these studies, we opted to calculate a library of MTM for modeling k_s, one for each state of the daily clearness coefficient k_d.

It must be noted that a synthetic time series of hourly k_s values can be obtained with this Markov model by setting an initial state of k_s and calculating the following state as a stochastic process with the associated probabilities contained in the row of P given by the initial state of k_s. This procedure will be explained in more detail in the following sections. However, before explaining the GHI estimation procedure, we describe the methodology to estimate the MTMs, for both the k_d and k_s.

2.4. Construction of the Markov Transition Matrices

2.4.1. First-Degree Markov Transition Matrix (MTM) for k_s

As it was explained in Section 2.3, a first-degree Markov transition matrix contains the probability of the variable transitioning from a state i to a state j in one time step which can be expressed as P_{ij}. This probability P_{ij} can be approximated by counting the number of times the variables go from i to j in one time step, f_{ij} and then dividing this number by the total number of transitions the variable makes from state i to any other state, T_i.

$$P_{ij} = \frac{f_{ij}}{T_i} \quad (7)$$

with,

f_{ij}: Number of times the variable passes from i to j.
T_i: Number of times the variable departs from i.

For a given state z of the daily clearness coefficient (k_d), the corresponding k_s MTM is calculated as follows:

(1) Extract all the hourly GHI values from the days of the dataset that have a $k_d = z$. Keep the information regarding the hour and day of the year that corresponds to each hourly datum.
(2) Calculate the corresponding k_s values using Equations (1)–(4), and discretize them using Table 1.
(3) Using the discrete k_s data, calculate the corresponding f_{ij} and T_i values for all k_s discrete values. The i and j values are iterated over all the states of k_s.

(4) Calculate the P_{ij} probabilities with Equation (7), and position them in the MTM at the corresponding (i, j) position.

This procedure can be iterated over all k_d states to obtain a first-degree MTM for each state of the daily clearness coefficient.

2.4.2. Second-Degree MTM for k_d

Since the k_d is sensitive to atmospheric variations at synoptic scales, estimates of the daily clearness coefficient are obtained from a second-degree Markov chain calculated at daily resolutions (i.e., the future state of the k_d depends on the states of the previous two days). The construction of a second-degree MTM is performed in the same way as the first-degree MTM with the exception that the transition probability is calculated as:

$$P_{(ij)k} = \frac{f_{(ij)k}}{T_{(ij)}} \tag{8}$$

where (i, j) values correspond to the current and previous adjacent state of the variable respectively, and k corresponds to its future state. In this case, each row in the second-degree MTM corresponds to a (i, j) pair and so, a second-degree MTM has dimensions $(m^2 \times m)$.

In addition, atmospheric variability can be different for wet and dry seasons in the region of study. Therefore, two second-degree MTMs for k_d, one for wet season months (MTM_{wet}) and a second for dry season months (MTM_{dry}), are obtained and used to generate the estimates of the daily clearness coefficient state. Consequently, we used a two-part model based on Markov chains to estimate the hourly series of GHI in the site of study.

2.5. Simulations of Global Horizontal Irradiance (GHI) Using a Markov Chains Model

As previously mentioned, the statistical behavior of the k_s of a particular day depends on the state of the k_d of that day. This means that for each state of the k_d illustrated in Table 1, there is a corresponding MTM for k_s, as it was also explained in the previous section. Therefore, the state of the k_d should be forecasted before the diurnal series of k_s can be estimated. Consequently, the proposed Markov model has two parts: the first part estimates the state of the daily clearness coefficient (k_d) using a second-degree Markov chain and the second part uses this estimate as a decision-maker variable to choose the appropriate first-degree MTM for estimating k_s. This procedure results in an estimated hourly series of k_s for one day-ahead. This process is illustrated in Figure 2.

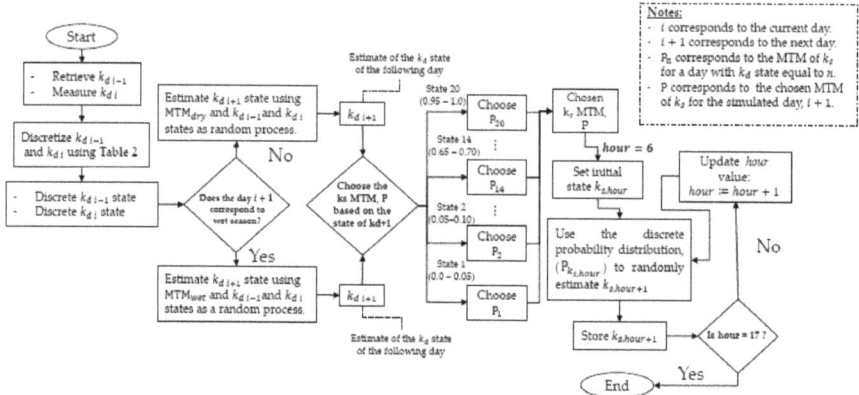

Figure 2. Simulation process of the normalized clearness coefficient, k_s using the two-part Markov model proposed in this work.

The procedure depicted in Figure 2 results in a series of hourly k_s states, one for each hour of the following day like $\{k_{s,6}, k_{s,7}, \ldots, k_{s,18}\}$, which vary from 1 to 20 (see Table 1). These values are then used to calculate the corresponding hourly GHI values. This is done as follows:

(1) Transform the estimates $\{k_{s,6}, k_{s,7}, \ldots, k_{s,18}\}$ from states (1 to 20) to extinction percentages (0.0–1.0). This is done by first assuming that for a state i of k_s, the continuous values inside the corresponding interval (see Table 1), follow a uniform distribution such that $k_s \sim \mathcal{U}((i-1) \times 0.05, i \times 0.05)$. Then, using a pseudo-random process a continuous value of k_s is picked from the given distribution.

(2) Once a continuous value of k_s is obtained and knowing the hour and the day for which it was estimated, the total hourly clearness coefficient, k_t must be calculated again in order to include the deterministic effects. This is done by using Equation (4) as:

$$k_{t,est} = k_{s,est} \times 1.031 \exp\left(\frac{-1.4}{0.9 + \frac{9.4}{A.M}}\right) + 0.1$$

(3) Now, given that the extraterrestrial radiation can be modeled as a deterministic variable, the Equation for the total clearness coefficient, k_t, can be used now to estimate the corresponding GHI value as:

$$GHI_{est} = k_t \times I_{ext}$$

This procedure is performed for each of the k_s hourly estimates to obtain the corresponding series of hourly GHI estimates.

The period March 2016 to February 2017 is used to calculate the MTMs for the model while the period May 2017 to May 2018 is used to evaluate the model performance. Since the stochastic model is computationally inexpensive, each day of the period between May 2017 and May 2018 was estimated 1000 times. With this information, it is possible to obtain the error distributions of the different realizations produced by the model, which allows a more general assessment of the skill of the model at estimating the GHI in Medellín to be performed.

2.6. The Weather Research and Forecasting (WRF) Model

We used the WRF model, a flexible, mesoscale model designed for atmospheric modeling, weather forecasting and climate simulations [10], as a benchmarking model for the proposed Markov model. WRF is a state-of-the-art, community-supported model that is subject of constant development and support by the National Center for Atmospheric Research (NCAR), as well as multiple contributions from the users' community. Thus, this model is under constant scrutiny and improvement [46]. Additionally, the WRF model has been used by meteorological and environmental agencies in Colombia, such as SIATA [47] and the Colombian Institute of Hydrology, Meteorology and Environmental Studies (IDEAM), as their choice for operational forecasting exercises for different regions of Colombia.

WRF has been used to estimate the incoming solar radiation for different regions of the world [8, 48–51]. It has also been modified to provide better estimates of the incoming solar radiation, like the case of the WRF-solar project [52] which introduced new parametrizations and modifications to the conventional WRF. These features were first included in version 3.6 of the WRF model and have been an integral part of this model since [53–56].

In order to compare the Markov model to the results obtained from WRF, we analyzed the simulations of 5 particular days corresponding to different cloud cover conditions in the site of study, as shown in Table 2. It must be noted that 5 simulated days do not represent a comprehensive validation period for comparing the WRF model against the proposed Markov model (May 2017 to May 2018). We limited the present analysis to a few days given the computational costs of the WRF simulations. These simulations are run at a relatively high resolution (including convective-permitting domains) and include different configurations for each case. In our case, convective-permitting simulations are important not only to better simulate the cloud field at smaller scales, but also to account for the orographic effects of the Tropical Andes, where Medellín is located. On the other hand, the use of different configurations of the model was needed for this study because, to the best of our knowledge, there is no peer-reviewed report of the skill of different microphysics schemes on the simulation of GHI for our region of interest. Thus, despite the limited number of simulated days with WRF, the comparison between our WRF simulations and the Markov model can still be useful for benchmarking the performance of the Markov model, and for gaining insights on the aspects of the model that should be improved.

Each 24-h WRF simulation started at 00:00:00 UTC (19:00:00 Local Time (LT)) of the day before and ended at 00:00:00 UTC (19:00:00 LT) of the day of interest. The first 6 h were considered as a spin-up period for each simulation. The integration time-step was set to 45 s and the shortwave radiation scheme was called every 15 min. The WRF outputs were also saved each 15 min. The resulting solar radiation series were averaged around each hour of the diurnal cycle; thus, an hourly temporal resolution was achieved.

Table 2. Particular days considered to analyze the performance of the clearness coefficient simulations obtained with the Markov chains-based model and the weather research and forecasting (WRF) model.

Date	Daily Clearness Coefficient (k_d)	Category
1 September 2017	0.72	Clear sky
23 December 2017	0.54	Broken clouds
5 June 2017	0.48	Cloudy
24 November 2017	0.39	Very cloudy (local conditions)
19 August 2017	0.18	Very cloudy (Synoptic conditions)

WRF was run using two nested domains. The outer domain includes Colombia, part of the Pacific Ocean and the Caribbean Sea and has a spatial resolution of 12 km (Figure 1a). The inner domain includes the region containing the city of Medellín and the municipalities of the Aburrá Valley and has a spatial resolution of 4 km. Because WRF numerically solves the Equations of the atmosphere, it requires initial and boundary conditions. These conditions were obtained from the Global Forecast System (GFS) final analysis (FNL), which provides information about the state of the atmosphere every 6 h.

We used five WRF configurations for each simulated day, each with different combinations of microphysics and cumulus schemes. The remaining parameterizations are common to all simulations. For the planetary boundary layer, we used the Mellor–Yamada–Janjic [57] whereas for shortwave and longwave radiation we used the rapid radiative transfer model for general circulation models (RRTMG). We selected RRTMG as the radiation scheme based on its reported performance for estimating GHI, as well as the fact that is currently the only radiation scheme capable of coupling with the microphysics schemes of [58,59], which tackles the microphysics-radiative inconsistency present in WRF [37]. These configurations are variations of the WRF-solar configuration [52].

In this work, a set of six experiments using different options of the cumulus and microphysics schemes were undertaken. Table 3 shows the different configurations used in this work. The Kain-Fritsch/Thompson-Eidhammer (KF-TE) experiment was considered here as the control experiment, being a slight variation of the default WRF-solar configuration. In order to explore the role of the cumulus parameterization at the edge of what is commonly considered the convective-resolving resolution (4 km), we included an additional configuration (Experiment KF-TE-02), in which the cumulus scheme is active in the inner domain.

The Thompson & Eidhammer scheme was one of the selected microphysics schemes because: (i) it provides the radiative effective radii of the hydrometeors to be used in the RRTMG scheme; and (ii) it takes into account the effects of aerosols over clouds, which subsequently affects incoming radiation [52]. The Morrison microphysics scheme [60] calculates the size distribution for more hydrometeors than the Thompson and Eidhammer scheme and has had positive performance in the estimate of the GHI in other studies [55]. The Grell cumulus scheme can reflect the effect of unresolved clouds over the incoming radiation. Schemes such as the Grell-3D scheme [61] allow the activation of the Deng mass flux scheme [62], which triggers the radiative feedback on both deep and shallow cumulus.

Considering that the effective resolution of a grid-point NWP model is larger than one grid cell, we averaged GHI values from WRF over a region around the location of the pyranometer corresponding to 3 × 3 grid boxes of the larger domain, (i.e., 36 km × 36 km). Model output is available every 15 min. The resulting series were hourly averaged, so they could be compared to the in-situ measurements via different error metrics as discussed in next section.

Table 3. Experiments scheme configurations. The Deng scheme is also referred as Deng's mass-flux-scheme.

Experiments	Cumulus		Shallow Convection		Microphysics	
	d01	d02	d01	d02	d01	d02
KF-TE	Kain-Fritsch [63]	Off	Off	Off	Thompson and Eidhammer	Thompson and Eidhammer
KF-TE-02	Kain-Fritsch	Kain-Fritsch	Off	Off	Thompson and Eidhammer	Thompson and Eidhammer
KF-MO	Kain-Fritsch	Off	Off	Off	Morrison	Morrison
GR-TE	Grell 3D	Off	Off	Off	Thompson and Eidhammer	Thompson and Eidhammer
GR-TE-DE	Grell 3D	Off	Deng	Off	Thompson and Eidhammer	Thompson and Eidhammer
GR-MO-DE	Grell 3D	Off	Deng	Off	Morrison	Morrison

2.7. Persistence-Markov Model

For further assessment of the proposed Markov model, a persistence-Markov model was also used in this work as a benchmarking model. Unlike the two-part Markov model previously described, this model assumes that the state of the k_d of the next day is the same as the k_d state of the current day. The process for obtaining the hourly series of GHI for the next day using the persistence-Markov model is the same as the one used for the proposed Markov model. This procedure is depicted in Figure 3.

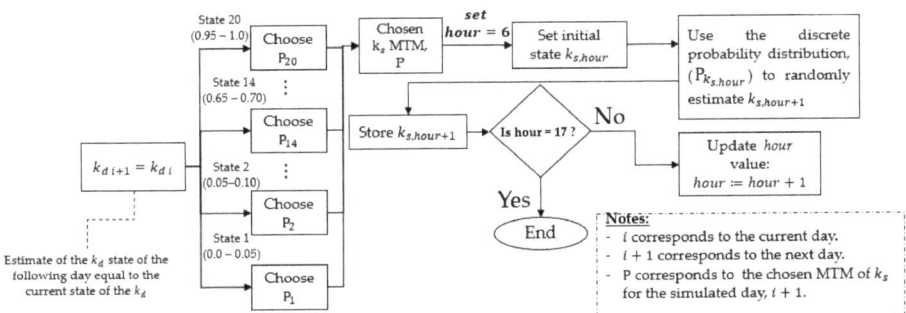

Figure 3. Simulation process of the normalized clearness coefficient, k_s using the persistence-Markov model proposed in this work.

The persistence-Markov model was used to simulate the same validation period of the proposed Markov model (i.e., May 2017–May 2018).

2.8. Error Metrics

The metrics used for evaluating the hourly estimates of the stochastic model and the WRF model are the root mean square error (RMSE; in W/m^2), the normalized root mean square error (nRMSE; in %), the mean bias error (MBE; in W/m^2), and the normalized mean bias error (nMBE; in %), as follows:

$$RMSE = \sqrt{\frac{1}{N}\sum_{i=1}^{N}(GHI_{est,i} - GHI_{meas,i})^2} \quad nRMSE = \frac{RMSE}{\overline{GHI}_{meas}} \quad (9)$$

$$MBE = \frac{1}{N}\sum_{I=1}^{n}(GHI_{est,i} - GHI_{meas,i}) \quad nMBE = \frac{MBE}{\overline{GHI_{meas}}} \tag{10}$$

- N corresponds to the number of hourly GHI values during the day.
- $GHI_{est,i}$ is the estimated average GHI value for the hour i.
- $GHI_{meas,i}$ is the measured average GHI value for the hour i.
- $\overline{GHI_{meas}}$ is the daily mean of the hourly GHI measured values.

The estimation errors of the daily clearness coefficient were calculated as the difference between the estimated k_d and the measured k_d, as follows:

$$\Delta k_d = k_{destimated} - k_{dmeasured} \tag{11}$$

3. Results

3.1. Characterization of the GHI

The assessment of the solar resource for the period March 2016 to February 2017 is performed with only daytime GHI values, from 6 LT to 18 LT, since these are the limits of the interval in which finite values of GHI (i.e., non-zero and non-missing) are measured by the pyranometers. The distribution of hourly GHI data throughout the period considered here is shown in Figure 4. Table 4 presents the statistical summary of the GHI for the period of March 2016–February 2017. Aside from the main, median and standard deviation of the period considered, Table 4 also presents the interquartile range (IQR) of the data distribution which is defined as the difference Q3–Q1. From Table 4, 50% of the radiation values measured by the SIATA pyranometer range from 90 W/m² to 643 W/m². The daytime values are skewed, with higher frequencies of the lowest radiation values, especially in the 0–100 W/m² range.

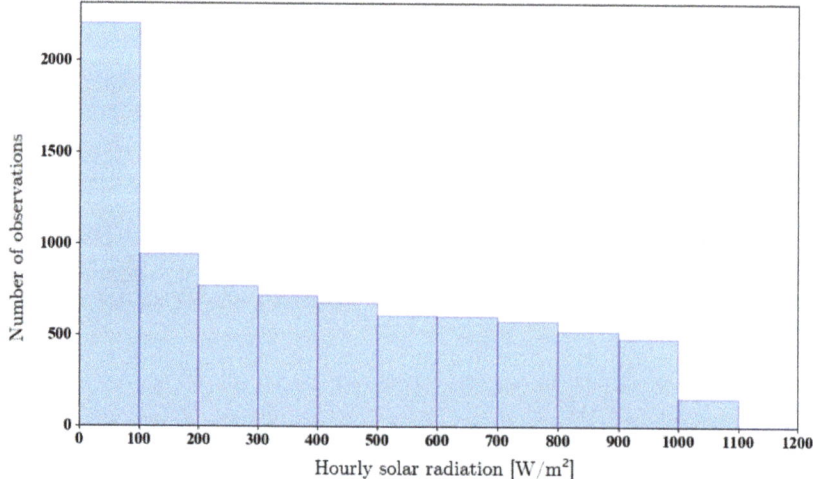

Figure 4. Histogram of hourly global horizontal irradiance (GHI) from the SIATA station recorded during the period 1 March 2016–28 February 2017.

Table 4. Statistical summary of the hourly GHI from the SIATA station recorded during the period 1 March 2016 to 28 February 2017.

Total Size of the Sample with Hourly Resolution = 4658				Percentiles (W/m²)				
Mean (W/m²)	Median (W/m²)	Standard Deviation (W/m²)	IQR (W/m²)	0%	25%	50%	75%	100%
385	332	315	553	0	90	332	643	1217

Figure 5 presents the GHI values corresponding to each hour of the day and each month during the period March 2016–February 2017. The highest values of GHI are observed from 11 to 13 h local time (LT). The months with the highest radiation values occurring during longer periods are March 2016, July 2016 and August 2016, and January 2017 and February 2017, reaching GHI values higher than 900 W/m². April 2016, May 2016 and late November 2016 are the periods where more GHI variability is observed, with April 2016 exhibiting several days with solar radiation below 300 W/m² at any time of the day.

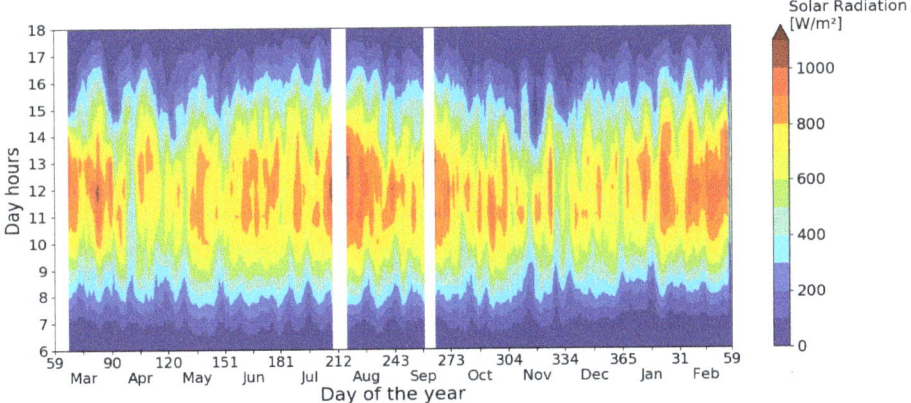

Figure 5. Hourly GHI values from SIATA station for all days of the period March 2016–February 2017. The color scale represents the intensity of solar radiation in W/m². White gaps correspond to missing values.

Reference [64] shows a distribution of the annual cycle of hourly radiation for Sevilla, Spain, similar to that shown in Figure 5, with an approximate mean of 383 W/m², suggesting a similar annual behavior of the GHI. However, measurements for both sites (Sevilla and Medellín) exhibit seasonal changes with rather different amplitudes. The 2016–2017 annual GHI series obtained from the SIATA pyranometer (Figure 5) shows a rather uniform pattern when compared to the larger contrasts between summer and winter seasons at higher latitudes that would alter significantly the daylight hours and the intensity of incoming radiation (see [64]). In spite of lacking the strong seasonal effects of middle latitudes, the SIATA pyranometer data suggests two periods of reduced solar radiation: April to early May 2016, and late October to November 2016. During November 2016, the time interval when solar irradiance exceeds 200 W/m² is reduced from ~10 h to ~7.5 h.

Figure 6 shows the violin plots of monthly GHI for 3 segments of the day-time. These time windows correspond to "morning" (6–10 LT), "noon" (10–14 LT), and "afternoon" (14–18 LT). The violin plots show the empirical probability distributions of the hourly GHI for each segment considered, as colored regions, which are drawn mirrored around the middle black lines. These lines present the same information as a boxplot, with the white dots indicating the position of the median, the horizontal small lines indication the quartile 1 (Q1) and quartile 3 (Q3). The thinner black lines that extend beyond these points reach the 5th percentile (p5) and the 95th percentile, respectively.

In the region of study, located in Central Colombia, precipitation shows two wet seasons, one during March–April–May and second one in September–October–November. By contrast, two dry seasons are observed in June–July–August and December–January–February [29,31,32,65]. For the period of study, the distributions in Figure 6 show that the GHI exhibited larger variability during the afternoon compared to the morning hours. February 2017 was the month with less variability at noon hours, most probably due to the persistence of clear skies during that time of the year (which is part of the dry season). The noon distributions for months like August 2016, January 2017, February 2017 presented skewed distributions towards high GHI values, while distributions at the same segment for months like April 2016 or November 2017 exhibited more uniform distributions of GHI, indicating a higher occurrence of low GHI values compared to the former mentioned months. This behavior is found to be characteristic for both clear sky and overcast conditions.

The potential occurrence of high radiation values at the surface is related to the magnitude of the extraterrestrial radiation, which in turn, for our region of interest (~6.25° N) exhibits some of its largest values between March and May (see e.g., Fig. 2.8 in [66]). However, more frequent cloud formation is observed over Medellín during this time of the year (first wet season), which corresponds to the early first rainy season for this region [29,31,32,65].

Figure 6 shows that the months within the dry season (e.g., August 2016 and February 2017) exhibit lower variability before noon, and higher variability in the afternoon hours, for the period considered. This behavior is due to the high number of clear sky days during these months. Months within the wet season (e.g., April 2016 and November 2016) exhibit larger variability throughout the day and GHI distributions more symmetrical around noon, unlike the dry season distributions of the period of study. Even though there is variability of hourly GHI during dry months, most of the hourly radiation records shown in Figure 6 are centered on higher values during the afternoon hours compared to other months of the year.

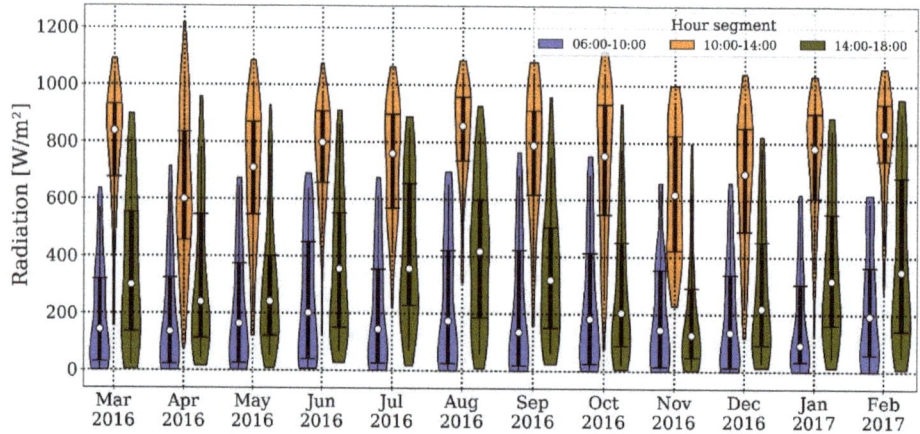

Figure 6. Daytime violin plots of GHI during the period March 2016–February 2017. Three hour segments are considered: 6–10 LT, 10–14 LT and 14–18 LT. The colored surfaces of each segment correspond to the empirical probability density function (PDF) of the GHI data. The black lines inside each distribution correspond to the boxplots of the GHI values and the white dots correspond to the median values of the distributions.

Table 5 shows the hourly mean coefficient of variation (CV) for each month of the period considered, along with the average values of daily solar energy the mean k_d values. The CV can be defined as:

$$CV = \frac{\mu}{\sigma} \qquad (12)$$

Table 5 shows that the dry season months (i.e., June 2016, July 2016, August 2016 and February 2017) exhibit the lowest relative variability during the day (less than 34%). Table 5 also shows the monthly mean daily solar energy values in kWh/m², commonly used in power systems and energy budget. August 2016 and February 2017 have the highest values of mean daily incoming energy and the lowest average of hourly CV. On the other hand, wet season months (i.e., April, May, October, November and December 2016) show the highest variability with CV values between 43% and 50% of the hourly GHI averages. April and November 2016 are the months with the highest hourly variability values (46% and 50%, respectively). Consistent with the mean daily values of solar energy, the mean k_d per month show that during the months of August 2016, January 2017 and February 2017, there was an overall low level of cloudiness, hence the high k_d values.

Table 5. Monthly mean hourly coefficient of variation (CV) and monthly mean daily radiation (in energy units of kilowatts-hour) from SIATA station during the period March 2016–February 2017.

Month	Hourly Mean CV (%)	Month Mean Daily Solar Energy (kWh/m²)	Mean k_d
March 2016	39.7	5.4	0.52
April 2016	46.2	4.7	0.45
May 2016	40.7	4.8	0.48
June 2016	29.9	5.6	0.56
July 2016	33.5	5.6	0.55
August 2016	31	6	0.58
September 2016	37.2	5.4	0.52
October 2016	45.7	5.0	0.50
November 2016	49.7	4.0	0.43
December 2016	43	4.6	0.51
January 2017	36.3	5.3	0.57
February 2017	32.9	5.9	0.60

3.2. Clearness Coefficient

We calculated the normalized clearness coefficient (k_s) for each hour throughout the period March 2016–February 2017, according to the SIATA records. Using these values, the empirical PDFs of k_s were calculated for each month during the period considered We calculated the normalized clearness coefficient (k_s) for each hour throughout the period March 2016–February 2017, according to the SIATA records. Using these values, the empirical probability density functions (PDFs) of k_s were calculated for each month during the period considered (Figure 7). Two main groups of PDFs are distinguished: one group including the months of June–July–August 2016 and January–February 2017, and a second group including the months of April, May, October, and November 2016. The first group corresponds to dry season months whereas the second group corresponds to wet season months [29,31,32,65].

Figure 7 shows that the distributions of the wet season months have peaks around lower values of k_s (~0.4), whereas the distributions for the dry season months exhibit larger frequencies for larger values of k_s (~0.6–0.8). Transition months like September and December 2016 show more uniform distributions of k_s.

The daily clearness coefficient, k_d, was calculated as the ratio of the daily mean GHI to the daily mean of the corresponding extraterrestrial radiation. This coefficient reflects the sky condition of the whole day in terms of how much solar radiation reaches the surface as a percentage of how much would have reached the surface, given that there was no atmosphere. Previous studies have shown that there is a statistical relationship between the value of the daily clearness coefficient, k_d, and the daily statistical distribution of the normalized hourly clearness coefficient, k_s, [9,45,64].

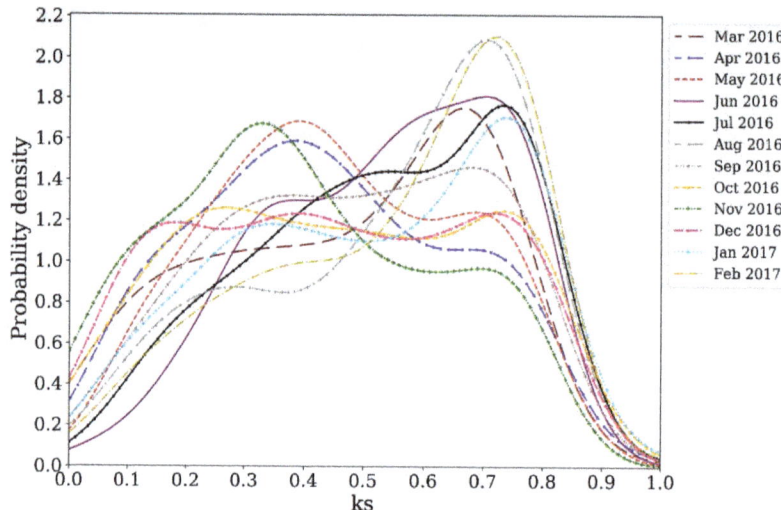

Figure 7. Empirical PDFs of k_s for each month between March 2016 and February 2017, according to SIATA records. The y-axis values can be larger than 1.0 since this is a distribution plot, thus meaningful information can only be obtained from its integral over an interval rather than from a single point. This integral is equal or less to 1.

Figure 8 shows the empirical probability distributions of the normalized hourly clearness coefficient grouped according to their corresponding daily clearness coefficient value. Only four ranges are shown to illustrate this relationship. Each colored line in Figure 8 denotes the empirical PDF of the hourly k_s values of a day that a has daily clearness coefficient contained in the interval shown in the corners of each panel, (e.g., every line in Figure 8a corresponds to the PDF of the k_s values of a day with a k_d between 0.25 and 0.30). In this way, what Figure 8 indicates is that the empirical PDFs of the k_s for days with the same state of k_d behave similarly.

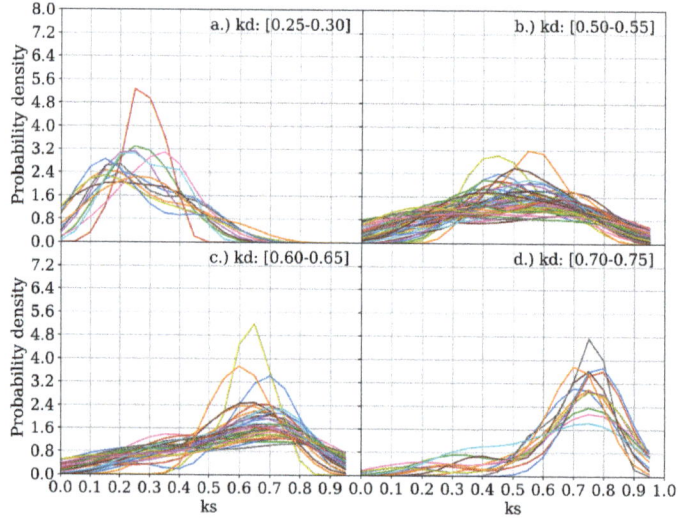

Figure 8. Empirical PDFs for four types of days with k_d in different percentages of daily mean extraterrestrial radiation reaching the surface: (**a**) 25%–30%, (**b**) 50%–55%, (**c**) 60%–65%, and (**d**) 70%–75%.

For days with low k_d, the corresponding k_s values are centered around lower values (Figure 8a). The days with higher k_d, on the other hand, present k_s values that are centered around higher values (Figure 8b). Both cases exhibit less k_s variability than the cases of days with intermediate k_d values (Figure 8b,c). Due to these observed differences between the k_s distribution according to the k_d, we considered a two-part Markov model.

3.3. Clearness Coefficient and GHI Estimates

3.3.1. Daily Clearness Coefficient Estimates

Figure 9 shows, for each of the considered days, the estimates of the daily clearness coefficient (k_d) obtained with each of the considered models (i.e., the first part of the proposed Markov model, persistence model, and the WRF experiments). Figure 9 presents these estimates as bar plots, with each bar representing a different model and its height representing the corresponding estimated value of k_d. The black horizontal lines in each of these panels indicates the corresponding k_d measured values. The blue bars in Figure 9 correspond to the mean k_d values estimated by the proposed Markov model, for each day.

For 1 September 2017 (Figure 9a), all WRF experiments overestimated the daily clearness coefficient for the outer domain and underestimated it for the inner domain. The persistence model underestimated the k_d by ~ 0.2 and the propose Markov model underestimated the k_d by ~ 0.05.

Figure 9b,c show the case of 23 December 2017and 5 June 2017respectively. For these cases, the Markov model and the Persistence model overestimated the measured k_d values, although the Persistence model had smaller overestimation than the proposed Markov model. For 23 December 2017 (Figure 9b), the KF-MO experiment for the inner domain estimated the k_d with an error of 0.01, being the closest estimate to the real k_d. In the case of 5th June 2017, only the KF-MO and GR-MO-DE experiments, both for the inner domain, underestimated the daily clearness coefficient, k_d. In the rest of the cases the k_d was overestimated.

Figure 9d corresponds to the case of 24th November 2017. This day presented a case where local-scale events seem to be responsible for the high extinction values over the incoming GHI. For this day, the Markov model shows a better performance than the Persistence model at estimating the k_d. In this case the Δk_d of the proposed Markov model is $\Delta k_d = 0.01$, while the persistence model exhibited a $\Delta k_d \approx 0.52$. For this day, the WRF experiments also overestimated the measured daily clearness coefficient, k_d consistently. Since the Markov model has a better performance at estimating transitions that are frequently found in the variable records used to train it, it is possible to assume that the events corresponding to the high levels of cloudiness for 24 November 2017, are events that frequently occurred during the March 2016–February 2017 period.

Figure 9e shows the simulations for 19 August 2017. This day, in contrast with 24 November 2017, was a case characterized by the presence of a synoptic scale event over the Caribbean Sea. Although all experiments reproduce high levels of GHI extinction at the location of the event, the GHI series reproduced over Medellín greatly vary from experiment to experiment. As can be seen from Figure 9e, the k_d estimates corresponding to GR-MO-DE-d01 and KF-MO-d01 are those closest to the measured value, with the GR-MO-DE-d01 estimate having the lowest error of 0.05. For this day, the Markov model reproduced an average k_d of 0.58 and the persistence model an average k_d of 0.53, which are values that lie far from the observed value of 0.18. This indicates that this type of events, which are not so commonly observed over the study region and are caused by events that occur beyond local scales, are hardly captured by the Markov model.

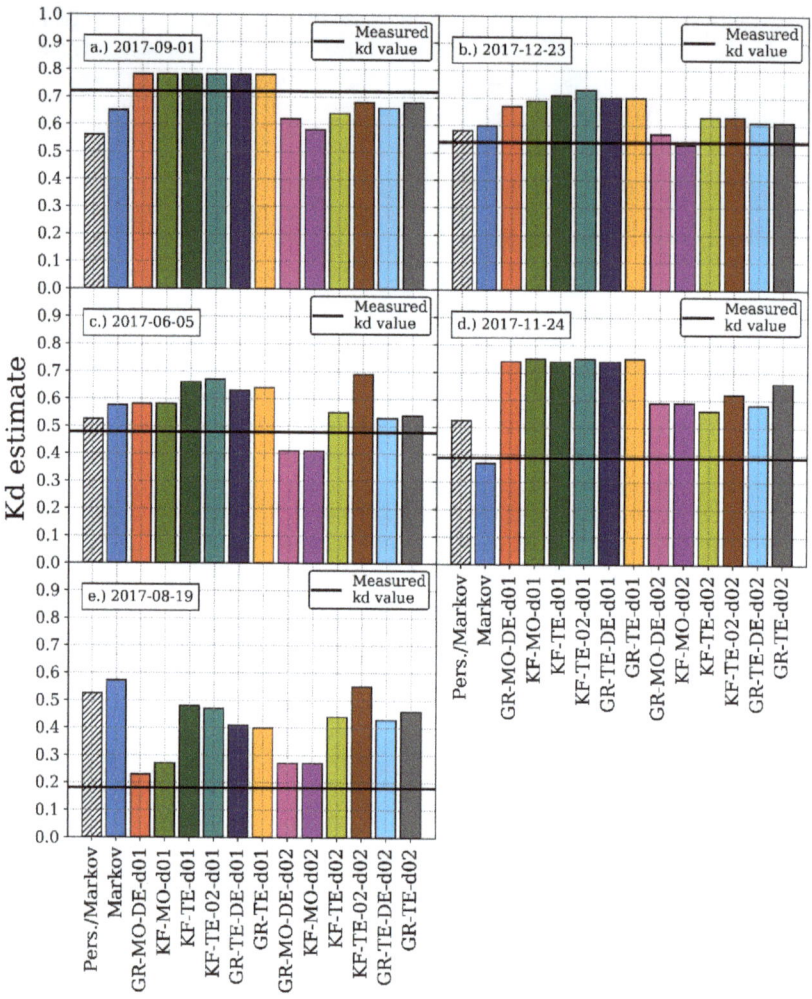

Figure 9. Bar plots of the k_d estimates Distributions of the daily clearness coefficient estimated by WRF, the Markov model and the persistence model. (**a**) k_d estimates distributions for 1 September 2017, measured $k_d = 0.72$. (**b**) k_d estimates distributions 23 December 2017. Measured $k_d = 0.54$. (**c**) k_d estimates distributions 6 June 2017. Measured $k_d = 0.48$. (**d**) k_d estimates distributions for 24 November 2017 measured $k_d = 0.39$. (**e**) k_d distributions for 19 August 2017, measured $k_d = 0.18$. The black lines correspond to the measured k_d values.

In general, the Markov model and the persistence model showed a similar performance at estimating the k_d, except for 1 September 2017 and 24 November 2017, were the propose Markov model had a better performance. It is also noted that, in general, the WRF experiments for the inner domain tended to produce lower k_d estimates than their counterpart in the outer domain.

Additionally, although the proposed Markov k_d estimates presented in Figure 9 are averaged values and the distribution of the estimates is not shown, we must indicate that the Markov model is only capable of producing 2 to 3 different states for each day presented in Figure 9.

3.3.2. Hourly Estimates of GHI

We evaluated the skill of the WRF model and the proposed Markov model at simulating the GHI at Medellín, Colombia, for 5 non-consecutive days with different cloud cover conditions (Table 2). Although only 5 simulation days do not represent a comprehensive simulation period, the comparison between the proposed Markov model and the WRF model at estimating the GHI for these 5 days still represents a good indicator of the performance of the proposed Markov model. The chosen days were selected based on their daily clearness coefficient value. For each particular day, we ran six simulations of the WRF model with different combinations of microphysics and cumulus schemes (Table 3). These simulations were later compared against the simulations obtained with the Markov model for the same 5 days.

Since the Markov model produces estimates in a stochastic way, each simulation results in a different hourly series of GHI, meaning that several realizations of a single day could be obtained in order to observe the overall behavior of the proposed Markov model. For this reason, and noting that the model is computationally inexpensive, each day in Table 2 is simulated 1000 times using the proposed Markov model. The resulting series for each day are plotted as violin plots, which are discriminated by each hour of the day, and are presented along with the estimate GHI series obtained with the WRF experiments in Figure 10. Violin plots are useful in this case because they include the approximate probability distribution of the estimates and also have information about the statistical metrics of the distribution such as the interquartile range and the median of the series. As in Figure 6, the white dots in the violin plots indicate the median value of the distributions and the height of the boxes inside each violin plot indicates the position and magnitude of the IQR of each distribution calculated as the difference between the Q3 and Q1 values, respectively.

Figure 10 presents the GHI hourly simulations obtained with the Markov model and the WRF model for each of the 5 selected days. The blue distributions correspond to the violin plots of the Markov estimates for each hour of each day and the colored lines correspond to the WRF GHI series for each experiment and for each domain considered (i.e., outer domain, d01 and inner domain, d02). The dashed black lines correspond to the measured hourly series of GHI for each day.

Figure 10a presents the hourly distributions of GHI for a clear sky day (1 September 2017, $k_d = 0.72$). For this day, the measured values of GHI are close to the IQR of the Markov estimate distributions. In the case of the WRF estimates, it is observed that series have a similar shape to the measured series, with the estimates of the KF-MO experiment presenting an RMSE increase of 77 W/m^2 from the outer domain (d01) with respect the inner domain (d02). In the case of the inner domain, d02, the WRF estimates consistently underestimated the measured GHI series.

Figure 10b presents the estimates for 23 December 2017. For this day, the measured series presented higher atmospheric extinction values during the morning hours than during the evening hours. For this day, the Markov model mostly overestimated the GHI during the morning hours, with only some extreme estimates falling near the measured values. This behavior is similar to that exhibited by most of the WRF experiments for this day. During the first evening hours (i.e., 12LT–15LT), the Markov model produced GHI values that were, in general, closer to the measured GHI values than the estimates produced by the WRF series for the inner domain (d02). In the case of the WRF estimates for the outer domain (d01), WRF produced GHI values that were very similar to the measured values and to the median values of the Markov distributions. During the last evening hours (i.e., 16LT–18LT), both models consistently overestimated the measured GHI values.

In the case of 5 June 2017 (Figure 10c), the Markov model had a satisfactory performance at estimating the GHI values between 9 LT–12 LT. However, it mostly overestimated the GHI during the afternoon hours as the measured values are near the lower tails of the estimate distributions. This matches the behavior of most of the WRF experiments, which also overestimated the GHI during the evening hours. However, the WRF experiments that used the Thompson and Eidhammer scheme, for the inner domain (i.e., KF-TE-d02, GR-TE-d02 and GR-TE-DE-d02), underestimated the clearness coefficient during the morning hours, unlike the rest of the experiments.

For the case of 24 November 2017 (Figure 10d), the Markov model consistently forecasted k_d states that were closer to the measured k_d, which caused the hourly GHI estimates of the Markov model to be closer to the hourly measured GHI values than the estimates obtained with the WRF experiments. This can be corroborated in Figure 9d. Although the Markov model had a better performance at estimating the hourly GHI than the WRF experiments for this day, both models consistently overestimated the measured GHI values for most of the evening hours (i.e., 14LT–18LT).

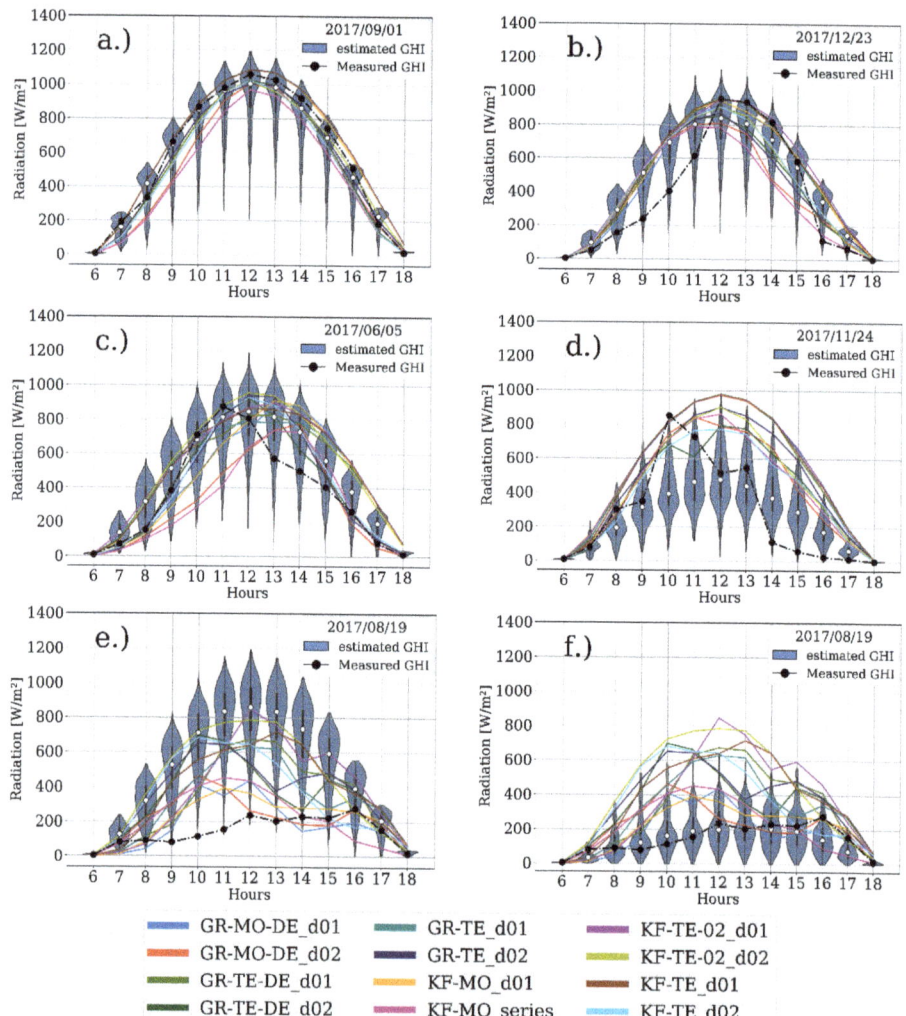

Figure 10. Hourly distributions of the GHI estimates produced by the Markov model and the WRF model for five particular days (Table 2). (**a**) Corresponds to 1 September 2017, with a $k_d = 0.72$. (**b**) Corresponds to 23 December 2017, with a measured $k_d = 0.54$. (**c**) Corresponds to 6 June 2017, with a measured $k_d = 0.48$. (**d**) Corresponds to 24th November 2017, with a measured $k_d = 0.39$. (**e**) Corresponds to 19 August 2017, with a measured $k_d = 0.18$. (**f**) Also corresponds to 19 August 2017, but in this case the estimated k_d was forced to be equal to the measured $k_d = 0.18$. The black dashed lines correspond to the measured hourly GHI values for each day.

The 19 August 2017 (Figure 10e) exhibits the lowest daily clearness coefficient value among the simulated days ($k_d = 0.18$). As it was previously mentioned, the high levels of cloudiness observed on this day are mainly associated to a tropical depression crossing the Caribbean and that would later transform into Hurricane Harvey [67]. For this day, the Markov model consistently overestimated the k_d for each of 1000 simulations performed (Figure 9e). These overestimations lead to an incorrect selection of the hourly MTM, which in turn, causes the consistent overestimations of the hourly values of GHI. This case is an example of how the estimates of the first part of the model largely affect the performance of the second part of the model. For this day, some of the WRF experiments exhibited a better performance at estimating the GHI than the Markov model. This was the case of the experiments that considered Morrison as the microphysics scheme. It was also observed that the outer domain (d01) estimates showed a better agreement with the measured GHI values than the nested domain (d02) estimates. Given that for 19 August 2017, the first part of the Markov model is not able to correctly simulate the measured state of the k_d in any of the 1000 simulations, an extra set of 1000 simulations were performed to observe the behavior of the second part of the model alone (i.e., the simulation of the hourly GHI). In order to do this, for these extra set of simulations, the state of k_d used as input for the second part of the Markov model was forced to be equal to the measured value (i.e., $k_d = 0.18$). The simulations obtained in this way for the Markov model are presented in Figure 10f. Most of the measured values of the GHI now fall inside the IQR of the Markov estimate distributions. In this case, when the base state of the series, which can be represented by the k_d, is correctly estimated, the model frequently reproduces GHI values that are closer to the measurements than in the opposite case (Figure 10e).

In order to assess the general performance of the Markov model at estimating the hourly GHI for the 5 selected days, Table 6 presents the summary of the RMSE errors of the hourly estimates of GHI obtained with the Markov model for each of the simulated days. The last row of Table 6 corresponds to the case where the estimated k_d was set equal to the measured k_d. The RMSE values presented for the Markov model correspond to the median values of the RMSE distributions. Table 6 also shows the summary of the RMSE errors of the hourly estimates of GHI obtained with the GR-MO-DE experiment since it has one of the highest performances at estimating the hourly GHI. The RMSE values presented for this experiment correspond to the mean value between the RMSE error for the outer domain, d01 and the inner domain, d02. Additionally, the RMSE median value of the persistence-Markov model is also presented in order to provide further benchmarking for the proposed Markov model.

Table 6. Summary of hourly GHI estimates produced by the Markov model and by the WRF experiment, GR-MO-DE.

Simulated day	Markov	Persistence-Markov	WRF: GR-MO-DE
	RMSE (Median)	RMSE (Median)	RMSE
01/09/2017 $k_d = 0.72$	144 W/m^2 (26%)	190 W/m^2 (33%)	116 W/m^2 (21%)
23/12/2017 $k_d = 0.54$	177 W/m^2 (47%)	171 W/m^2 (45%)	174 W/m^2 (46%)
05/06/2017 $k_d = 0.48$	165.1 W/m^2 (44%)	158 W/m^2 (43%)	204 W/m^2 (55%)
24/11/2017 $k_d = 0.39$	209 W/m^2 (76%)	233 W/m^2 (85%)	288 W/m^2 (104%)
19/08/2017 $k_d = 0.18$	415.2 W/m^2 (288.4%)	372 W/m^2 (258%)	133 W/m^2 (92%)
19/08/2017 $k_d = 0.18$ (k_d corrected)	97.5 W/m^2 (68%)	372 W/m^2 (258%)	133 W/m^2 (92%)

According to Table 6, for 1st September 2017, the Markov model exhibits an RMSE error of 144 W/m², which is lower than the RMSE value for the Persistence-Markov model but higher than the mean RMSE produced by the WRF experiment. This indicates a lower performance at estimating the hourly GHI for this day compared to the WRF experiment but an improvement with respect the persistence-based model. For 23 December 2017, the Markov model, the persistence-Markov model and the WRF experiments presented a similar performance at estimating the hourly GHI, all of them exhibiting nRMSE values of 47%, 45% and 46%, respectively. For the case of 6 June 2017, the performance of the Markov model increases with respect to the performance of the WRF experiments at estimating the GHI, however, the persistence-Markov model presents an improvement with respect the proposed Markov model. For the case of 24 November 2017, it can be seen that the proposed Markov model has a better performance than the persistence-Markov mode and the WRF model, presenting a lower RMSE than the other two models. This shows that for this particular day, neither the persistence-based model nor the WRF model were capable of simulating the correct state of the k_d, while the proposed Markov model did. Finally, for the case of 19 August 2017, the WRF presents a better performance at reproducing the effects of the larger scale event over the region of study, while the proposed Markov model and the persistence-Markov model fail to reproduce these effects over the GHI and thus, result in estimations with high RMSE values.

In general, the Markov model exhibits a lower RMSE values at estimating the hourly GHI in Medellín than the WRF experiment except for 1 September 2017 and 19 August 2017. For the latter simulation day, the Markov model is not able to reproduce the effects of the large-scale event over the region of study, while the WRF experiment GR-MO-DE is capable of producing large atmospheric extinction levels over the hourly GHI estimates.

As an additional assessment of the performance of these models for the 5 days selected, Table 7 shows the median of the RMSE values corresponding to each model. It can be observed that the proposed Markov model had a better performance at estimating the GHI than the persistence-Markov model. Also, it can be seen that the WRF model had a similar performance to the proposed Markov model with a small improvement of 3 W/m².

Table 7. Overall performance of the Markov model for the 5 chosen simulation days.

Error Metric	Markov	Persistence-Markov	WRF:GR-MO-DE
Median RMSE (W/m²)	177	190	174

3.4. Daily Forecasts of GHI for the Validation Period of May 2017–May 2018

Figure 11 shows the histogram of frequencies of the Δk_d errors obtained for the period of May 2017–May 2018. Most of the Δk_d errors are between 0.0 and 0.15, being the values between 0.05 and 0.15 the most frequent ones. This shows that the first part of the model mostly overestimates the k_d, producing k_d values that usually are 1 state or 3 states above the measured values (see Table 1). On the other hand, the values corresponding to the correct estimations of the k_d ($\Delta k_d = 0$) are also highly frequent.

Since the Markov-based model formulated here is computationally inexpensive, it was possible to perform 1000 simulations of each day during the period May 2017–May 2018. From this procedure it is possible to analyze the distribution of the errors of the different realizations produced by the model. Figure 12 show the distributions of the RMSE (Equation (9)) vs the estimation error Δk_d (Equation (11)), but this time presented as boxplots. In these plots, the lower whisker of the boxplots corresponds to Q1 − 1.5 × IQR, while the upper whisker corresponds to Q3 + 1.5 × IQR. The colored boxes enclose the IQR with the lines drawn within the boxes representing the median value. The colors of the boxes indicate the period in which the estimations were made.

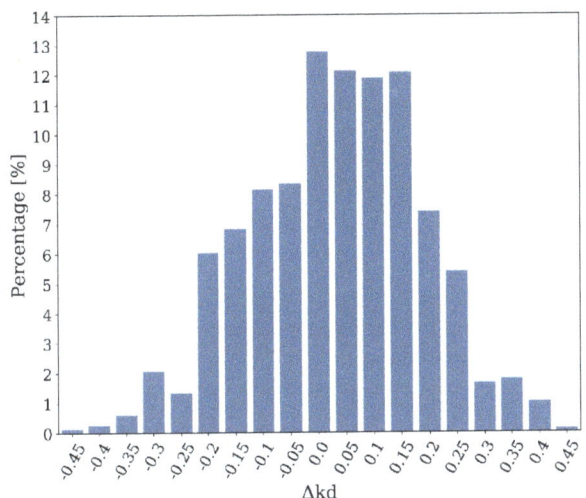

Figure 11. Frequency histogram of the Δk_d of the estimates performed during the period May 2017–May 2018.

Figure 12 illustrates how the estimation error of k_d (Δk_d) affects the RMSE distributions of the GHI estimates. Figure 12 also shows the RMSE distributions for wet and dry seasons.

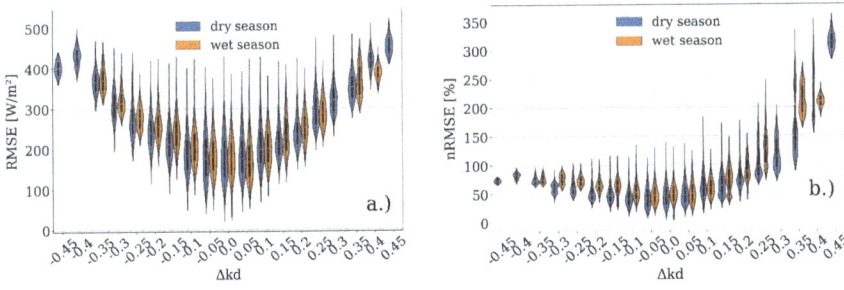

Figure 12. (a) Distributions of the root mean square error (RMSE) vs Δk_d for the period May 2017–May 2018. (b) As in (a) but for nRMSE. Orange boxes correspond to the wet season estimates. Blue boxes correspond to the dry season estimates.

Figure 12a,b shows the impact that the estimation of the k_d (first part of the Markov model) has on the hourly estimates of GHI (second part of the Markov model). However, Figure 12a,b do not allow evaluating the hourly estimates without being affected by the errors of the k_d. From this analysis is neither possible to observe the performance of the second part of the model alone at estimating the hourly GHI during days with different sky conditions (k_d). One way to do this is by simulating the period of May 2017 to May 2018 using the observed k_d values of each day as input for the second part of the Markov model. If this procedure is repeated a large number of times, it would be possible to obtain the error distributions of the GHI estimates for different sky conditions, which are given by the values of the observed k_d.

Finally, the proposed Markov model is compared against the Persistence-Markov model for the validation period of May 2017–May 2018. Table 8 shows that the proposed Markov model has a small improvement with respect the persistence-Markov model when evaluated during the period May 2017–May 2018. However, it is observed that during this period, the proposed Markov model exhibited as larger MBE than the persistence-Markov model. This is consistent with the overestimations in the k_d produced by the first part of the proposed Markov model, as can be observe in Figure 11.

Table 8. Overall error for the validation period of May 2017–May 2018.

Error Metric	Markov	Persistence-Markov
Median RMSE (W/m^2)	214	217
Median MBE (W/m^2)	33.9	7.21
Δk_d	0.076	0.003

The bias exhibited by the proposed Markov model at estimating the k_d reflects the fact that more than one year of data for training the model is necessary to train the model in order to produce a satisfactory estimator of the k_d. It is also worth noting that the persistence-Markov model displays a very low value of MBE, and not just when compared against the proposed Markov model.

4. Concluding Remarks

4.1. Solar Assessment

In this work, we analyzed global horizontal irradiance data from a pyranometer station located in Medellín, Colombia, with records during the period March 2016 to February 2017. Because of the low percentage of missing values, hourly and daily averages were computed without special interpolation algorithms. Although Medellín corresponds to a mountainous tropical city with months exhibiting a large number of cloudy days during the period considered in this work, most of the days during this period exhibited around 4 to 5 h with radiation values above 650 W/m^2, typically surpassing the 800 W/m^2 between 11 and 13LT. Additionally, the measured data showed that the average daily solar energy that reached the surface in the region during March 2016 to February 2017 was of ~5 kWh/m^2, which corresponds to 5 h of equivalent peak sun hours (PSH). The hourly measurements showed that this value would increase to an average of 5.5 PSH during the dry season months. This resource availability is important since it is comparable or even larger than what has been observed in regions that are referents regarding the use of solar radiation for energy production, like California, USA, with average values ranging from 3.8 to 6.3 PSH [68], or cities in Germany, with average PSH values that can range from 2.9 PSH to 3.6 PSH [69].

Additionally, during the dry season months, when the GHI values tend to be higher, solar variability decreases, which makes the use of the solar resource even more significant considering that during these months, especially when El Niño events occur, the main energy generation resource in Colombia (i.e., hydro energy) is heavily affected [30,70].

This is particularly important for Colombia since the country depends energetically on hydroelectric plants, which account for the 68.31% of the 17.35 GW of total installed capacity [71]. For this reason, it is essential to have complementary energy sources during dry periods that can reduce the risk of energy supply cuts.

4.2. Two-Part Markov Chain Model

In this work, a two-part model based on Markov chains was proposed. Additionally, a Numerical Weather Prediction model (WRF) was used as a benchmark for the estimates obtained with the Markov model. In general, the Markov model exhibited typical Δk_d ranging from 0.0 to 0.15. Also, regarding the k_d estimates, the estimation errors for all the WRF experiments ranged from −0.03 to 0.43. These results show that, in general, the first part of the Markov model (i.e., the part that estimated the k_d), has a comparable performance to the WRF model at estimating the daily clearness coefficient. Both models showed a consistent positive bias at estimating the k_d, even the WRF experiment of GR-MO-DE, which presented an average $\Delta k_d = 0.13$, which is also comparable to the typical Δk_d values from the Markov model. On the other hand, the persistence-Markov model exhibited a lower Δk_d for the validation period of May 2017–May 2018, but with a similar performance to the proposed Markov model in terms of the RMSE.

Although further inspection would be required to determine the origin of this positive bias in the Markov model it is clear that more than one year of data is necessary to produce more generalizable estimates of the k_d, and therefore, of the hourly GHI. This issue was also observed for the case of the 5 selected days (Figure 9), where the Markov model was only capable of estimating 2 to 3 different states for each day considered in this work.

In addition, the positive bias in our Markov model could be partially due to anomalous climate conditions during the training period.

During 2015/2016 a particularly strong *El Niño* event took place [72], which greatly affected the presence of clouds during the beginning of 2016, potentially producing larger values of k_d than in a normal year. Again, this shows how peculiarities of one year may cause biased estimates of the clearness coefficient, and therefore, longer periods for training the model are required.

The performance at estimating the hourly GHI of the Markov model and the WRF experiment, GR-MO-DE is presented in Table 7. For the five specific days considered in this study, the Markov model had a performance comparable to the performance of the GR-MO-DE WRF experiment, except for the case of the synoptic event, where the WRF experiment exceeded the skill of the Markov model at estimating the GHI. However, it is worth noting, as well, that for the overcast day associated to a more frequently measured type of event (i.e., 24 November 2017), the Markov model had a better performance at reproducing the hourly GHI than the WRF experiment.

When a more general assessment of the Markov is performed during the period May 2017–May 2018, RMSE errors inside the IQR of the hourly estimates produced by the Markov model for clear sky days ranged from ~57 W/m² to ~171 W/m² (~16%–~38%). For broken cloud sky conditions, the RMSE values ranged from ~149 W/m² to ~250 W/m² (~32%–~81%) while for overcast sky conditions, RMSE values ranged from ~80 W/m² to ~170 W/m² (~50%–~110%). In general, these results are in agreement with the errors found in similar studies regarding the hourly GHI forecasting for one day-ahead [56,73,74]. Although not frequently, the Markov model managed to produce GHI hourly estimates with very low RMSEs, even for highly overcast conditions. These values ranged from ~30 W/m² to ~80 W/m² (~20%–~50%). Even though these values are not typical among the estimates obtained in this work, they indicate that the model has the potential to produce series that are closer to the measured series for overcast condition than what is typically found in other works [8,50,51,73,75–77]. This improvement can be achieved by a further characterization of the hourly GHI series that could return typical intervals in which the hourly GHI usually lies for each hour of the day. Based on these intervals, the simulated series can be evaluated after they are produced and discarded if necessary. Additionally, because of how the Markov model was trained in this study, the hourly simulations of GHI did not take into account the dependency on the time of the day and the atmospheric mechanisms that could affect cloudiness. As a way of correcting this, the transition probabilities stored inside the Markov transition matrices could be obtained considering the time of the day.

We found that the estimates of the daily clearness coefficient, k_d, are fundamental for achieving the lowest RMSE values of the hourly series of GHI using the Markov model. Overestimations and underestimations of the k_d lead to error distributions that increase in magnitude with the bias of the k_d estimates. We also found that combining a persistence model for estimating the k_d with the second part of the proposed Markov model (i.e., the part that produces the hourly GHI estimates), results in a low bias model that could serve as an initial alternative for modeling the hourly GHI in the site of Medellín, Colombia.

In general, we believe further studies could start by improving the k_d prediction, which depends on meteorological process at synoptic scale, and so models like the weather prediction models could be a plausible alternative to simulate the k_d of the next day and which can later be used as an input for the second part of the Markov model proposed in this work. We are currently assessing the possibility of coupling the WRF model with the proposed Markov model.

The present work, despite the short time period of study, represents a novel contribution in terms of a detailed diagnostic of the solar resource and the performance of a variety of modeling tools for its day-ahead estimation for our region of interest. As such, this study is a first needed step for the assessment and modeling of solar energy for the city of Medellín, located within the tropical Andes.

Author Contributions: Conceptualization, J.U.-O., J.A.M. and P.A.A.; methodology, J.U.-O., J.A.M. and P.A.A.; software, J.U.-O.; validation, J.U.O.; formal analysis J.U.-O.; investigation, J.U.-O., J.A.M. and P.A.A.; resources, J.U-O; data curation, J.U.-O.; writing—original draft preparation, J.U.-O.; writing—review and editing, J.U.-O., J.A.M., P.A.A. and Á.J.-D. visualization, J.U.-O.; supervision, J.A.M. and P.A.A.; project administration, J.A.M. and P.A.A.; funding acquisition, J.U.-O., J.A.M. and P.A.A.

Funding: The authors gratefully acknowledge the financial support provided by the Program "Colombia Científica" within the framework of the called *Ecosistema Científico* (Contract No. FP44842-218-2018). We also acknowledge Sistema de Alertas Tempranas de Medellín for gently providing the GHI data used in this study. Julián Urrego-Ortiz was supported by the scholarship "Estudiante Instructor" funded by Universidad de Antioquia–Colombia.

Conflicts of Interest: The authors declare no conflict of interest.

References

1. Krakauer, N.; Cohan, D. Interannual variability and seasonal predictability of wind and solar resources. *Resources* **2017**, *6*, 29. [CrossRef]
2. Salgueiro, V.; Costa, M.J.; Silva, A.M.; Bortoli, D. Effects of clouds on the surface shortwave radiation at a rural inland mid-latitude site. *Atmos. Res.* **2016**, *178*, 95–101. [CrossRef]
3. Vignola, F.; Grover, C.; Lemon, N.; McMahan, A. Building a bankable solar radiation dataset. *Sol. Energy* **2012**, *86*, 2218–2229. [CrossRef]
4. CREG. *Resolución No. 30 de febrero de 2018*; CREG: Bogotá, Colombia, 2018; p. 27.
5. Energías Renovables. Available online: http://www1.upme.gov.co/Paginas/Energias-renovables.aspx (accessed on 22 July 2019).
6. Hidroituango, el Megaproyecto de Ingeniería en Colombia, Tiene Otro Problema: Un Socavón a 40 Metros CNN. Available online: https://cnnespanol.cnn.com/2019/01/14/hidroituango-el-megaproyecto-de-ingenieria-en-colombia-tiene-otro-problema-un-socavon-a-40-metros/#0 (accessed on 22 July 2019).
7. Al-Kayiem, H.; Mohammad, S. Potential of renewable energy resources with an emphasis on solar power in Iraq: An outlook. *Resources* **2019**, *8*, 42. [CrossRef]
8. Lauret, P.; Diagne, M.; David, M. A neural network post-processing approach to improving NWP solar radiation forecasts. *Energy Procedia* **2014**, *57*, 1044–1052. [CrossRef]
9. Ngoko, B.O.; Sugihara, H.; Funaki, T. Synthetic generation of high temporal resolution solar radiation data using Markov models. *Sol. Energy* **2014**, *103*, 160–170. [CrossRef]
10. Skamarock, W.C.; Klemp, J.B.; Dudhi, J.; Gill, D.O.; Barker, D.M.; Duda, M.G.; Huang, X.-Y.; Wang, W.; Powers, J.G. *A Description of the Advanced Research WRF Version 3*; NCAR: Boulder, Colorado, USA, 2008.
11. Perez, R.; Lorenz, E.; Pelland, S.; Beauharnois, M.; Van Knowe, G.; Hemker, K.; Heinemann, D.; Remund, J.; Müller, S.C.; Traunmüller, W.; et al. Comparison of numerical weather prediction solar irradiance forecasts in the US, Canada and Europe. *Sol. Energy* **2013**, *94*, 305–326. [CrossRef]
12. Perez, R.; Kivalov, S.; Schlemmer, J.; Hemker, K.; Renné, D.; Hoff, T.E. Validation of short and medium term operational solar radiation forecasts in the US. *Sol. Energy* **2010**, *84*, 2161–2172. [CrossRef]
13. Doorga, J.R.S.; Rughooputh, S.D.D.V.; Boojhawon, R. Modelling the global solar radiation climate of Mauritius using regression techniques. *Renew. Energy* **2019**, *131*, 861–878. [CrossRef]
14. Ibrahim, S.; Daut, I.; Irwan, Y.M.; Irwanto, M.; Gomesh, N.; Farhana, Z. Linear regression model in estimating solar radiation in perlis. *Energy Procedia* **2012**, *18*, 1402–1412. [CrossRef]
15. Gairaa, K.; Bakelli, Y. A Comparative study of some regression models to estimate the global solar radiation on a horizontal surface from sunshine duration and meteorological parameters for Ghardaïa Site, Algeria. *ISRN Renew. Energy* **2013**, *2013*, 1–11. [CrossRef]
16. Paoli, C.; Voyant, C.; Muselli, M.; Nivet, M.L. Forecasting of preprocessed daily solar radiation time series using neural networks. *Sol. Energy* **2010**, *84*, 2146–2160. [CrossRef]
17. Ji, W.; Chee, K.C. Prediction of hourly solar radiation using a novel hybrid model of ARMA and TDNN. *Sol. Energy* **2011**, *85*, 808–817. [CrossRef]

18. David, M.; Ramahatana, F.; Trombe, P.J.; Lauret, P. Probabilistic forecasting of the solar irradiance with recursive ARMA and GARCH models. *Sol. Energy* **2016**, *133*, 55–72. [CrossRef]
19. Notton, G.; Voyant, C. Forecasting of intermittent solar energy resource. In *Advanced in Renewable Energies and Power Technologies*; Yahyaoui, I., Ed.; Elsevier Science: Amsterdam, The Netherlands, 2018; pp. 78–114. ISBN 9780128129593.
20. Aguiar, R.; Collares-Pereira, M. TAG: A time-dependent, autoregressive, Gaussian model for generating synthetic hourly radiation. *Sol. Energy* **1992**, *49*, 167–174. [CrossRef]
21. Bouabdallah, A.; Olivier, J.C.; Bourguet, S.; Machmoum, M.; Schaeffer, E. Safe sizing methodology applied to a standalone photovoltaic system. *Renew. Energy* **2015**, *80*, 266–274. [CrossRef]
22. Poggi, P. Stochastic study of hourly total solar radiation in. *Int. J. Climatol.* **2000**, *1860*, 1843–1860. [CrossRef]
23. Hocaoglu, F.O.; Serttas, F. A novel hybrid (Mycielski-Markov) model for hourly solar radiation forecasting. *Renew. Energy* **2017**, *108*, 635–643. [CrossRef]
24. Shakya, A.; Michael, S.; Saunders, C.; Armstrong, D.; Pandey, P.; Chalise, S.; Tonkoski, R. Solar irradiance forecasting in remote microgrids using markov switching model. *IEEE Trans. Sustain. Energy* **2017**, *8*, 895–905. [CrossRef]
25. Li, S.; Ma, H.; Li, W. Typical solar radiation year construction using k-means clustering and discrete-time Markov chain. *Appl. Energy* **2017**, *205*, 720–731. [CrossRef]
26. Munkhammar, J.; Widén, J. A spatiotemporal Markov-chain mixture distribution model of the clear-sky index. *Sol. Energy* **2019**, *179*, 398–409. [CrossRef]
27. Hocaoğlu, F.O. Stochastic approach for daily solar radiation modeling. *Sol. Energy* **2011**, *85*, 278–287. [CrossRef]
28. Eugster, W. Mountain Meteorology: Fundamentals and Applications. *Mt. Res. Dev.* **2001**, *21*, 200–201. [CrossRef]
29. Bedoya-Soto, J.M.; Aristizábal, E.; Carmona, A.M.; Poveda, G. Seasonal shift of the diurnal cycle of rainfall over medellin's valley, central andes of Colombia (1998–2005). *Front. Earth Sci.* **2019**, *7*, 92. [CrossRef]
30. Poveda, G.; Mesa, O.J.; Waylen, P.R. *Nonlinear Forecasting of River Flows in Colombia Based Upon ENSO and Its Associated Economic Value for Hydropower Generation*; Springer: Dordrecht, The Netherlands, 2003; pp. 351–371.
31. Poveda, G.; Waylen, P.R.; Pulwarty, R.S. Annual and inter-annual variability of the present climate in northern South America and southern Mesoamerica. *Palaeogeogr. Palaeoclimatol. Palaeoecol.* **2006**, *234*, 3–27. [CrossRef]
32. Poveda, G. La hidroclimatología de Colombia: una síntesis desde la escala inter-decadal hasta la escala diura por ciencias de la tierra. *Rev. Acad. Colomb. Cienc. Exactas, Físicas y Nat.* **2004**, *28*, 201–222.
33. Badescu, V. *Modeling Solar Radiation at the Earth's Surface*; Springer: Berlin Heidelberg, Germany, 2008; p. 53. ISBN 978-3-540-77454-9.
34. Kasten, F. A new table and approximate formula for relative optical air mass. *Archiv für Meteorologie, Geophysik und Bioklimatologie* **1966**, *14*, 206–223. [CrossRef]
35. Kasten, F.; Young, A.T. Revised optical air mass tables and approximation formula. *Appl. Opt.* **2000**, *28*, 4735–4738. [CrossRef]
36. Liu, B.Y.H.; Jordan, R.C. The interrelationship and characteristic distribution of direct, diffuse and total solar radiation. *Sol. Energy* **1960**, *4*, 1–19. [CrossRef]
37. Thompson, G.; Tewari, M.; Ikeda, K.; Tessendorf, S.; Weeks, C.; Otkin, J.; Kong, F. Explicitly-coupled cloud physics and radiation parameterizations and subsequent evaluation in WRF high-resolution convective forecasts. *Atmos. Res.* **2016**, *168*, 92–104. [CrossRef]
38. Iqbal, M. *An Introduction to Solar Radiation*; Elsevier Inc.: Vancouver, BC, Canda, 1983; ISBN 978-0-12-373750-2.
39. Perez, R.; Ineichen, P.; Seals, R.; Zelenka, A. Making full use of the clearness index for parameterizing hourly insolation conditions. *Sol. Energy* **1990**, *45*, 111–114. [CrossRef]
40. Palomo, E. Hourly solar radiation time series as first-order Markov chains. In Proceedings of the Actes du International Solar Energy Society Solar World Congress, Kobe, Japan, September 1989; pp. 2146–2150.
41. Munkhammar, J.; van der Meer, D.; Widén, J. Probabilistic forecasting of high-resolution clear-sky index time-series using a Markov-chain mixture distribution model. *Sol. Energy* **2019**, *184*, 688–695. [CrossRef]
42. Nguyen, B.T.; Pryor, T.L. A computer model to estimate solar radiation in Vietnam. *Renew. Energy* **1996**, *9*, 1274–1278. [CrossRef]
43. Aguiar, R.J.; Collares-Pereira, M.; Conde, J.P. Simple procedure for generating sequences of daily radiation values using a library of Markov transition matrices. *Sol. Energy* **1988**, *40*, 269–279. [CrossRef]

44. Graham, V.A.; Hollands, K.G.T. A method to generate synthetic hourly solar radiation globally. *Sol. Energy* **1990**, *44*, 333–341. [CrossRef]
45. Hollands, K.G.T.; Huget, R.G. A probability density function for the clearness index, with applications. *Sol. Energy* **1983**, *30*, 195–209. [CrossRef]
46. Diagne, M.; David, M.; Lauret, P.; Boland, J.; Schmutz, N. Review of solar irradiance forecasting methods and a proposition for small-scale insular grids. *Renew. Sustain. Energy Rev.* **2013**, *27*, 65–76. [CrossRef]
47. Hoyos, C.D.; Zapata, M.H. *Análisis del Impacto de la Interacción Suelo-Atmósfera en las Condiciones Meteorológicas del Valle de Aburrá Utilizando el Modelo WRF*; Universidad Nacional de Colombia: Bogotá, Colombia, 2015.
48. Diagne, M.; David, M.; Boland, J.; Schmutz, N.; Lauret, P. Post-processing of solar irradiance forecasts from WRF Model at Reunion Island. *Energy Procedia* **2014**, *57*, 1364–1373. [CrossRef]
49. Incecik, S.; Sakarya, S.; Tilev, S.; Kahraman, A.; Aksoy, B.; Caliskan, E.; Topcu, S.; Kahya, C.; Odman, M.T. Evaluation of WRF parameterizations for global horizontal irradiation forecasts: A study for Turkey. *Atmósfera* **2019**, *32*, 143–158. [CrossRef]
50. Zempila, M.M.; Giannaros, T.M.; Bais, A.; Melas, D.; Kazantzidis, A. Evaluation of WRF shortwave radiation parameterizations in predicting Global Horizontal Irradiance in Greece. *Renew. Energy* **2016**, *86*, 831–840. [CrossRef]
51. Lara-Fanego, V.; Ruiz-Arias, J.A.; Pozo-Vázquez, D.; Santos-Alamillos, F.J.; Tovar-Pescador, J. Evaluation of the WRF model solar irradiance forecasts in Andalusia (southern Spain). *Sol. Energy* **2012**, *86*, 2200–2217. [CrossRef]
52. Jimenez, P.A.; Hacker, J.P.; Dudhia, J.; Haupt, S.E.; Ruiz-Arias, J.A.; Gueymard, C.A.; Thompson, G.; Eidhammer, T.; Deng, A. WRF-SOLAR: Description and clear-sky assessment of an augmented NWP model for solar power prediction. *Bull. Am. Meteorol. Soc.* **2016**, *97*, 1249–1264. [CrossRef]
53. Arbizu-Barrena, C.; Ruiz-Arias, J.A.; Rodríguez-Benítez, F.J.; Pozo-Vázquez, D.; Tovar-Pescador, J. Short-term solar radiation forecasting by advecting and diffusing MSG cloud index. *Sol. Energy* **2017**, *155*, 1092–1103. [CrossRef]
54. Ruiz-Arias, J.A.; Arbizu-Barrena, C.; Santos-Alamillos, F.J.; Tovar-Pescador, J.; Pozo-Vázquez, D. Assessing the surface solar radiation budget in the WRF model: A spatiotemporal analysis of the bias and its causes. *Mon. Weather Rev.* **2016**, *144*, 703–711. [CrossRef]
55. De Meij, A.; Vinuesa, J.F.; Maupas, V. GHI calculation sensitivity on microphysics, land- and cumulus parameterization in WRF over the Reunion Island. *Atmos. Res.* **2018**, *204*, 12–20. [CrossRef]
56. Verbois, H.; Rusydi, A.; Thiery, A. Probabilistic forecasting of day-ahead solar irradiance using quantile gradient boosting. *Sol. Energy* **2018**, *173*, 313–327. [CrossRef]
57. Janjić, Z.I. The step-mountain Eta coordinate model: Further developments of the convection, viscous sublayer, and turbulence closure schemes. *Mon. Weather Rev.* **1994**, *122*, 927–945. [CrossRef]
58. Thompson, G.; Field, P.R.; Rasmussen, R.M.; Hall, W.D. Explicit forecasts of winter precipitation using an improved bulk microphysics scheme. Part II: Implementation of a new snow parameterization. *Mon. Weather Rev.* **2008**, *136*, 5095–5115. [CrossRef]
59. Thompson, G.; Eidhammer, T. A Study of aerosol impacts on clouds and precipitation development in a large winter cyclone. *J. Atmos. Sci.* **2014**, *71*, 3636–3658. [CrossRef]
60. Morrison, H.; Thompson, G.; Tatarskii, V. Impact of cloud microphysics on the development of trailing stratiform precipitation in a simulated squall line: Comparison of one- and two-moment schemes. *Mon. Weather Rev.* **2009**, *137*, 991–1007. [CrossRef]
61. Grell, G.A.; Dévényi, D. A generalized approach to parameterizing convection combining ensemble and data assimilation techniques. *Geophys. Res. Lett.* **2002**, *29*, 38-1–38-4. [CrossRef]
62. Deng, A.; Gaudet, B.; Dudhia, J.; Alapaty, K. Implementation and evaluation of a new shallow convection scheme in WRF. In Proceedings of the 26th Conference on Weather Analysis and Forecasting/22nd Conference on Numerical Weather Prediction, Atlanta, GA, USA, 2–6 February 2014; pp. 2–6.
63. Kain, J.S. The kain–fritsch convective parameterization: An update. *J. Appl. Meteorol.* **2004**, *43*, 170–181. [CrossRef]
64. Moreno-Tejera, S.; Silva-Pérez, M.A.; Lillo-Bravo, I.; Ramírez-Santigosa, L. Solar resource assessment in Seville, Spain. Statistical characterisation of solar radiation at different time resolutions. *Sol. Energy* **2016**, *132*, 430–441. [CrossRef]

65. Bowman, K.P.; Fowler, M.D. The diurnal cycle of precipitation in tropical cyclones. *J. Clim.* **2015**, *28*, 5325–5334. [CrossRef]
66. Liou, K.-N. *An Introduction to Atmospheric Radiation*, 2nd ed.; Elsevier Science: Amsterdam, The Netherlands, 2002; ISBN 0124514510.
67. US Department of Commerce, NOAA, N.W.S. Hurricane Harvey Info. Available online: https://www.weather.gov/hgx/hurricaneharvey (accessed on 22 September 2019).
68. Roberts, B.J. Solar Maps | Geospatial Data Science | NREL. Available online: https://www.nrel.gov/gis/solar.html (accessed on 13 June 2019).
69. HotSpot Energy Solar Sun Hours | Average Daily Solar Insolation | Europe. Available online: https://www.hotspotenergy.com/DC-air-conditioner/europe-solar-hours.php (accessed on 13 June 2019).
70. World Bank. *Colombia Recent Economic Developments in Infrastructure II*; World Bank: Washington, DC, USA, 2004.
71. Capacidad Efectiva Por Tipo de Generación. Available online: http://paratec.xm.com.co/paratec/SitePages/generacion.aspx?q=capacidad (accessed on 22 July 2019).
72. Stockdale, T.; Balmaseda, M.; Ferranti, L. The 2015/2016 El Niño and Beyond. Available online: https://www.ecmwf.int/en/newsletter/151/meteorology/2015-2016-el-nino-and-beyond (accessed on 22 October 2019).
73. Aryaputera, A.W.; Yang, D.; Walsh, W.M. Day-ahead solar irradiance forecasting in a tropical environment. *J. Sol. Energy Eng.* **2015**, *137*, 051009. [CrossRef]
74. Verbois, H.; Huva, R.; Rusydi, A.; Walsh, W. Solar irradiance forecasting in the tropics using numerical weather prediction and statistical learning. *Sol. Energy* **2018**, *162*, 265–277. [CrossRef]
75. Husein, M.; Chung, I.Y. Day-ahead solar irradiance forecasting for microgrids using a long short-term memory recurrent neural network: A deep learning approach. *Energies* **2019**, *12*, 1856. [CrossRef]
76. Lan, H.; Yin, H.; Hong, Y.Y.; Wen, S.; Yu, D.C.; Cheng, P. Day-ahead spatio-temporal forecasting of solar irradiation along a navigation route. *Appl. Energy* **2018**, *211*, 15–27. [CrossRef]
77. Lan, H.; Zhang, C.; Hong, Y.Y.; He, Y.; Wen, S. Day-ahead spatiotemporal solar irradiation forecasting using frequency-based hybrid principal component analysis and neural network. *Appl. Energy* **2019**, *247*, 389–402. [CrossRef]

© 2019 by the authors. Licensee MDPI, Basel, Switzerland. This article is an open access article distributed under the terms and conditions of the Creative Commons Attribution (CC BY) license (http://creativecommons.org/licenses/by/4.0/).

Article

Energy Saving Potential of Industrial Solar Collectors in Southern Regions of Russia: The Case of Krasnodar Region

Svetlana Ratner [1,2], Konstantin Gomonov [1,*], Svetlana Revinova [1] and Inna Lazanyuk [1]

1. Department of Economic and Mathematical Modelling, Peoples' Friendship University of Russia (RUDN University), 6 Miklukho-Maklaya Street, Moscow 117198, Russian; ratner-sv@rudn.ru (S.R.); revinova-syu@rudn.ru (S.R.); lazanyuk-iv@rudn.ru (I.L.)
2. Economic Dynamics and Innovation Management Laboratory, V.A. Trapeznikov Institute of Control Sciences, Russian Academy of Sciences, 65 Profsoyuznaya Street, Moscow 117997, Russian
* Correspondence: gomonov-kg@rudn.ru; Tel.: +7-495-433-4065

Received: 14 January 2020; Accepted: 10 February 2020; Published: 17 February 2020

Abstract: Industrial low-temperature processes are a promising sector for the introduction of solar collectors, which can partially, and in some cases, completely, replace traditional heat supply technologies. In Krasnodar Region (Russia), it is shown that the energy-saving potential when introducing industrial solar collectors only at food industry enterprises can make up 16%–17% of the total amount of thermal energy produced in the region annually. The global market of industrial solar collectors is currently developing almost without any government incentives, only due to market mechanisms, which indicates the commercial attractiveness of the technology. According to the predicted estimates, levelized cost of energy produced by industrial solar collectors in the southern regions of Russia may amount to 3.8–6.6 rubles per kWh. Even though the forecast estimates are higher than current tariffs, the economic feasibility of using solar collectors in the industry increases significantly if it is not possible to connect to centralized heating networks, as well as in the case of the seasonal load of industrial facilities. As a measure of state incentives for the development of industrial solar collectors in Russia, we offer state co-financing of demonstration projects of Russian manufacturers. This will increase the level of awareness of the population and businesses about the capabilities of this technology. Also, it will increase the technical competencies and innovative potential of companies involved in the production and installation of solar collectors.

Keywords: heat supply of industrial processes; renewable energy; solar collectors; economic efficiency

1. Introduction

In recent decades, renewable energy has developed enormously throughout the world. The transition to renewable energy sources is considered both in academic and business societies as an essential step towards the formation of a circular economy and achieving sustainable development goals [1,2].

The average annual growth rate of installed capacity of renewable energy sources (RES) in the period from 2009-2018 amounted to 8.4% [3], and, starting from 2015, net capacity additions for renewable power are higher than for fossil fuels and nuclear all together [4]. The average annual growth rate of energy generation based on renewable energy sources in the period from 2009–2017 amounted to almost 6% (Figure 1). At the end of 2017, investments in renewable energy-based electricity generation for the first time in history exceeded investments in traditional types of electricity generation (including nuclear energy), most of which came from countries with developing economies [5]. In 2018 global investment in RES (including large hydropower plants) reached USD 288.9 billion. Despite an

11% decrease compared to 2017, that was the fifth year in a row that investment exceeded USD 230 billion [4].

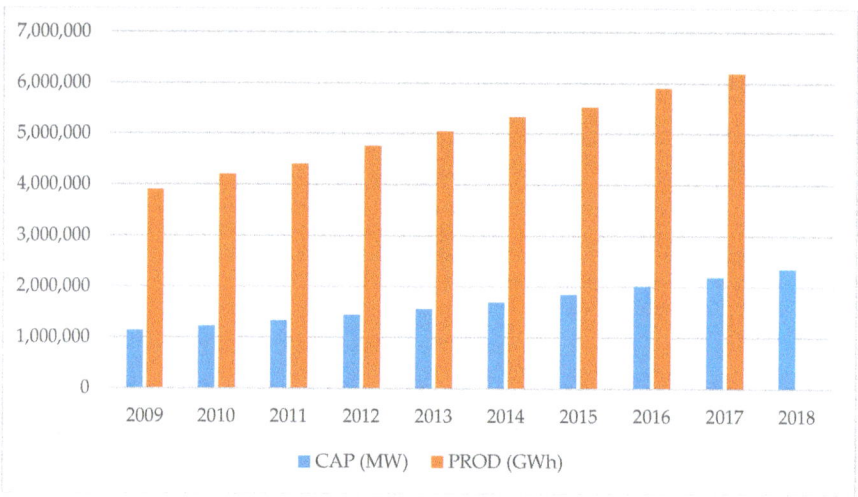

Figure 1. Total renewable energy capacity and production in the world. Source: authored based on The International Renewable Energy Agency data [3].

Today, almost all countries of the world have goals at the state level for the development of renewable energy [5–8]. However, despite significant progress in electricity generation, the introduction of renewable energy technologies in the heat supply and heat generation sector, including for industrial needs, is still slow, despite the fact that these types of final energy consumption have the most significant shares in the global energy balance. Thus, according to REN21 statistics of 2016, 51% of all energy consumed in the world is spent on heat supply, and this sector contributes nearly 40% of global energy-related CO_2 emissions [4]. The share of modern renewables in final heat consumption globally is only 9.8% and distributed between the following modern technologies [5]:

1. *Biogeneration:* boilers using solid biomass; the use of biogas in central heating systems; the addition of biogas to the gas supply grid; the direct use of biogas for cooking.

2. *Solar collectors:* used for heating water and, to a lesser extent, heating buildings. In recent years, the scale of use in central heating systems and industry has increased significantly [5].

3. *Geothermal energy:* used in central heating systems, for swimming pools, greenhouses, as well as in industry. All three technologies together contribute 8% in final heat consumption.

4. *Heating with renewable electricity:* the use of electricity generated by solar panels, wind farms, etc., for the operation of heat pumps in the residential, commercial, and industrial sectors. In 2016, this sector contributed 1.8% of the final heat consumption.

The use of solar energy in the heat supply of buildings in the residential and commercial sectors has a rather long history and is well studied in the literature [9–11], while the use of solar energy in industrial production is currently only developing. The primary constraint so far is the impossibility of providing round-the-clock heat supply to the production process using solar energy. To overcome this technical barrier, it is necessary to install additional equipment like heat storage systems, which significantly increase the cost of the entire solar installation [11–13]. The high initial cost of acquiring and installing equipment (solar collectors and heat storage systems) is the second most crucial constraint, which is especially essential for small and medium enterprises that do not have a sufficiently large volume of current assets [12]. At the same time, government support measures for the development of this type of renewable energy sources are not yet widespread. So, in 2018, according to REN21 [4], already 135 countries of the world carried out various government policy measures aimed at supporting renewable energy in the electricity generation sector, while incentive measures for renewable energy technologies in the heating sector were introduced in only 20 countries.

Currently, the use of solar energy is most developed in the food industry [14], primarily because most of the production processes associated with food processing are low-temperature (Table 1). So, for example, processes such as various types of drying, cleaning, washing, heating water, pasteurization, and sterilization do not require temperatures above 250 °C, which can easily be achieved using various types of solar collectors. The second most common user of solar collectors is the textile industry, in which many production processes (such as cleaning, drying, washing, pressing) do not require high temperatures [15–17].

Table 1. Industrial processes are potentially suitable for the use of solar collectors as heat supply equipment [14–17].

Sector	Industrial Process	Temperature Range, °C
Chemicals	Biochemical reaction	20–60
	Distillation	100–200
	Compression	105–165
	Compression	110–130
Foods and beverages	Blanching	60–100
	Scalding	45–90
	Evaporating	40–130
	Cooking	70–120
	Pasteurization	60–145
	Smoking	20–85
	Cleaning	60–90
	Sterilization	100–140
	Tempering	40–80
	Drying	40–200
	Washing	30–80
Paper and paperboard manufacturing	Bleaching	40–150
	De-inking	50–70
	Cooking wood in a chemical solution	110–180
	Drying	95–200
Fabricated Metal	Chromating	20–75
	Degreasing	20–100
	Electroplating	30–95
	Phosphating	35–95
	Purging	40–70
	Drying	60–200
Rubber and Plastics	Drying	50–150
	Preheating	40–70
Textile industry	Bleaching	40–100
	Coloring	40–130
	Drying	60–90
	Washing	50–100
	Fixing	160–180
	Pressing	80–100
Wood industry	Steaming	70–90
	Pickling	40–70
	Compression	120–170
	Cooking	80–90
	Drying	40–150
Mining	Cleaning	~60
	Electro-winning	~50
	Other processes	~80
Agriculture	Drying	80
	Water heating	90
Automobile industry	Water heating	90
	Cleaning	120
	Other processes	~50

Depending on the required temperature level of the production process, various types of solar thermal collectors are used, from the most straightforward and cheapest air flat-plate collectors, suitable for temperatures up to 100 °C to the more complex Fresnel collector or parabolic trough collectors for temperatures up to 400 °C [14–17].

This study aims to review the current state of the world market of industrial solar collectors and assess the possibilities of their application in individual industrial sectors of the Russian Federation. The rest of the paper is organized as follows: in Section 2 we describe materials for the study and basic methodology; Section 3 gives a brief overview of the research background and particular main trends and status-quo of industrial solar collectors in the world and in Russia; in Section 4 we present the results of calculations for estimation of the expected economic efficiency of industrial solar collectors in the southern regions of Russia and estimation of the potential for their use in the Krasnodar Region; Section 5 discusses the results of the study and gives some policy recommendations; the final section concludes the study and discusses its added value for academic literature.

2. Methods

The information base of the study was the analytical materials of the project of the World Energy Agency "Integration of Solar Heat into Industrial Processes" (IEA SHC Task49 / IV SHIP), materials of the REN21 expert network and the analytical agency Solar and Wind Energy. The current state and the trends in the development of solar heat in Russia was studied based on the data of Austrian Institute for Sustainable Technologies (IFA Solar Heating and Cooling Program), and the data of Russian Litvinchuk HVAC Marketing Agency (http://www.litvinchuk.ru/), which specializes in research for heating, air conditioning, and cooling systems markets. The data for assessing the potential of using solar collectors in the industry of the Krasnodar Region were obtained from the statistical collection "Krasnodar Territory in Figures, 2016" [18] and open data from the Federal State Statistics Service, presented on the official website in the section Technological Development of Economic Sectors / Energy Efficiency (https://www.gks.ru/folder/11189).

The traditional approach is widely used to calculate the economic efficiency of industrial solar collectors in Russian scientific literature (see, for example, [11,19]). This approach is based on calculating the payback period of equipment T (years) through the cost of replaced energy, the cost of energy produced by the solar collector, and the coefficient of efficiency of conversion of solar energy into thermal energy (the conversion factor) by the formula:

$$T = \frac{C_{SC}}{S_p \eta C_{th}} \quad (1)$$

where

S_{SC}—cost of heat generated by the solar collector (rubles/m^2);
C_{th}—cost of replaced energy (rubles/kWh);
S_p—the total intensity of solar radiation in the plane of the solar collector (kWh/m^2);
η—the conversion factor of solar energy into heat.

This approach gives the most accurate results in the case of calculating the economic efficiency of a particular solar collector installed in a certain way in a specific geographic location but is poorly suited for predicting and assessing the economic potential of using solar collectors on a scale of the industrial sector of the region. Firstly, it does not take into account changes in the value of money over time (discount coefficient), and secondly, it requires data on the exact locations of all industrial facilities on which the installation of solar collectors is planned. When calculating the regional potential for such a region as the Krasnodar Territory, with an area of 76,000 square km and a length from north to south of more than 320 km, and from west to east of more than 350 km, this approach creates significant computational difficulties [20] and at the same time does not give any advantages over less accurate methods, since it still leads to the need for data averaging.

Therefore, in this study, we used an approach based on the construction of a linear regression model based on statistics on the performance of flat solar water collectors in different regions of the world presented in the source [15]. The explanatory variable (proxy) in the model is the level of solar insolation. Further, the calculated value of the productivity of the solar collector was substituted into the formula LCOE (levelized cost of energy) [21,22]:

$$LCOE = \dfrac{I_0 + \sum\limits_{t=1}^{T} A_t \cdot (1+r)^{-t}}{\sum\limits_{t=1}^{T} SE \cdot (1+r)^{-t}}, \qquad (2)$$

where

I_0—the unit cost of equipment, taking into account the installation (euro /m^2);

A_t—equipment maintenance cost in year t (according to [23] is assumed to be equal 0.25%–0.5% depending on the type of collector);

SE—the amount of energy produced in year t;

T—the life cycle of equipment (years);

r—discount rate, reflecting the change in the value of money over time (for calculations in euros, as a rule, it is assumed to be equal to 3%).

Values I_0 and T were taken as average for equipment of a similar class, A_t as the average value of labor costs in countries with a comparable standard of living and wages, and r as the average inflation rate in Russia over the past five years.

One can quickly notice that the advantage of our approach is, on the one hand, simplicity, and, on the other hand, taking into account essential factors affecting the economic efficiency of the solar collector, such as the costs of its installation and maintenance, the life cycle of the solar collector, and the change in the cost of money over time. Schematically, the logic of our study is reflected in Figure 2.

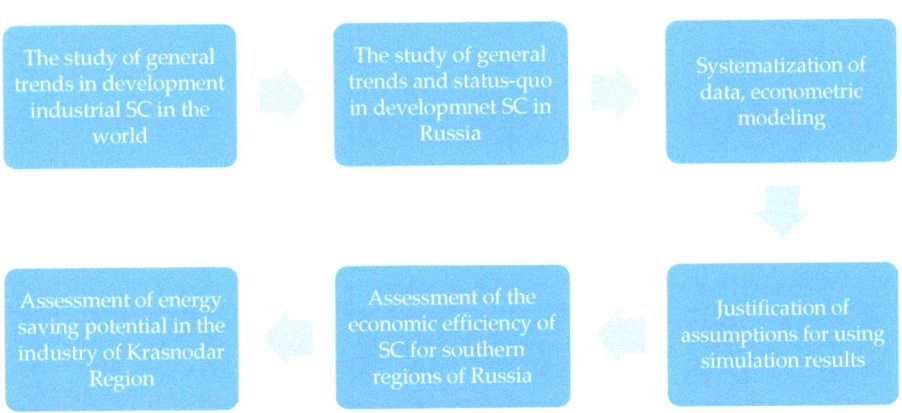

Figure 2. The general algorithm of the study.

Thus, to achieve the main goal of the study, we needed to solve the following two problems: 1) to analyze the structure of industrial production in the region and assess the volume of low-temperature industrial processes, and, possible demand for industrial solar collectors; 2) to determine under what conditions a transition of low-temperature industrial processes to solar energy can be economically feasible.

3. General Trends in Development Industrial Solar Collectors in the World and Russia (Research Background)

3.1. Industrial Solar Heat Worldwide

Solar Heat for Industrial Processes (SHIP) is a fast-growing new global market [15,23]. According to the data of Austrian Research Institute for Sustainable Technologies AEE INTE, which is at the current moment the leading European research center for hybrid heating systems, the number of industrial solar installations at the end of 2018 (the latest statistics) is estimated as 741 systems with a total collector area of more than 662,000 square meters. Moreover, if earlier the leaders in this market were technologically developed countries, in recent years several promising projects have been implemented in the territories of developing countries, from small demonstration plants to large-scale systems with a capacity of 100 MWth (Figure 3). In 2017, 124 systems were installed in the industrial sector with a total collector area of more than 190,000 square meters, and in 2018 another 108 industrial new solar systems with a total area of more than 54,000 square meters were commissioned (Table 2).

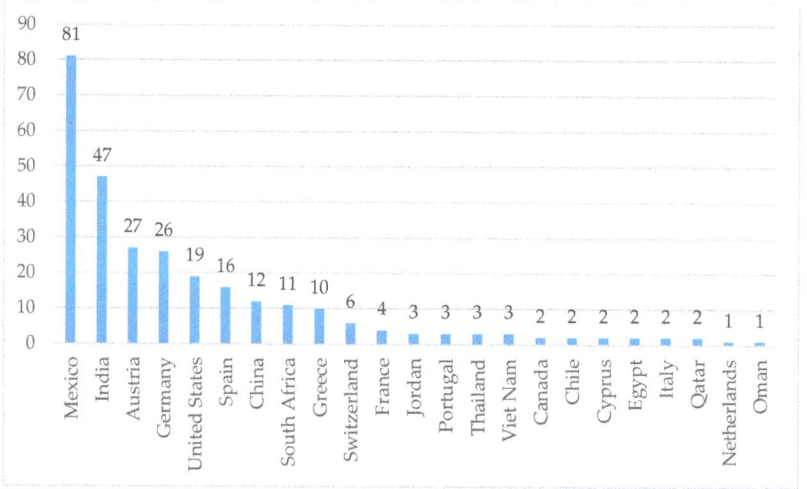

Figure 3. The number of industrial collectors with a capacity of at least 700 kW in different countries of the world in March 2019 [23].

Table 2. Statistics of installations of industrial solar collectors by country in 2017–2018, [15,23].

Country	2017			2018		
	Number of New Collectors	Gross Area, m²	The Average Area of Collector, m²	Number of New Collectors	Gross Area, m²	The Average Area of Collector, m²
Oman	1	148,000	148,000	-	-	-
Mexico	36	6411	178	51	6,898	135
India	36	15,313	425	10	3964	396
China	19	11,534	607	15	28,813	1921
Austria	2	1758	893	3	435	145
France	2	2052	1026	2	5,543	2772
Afghanistan	1	3260	3260	-	-	-
Jordan	1	1254	1254	-	-	-
Germany	-	-	-	9	1589	177
Spain	-	-	-	3	1218	406
Others	12	2971	114	15	5193	346
Total	124	192,580		108	53,654	

The information on the most significant industrial solar collectors is presented in Table 3.

Table 3. Largest industrial solar collectors in the world. Source: authored based on data from [15] and https://www.solarthermalworld.org.

Name of the System (Country)	Gross Area, 1000 m^2	Capacity, MWth	Type of Industrial Processes	Type of Collector
Miraah (Oman)	148	100	Heavy oil production	Parabolic through
Gaby Copper mine (Chile)	39.3	28	Copper mining	Flat plate
Qier Solar (China)	13	9	Dyeing fabrics	Flat plate
Prestage Foods Factory (USA)	7.8	5	Sanitation	Flat plate
Heli Lithium (China)	3.3	2.3	Lithium-ion Battery Manufacturing	Evacuated tube
Kabul Meat Factory (Afghanistan)	3.26	2.2	Meat processing	Parabolic through
Polyocean Algal Industry Group (China)	2.2	1.5	Seafood processing	Evacuated tube
Japan Tobacco International (Jourdan)	1.25	0.7	Cigarette production	Frenel

In February 2018, the world's largest 4-block solar power plant Miraah with a capacity of 100 MWth was commissioned. The heating system supplies steam (660 tons daily) to the Amal field in southern Oman. Steam is used in the production of viscous and heavy oil. The system consists of parabolic solar collectors placed in a greenhouse in order to protect against wind and sand. The greenhouse turned out to be a successful and economical solution, as it allows reduction of the cost of cleaning and washing the collectors, as well as makes them lighter and less resource-intensive [15]. The second-largest industrial solar thermal plant was installed in Chile in June 2013 near the copper ore mine. Its capacity is 27.5 MWth [23].

The largest 2.3 MWth solar power plant in China was commissioned in 2017 to supply steam to one of Heli Lithium Industry's plants (producing lithium-ion batteries for electric forklifts). Vacuum tube collectors with a total area of 3300 m^2 are installed at the power station. Another large power plant with a capacity of 1.5 MWth, also using tubular vacuum collectors (total area of 2200 m2), is located in Qingdao in Shandong province in eastern China and supplies heat to the seafood processing company Polyocean Algal Industry Group [15].

At the end of 2017, the first parabolic collector was installed to heat a meat processing plant in Afghanistan. The total area of the collector was 3260 m^2. In Jordan, a Fresnel collector with a capacity of 700 kWh (total area of 1254 m^2) for direct steam generation for the needs of the Japanese tobacco factory Japan Tobacco International has been installed [15].

Since the market for industrial solar collectors is still very young, a complete system of statistical accounting of its structure and dynamics has not yet been formed. The complete detailed information is currently collected in the framework of the project of the World Energy Agency "Integration of solar heat into industrial processes" (IEA SHC Task49 / IV SHIP), carried out jointly by experts from 16 countries during 2015–2018. During the interview, experts collected information on 308 industrial heat collectors out of the 741 known. Figure 4 shows the distribution of the recorded 308 objects by size [23]. The first group includes the two largest solar thermal power plants (more than 21 MWth), which are also described in detail in Table 3. To the second group belongs 33 heating plants with a capacity of 0.7 to 21 MWth (or a total area of 1000 to 2999 m^2), followed by a group of 57 heating systems with a capacity of 0.35 to 0.7 MW (area from 500 up to 1000 m^2). The most significant number, 139 heating systems with a capacity of less than 0.35 MW (or a total area of less than 500 m^2) represent the fourth group of industrial solar power plants; the fifth group includes small heating systems with an area of up to 100 m², which includes 77 systems.

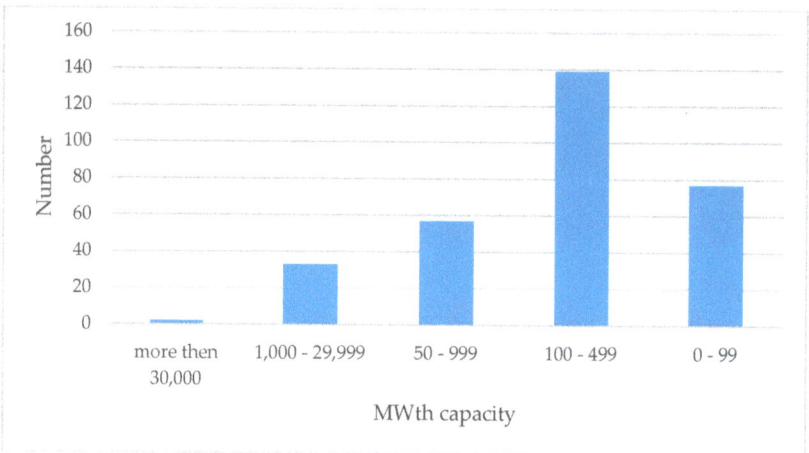

Figure 4. Distribution of industrial solar installations by size [23].

In the food industry 112 solar collectors are currently used, 31 in the beverage industry, and 24 in the textile industry (Figure 5). However, even though these industries are leading in the number of installations of solar collectors, the mining industry is the undisputed leader in the volume of their use (collector area and capacity) (Figure 6).

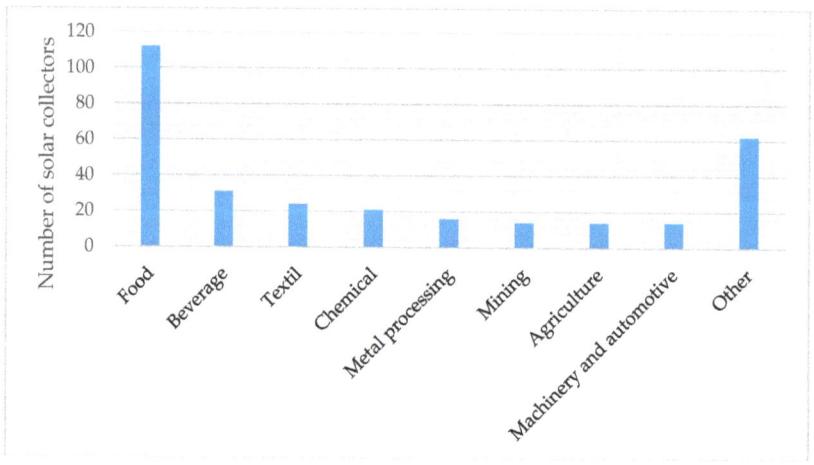

Figure 5. Distribution of industrial solar installations (numbers) by application [23].

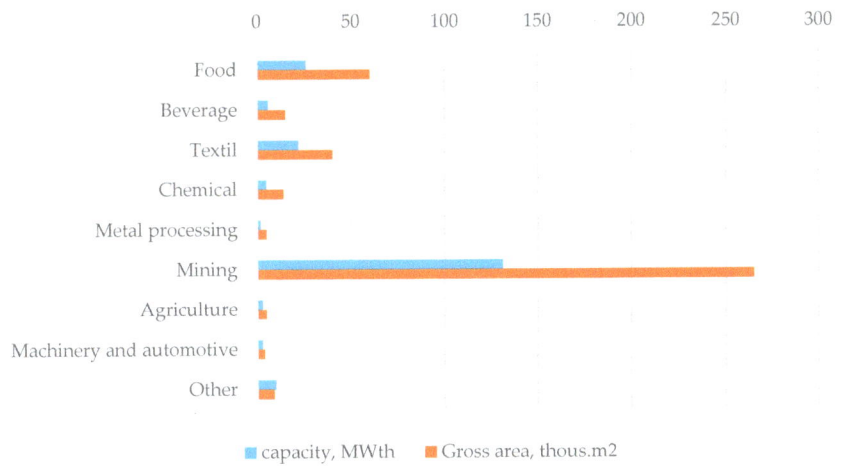

Figure 6. Distribution of industrial solar installations (capacity, gross area) by application [23].

Currently, the leaders in the cumulative area and capacity of installed industrial solar collectors are Oman, China, Chile, USA, Mexico, and India (Figure 7). Other countries are significantly inferior to them in the development of the SHIP market. With the exception of the United States, the leading countries in terms of the development of solar collectors in the industry are rapidly developing industrial countries in which the industrial sector mainly generates the growth in demand for thermal energy. For example, the growth of thermal energy consumption in the industrial sector of India in the period from 2010 to 2015 increased by more than 30% [24,25].

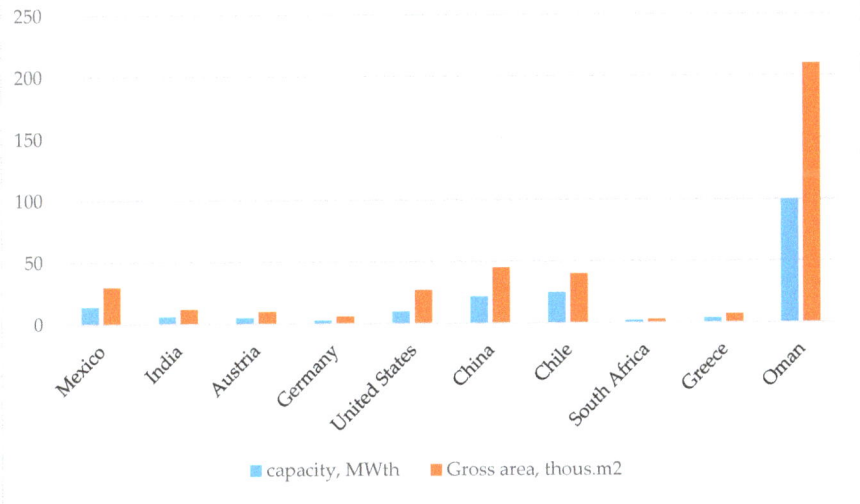

Figure 7. Cumulative area and cumulative power of installed solar collectors in the leading countries [23].

As for the types of commonly used solar collectors, they are flat plate collectors (139), followed by parabolic (58) and evacuated trough collectors (46). However, in terms of total area, parabolic collectors are superior to flat and vacuum (Figure 8).

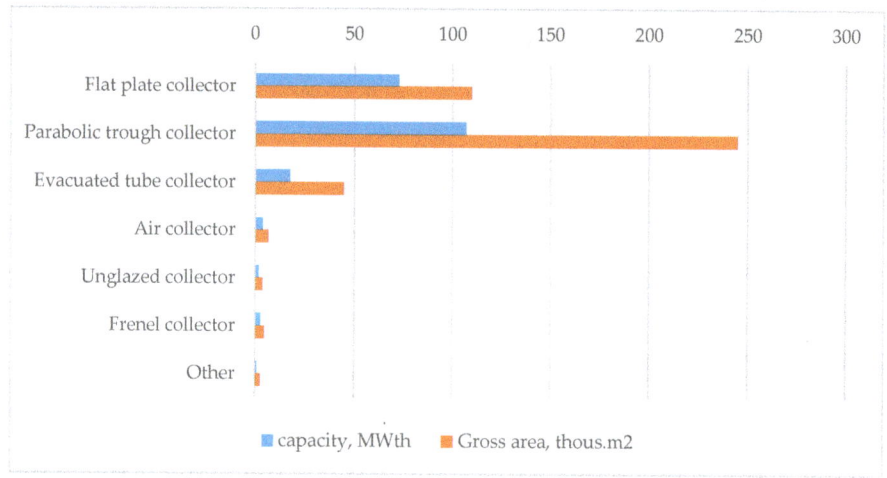

Figure 8. Distribution of industrial solar collectors by type [23].

The largest manufacturers of flat solar collectors are Chinese companies: Sunrain (annual turnover of more than 600 million US dollars), BTE Solar, Five Star, and others [26]. The list of leading companies among manufacturers of solar collectors in 2017 also includes the Austrian company Greenonetec and the German Bosch Thermotechnik (according to the data of analytical agency Sun&Wind Energy http://www.sunwindenergy.com/solar-thermal/2017-ranking-worlds-largest-flat-plate-collector-manufacturers. In addition to these countries, the production of solar collectors on a large scale is also established in Greece, Italy, Turkey, Australia, Mexico, Bulgaria, Poland, and India. In the production of solar collectors, relatively simple technologies are used that do not require special licensing and are relatively easy to copy; therefore many manufacturing companies co-finance projects with state support for the development of industrial solar collectors to expand their sales market [27,28].

In South Africa there is the system of incentives for demonstration projects on the use of high-power solar collectors in the food and textile industries, aimed at raising awareness about the possibilities of using this technology and developing the production of power equipment [5,29]. Industrial solar collectors have also received support in Tunisia as part of the Prosol industrial development program, launched in 2010 with financial support from the Italian Ministry of the Environment and the United Nations Environment Program. Benetton Textile Mill is a demonstration project, in which 1000 m^2 of flat plate solar collectors were installed on the roofs of production facilities in 2016. The success of the project allows for the replication of the technology in other sectors of industry to achieve the national goal of 14,000 m^2 solar collectors by the end of 2020 [5].

3.2. Status-Quo of Solar Heat in Russia

There are no official statistics on the solar heat supply in Russia; therefore, estimates of international organizations on solar collectors are based on an expert method based on a survey of leading Russian specialists in this field [11]. Dynamics of development of the solar heat supply in Russia in the period from 2013 to 2017 (last statistics) are presented in Figure 9. The predominance of flat plate solar water collectors (FPC water) in Russia is explained by the fact that the amount of heat generated by them per unit area in the winter is significantly higher than the amount of heat generated by vacuum collectors.

The fact that snow and frost on the surface of flat solar collectors can be removed faster than from vacuum collectors is also essential.

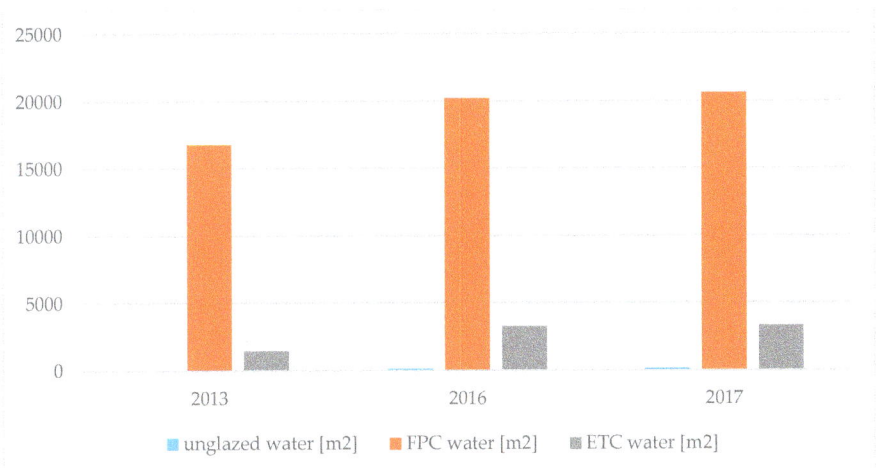

Figure 9. Gross area of solar collectors (for all applications) in Russia.

According to Austrian Institute for Sustainable Technologies estimates, at the end of 2017, the largest share of solar collectors was in the district heating sector (Figure 10). This is explained by lower unit costs for its construction and operation since equipment such as tanks, pumps, and chemical water treatment plants are already available in the regular boiler room. The same is true for well-qualified personnel. The share of solar collectors in the industry is only about 3.8% (about 900 m^2).

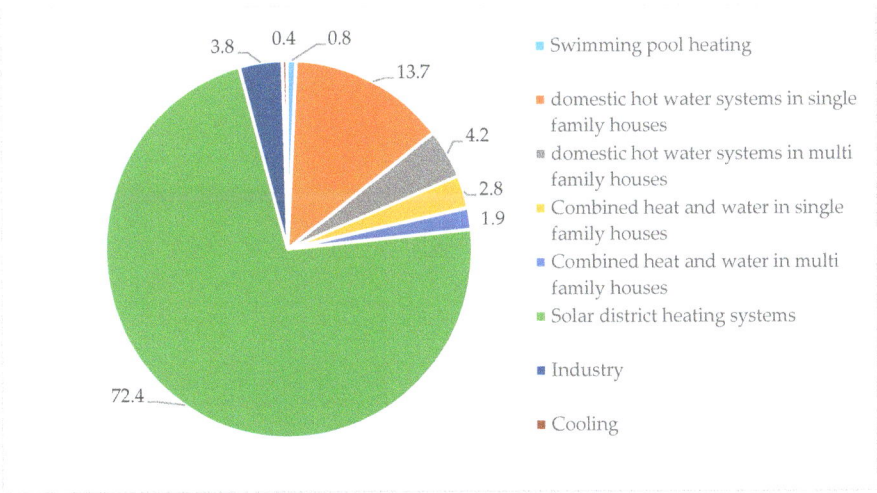

Figure 10. Distribution of solar collectors in Russia by application [11].

The largest number of solar collectors is operated in the southern regions of Russia: Krasnodar, Stavropol, Astrakhan, and Volgograd regions, as well as in the regions of the Far Eastern Federal District with a fairly high level of solar radiation and a high percentage of decentralized heat supply. They are the Republic of Buryatia, Khabarovsk, and Primorsky regions. In the Krasnodar Territory, as a region with a highly developed tourist infrastructure, more than half of all solar collectors are installed for the hot water supply of hotels and other tourist facilities. In Buryatia, most solar collectors are involved in industry. In the boiler house in the city of Narimanov (Astrakhan Region) the largest solar collector in Russia, with an area of 4400 m² and manufactured by Buderus (Germany), is used for hot water supply.

A significant obstacle to the wider distribution of solar collectors in Russia is the low awareness of the population about the capabilities of this technology, as well as the lack of service companies that provide turnkey solar collector installation services [30].

4. Results

4.1. Assessment of the Economic Efficiency of Industrial Solar Collectors in Russia

The economic efficiency of industrial solar collectors, as well as other solar energy conversion devices, primarily depends on the level of solar insolation (irradiation), which determines their performance (specific solar yield, kWh/m²-a). Other significant factors are the initial cost of the collector itself, the costs of its installation and maintenance, and the life cycle of the solar collector. All these factors, as well as the discount coefficient, are taken into account in the indicator levelized cost of energy, which is the most common in a comparative assessment of the economic efficiency of various energy technologies. In other words, it reflects the average cost of a unit of energy produced using the specific generating device for the entire period of operation of the equipment [21,22]. Despite some criticism [31], this metric is currently the most widely used in the literature [22,32,33].

According to IEA SHC Task49/IV SHIP experts, the LCOE value for large-scale solar collectors used for the hot water supply of residential houses ranges from 2 euro cents (in 2016 prices) per kWh of thermal energy in India to 14 euro cents in Austria, Denmark, Canada, and France [15]. This significant difference is determined mainly by the cost of labor in these countries. Assuming that the cost of the installed industrial solar collector, similar in quality and performance, is approximately equal to the cost of the collector used for hot water supply (from 200 to 1160 euros per m²), and the temperature of water heating in them is comparable to the temperature necessary to provide heat for low-temperature industrial processes, we can estimate expected the LCOE for industrial collectors in Russia. To this end, we will construct a model of paired linear regression based on the performance of large-scale solar collectors used to heat water at a certain level of solar radiation, given in the source [15] for 62 capitals of the world. We consider the level of solar insolation (X) as a factor, and as the dependent variable (Y) we take the performance of the average large-scale solar collector with a horizontal panel for hot water supply. Using the least-squares method implemented in STATISTICA 10.1, we obtain the model, presented in Table 4.

Table 4. Dependence of performance (specific solar yield, kWh/m²-a) of the average large-scale flat plate solar collector on horizontal irradiation.

Parameters of Linear Regression Model	Value
Solar insolation, kWh/m²-a	0.389 ***
const	41.719 **
R²	0.97
P(F-stat)	0.0001

***—1% significance level, **—5% significance level.

Given the high statistical quality of the constructed model, it can be used to predict the expected performance of the solar collector anywhere in the world. So, for example, for the level of solar insolation in Sochi (average annual horizontal irradiation of 1365.1 kWh/m²), we obtain the expected productivity of the solar collector 573.24 (kWh/m²). Substituting the predicted estimates of collector productivity in the model (Table 4, Figure 11), and using the lower estimates for the volume of initial specific investments in the purchase and installation of equipment from the source [15], we obtain LCOE estimates for the cost of energy produced using industrial solar collectors in the region of Sochi, in the range of 0.052–0.09 euros per 1 kWh or 3.8–6.6 rubles/kWh (when converting euros to rubles at the rate of 1 euro = 73 rubles).

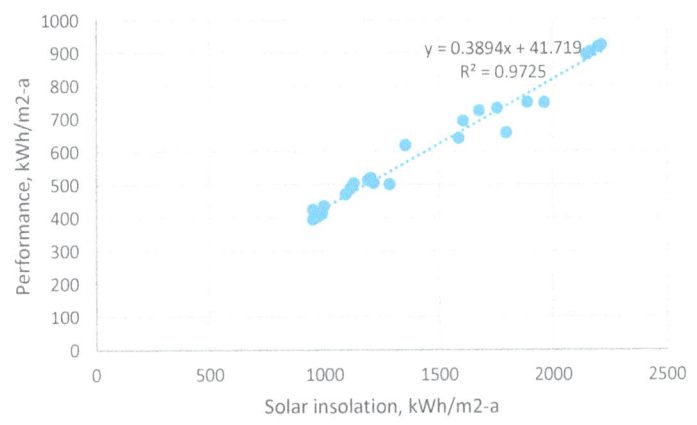

Figure 11. Linear regression model.

The use of lower estimates of specific investments is because the cost of labor in Russia is much lower than in developed European countries. If, as estimates of specific investments we take the values of these indicators as equal to France, Canada, or Denmark (as the highest), then the LCOE will range from 5.4 to 8.03 rubles/kWh.

Both the upper and lower estimates of the cost of thermal energy produced by solar collectors are higher than when using traditional hydrocarbon technologies for thermal energy production in the chosen region [34]. In the district heating zone, the tariff for thermal energy at the beginning of 2020 was only 1.5–2 rubles/kWh. However, with the rise in price of hydrocarbon sources and the introduction of taxes on greenhouse gas emissions (which is currently being discussed in the world expert community as a necessary measure to achieve the goals of the Paris Climate Agreement) [35–38], the commercial attractiveness of new technologies can significantly increase in those regions of Russia, where the average annual level of solar insolation is relatively high, and the need to provide energy for low-temperature industrial processes is quite large. Also, the economic feasibility of using solar collectors increases significantly if it is impossible to connect to centralized heating grids, as well as in the case of the seasonal load of industrial facilities (for example, in enterprises for the production of canned vegetables, sugar beet processing, etc.) One of these regions, where the development potential of industrial solar collector technology is large enough, is the Krasnodar Region.

4.2. Assessment of Energy-Saving Potential in the Industry of Southern Russia (for Example, the Krasnodar Region)

The southern regions of Russia not only have suitable natural and climatic conditions for the efficient use of solar collectors in various types of economic activities but are also regions with developed industrial sectors in which low-temperature production processes predominate. Therefore, for example, in the structure of thermal energy consumption in the industry of the Krasnodar Territory during 2010–2015 (Figure 12) the production of food, including drinks and tobacco, steadily occupies the fifth position, following only the chemical and metallurgical industries, the production of electricity, steam, and water, as well as the production of petroleum products [18].

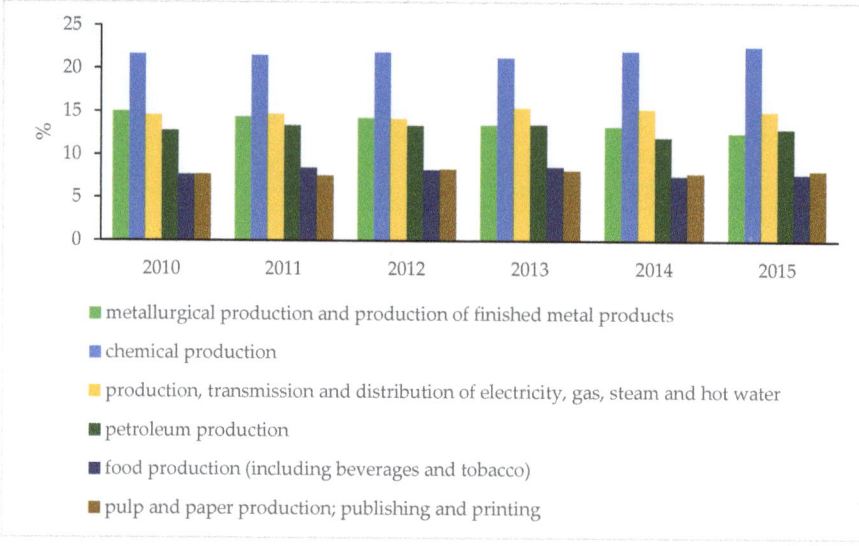

Figure 12. Industries in the Krasnodar Region leading in the consumption of thermal energy. Source: Statistical Digest Industrial "Production in Russia—2016", Part 5.4 "Consumption of certain types of fuel and energy resources by type of economic activity" https://gks.ru/bgd/regl/b16_48/Main.htm [18].

Note that at least two more of the above leading industries are also potential applications of solar collectors (chemical and petrochemical production).

However, even if we evaluate the energy-saving potential when introducing solar collectors only for food industry enterprises, we get an approximate estimate of 2,612,000 Gcal per year (detailed calculations are given in Table 5).

Table 5. An assessment of thermal energy consumption by food industry enterprises of the Krasnodar Region.

Goods	Specific Consumption of Thermal Energy, Thousand kcal per ton/Thousand per Deciliter) [1]	Production Volume, Thousand Tones/Thousand Deciliter [13]	Total Thermal Energy Consumption, Thousands kcal
Sugar form sugar beet	1376.3	1505.2	2,071,606,760
Bread and Bakery	285.7	303.4	86,681,380
Milk	250	309.5	77,375,000
Cheese and cheese Products	752	65	48,880,000
Vegetable oil	348	512	178,176,000
Butter	1365.6	10.1	13,792,560
Alcohol	1539.4	17,870.5	27,509,847.7
Beer	2156.9	20,200	43,569,380
Meat	155.7	76.6	11,926,620
Meat products	351.9	65.2	22,943,880
Compound feed	30.3	975.7	29,563,710
Total			2,612,025,138

[1] According to the Federal State Statistics Service http://www.gks.ru/wps/wcm/connect/rosstat_main/rosstat/ru/statistics/economydevelopment/#.

At the current moment, the production of thermal energy in the region is ensured by the operation of four thermal power plants and 2848 boiler houses. In 2017 5,316,500 Gcal of thermal energy were produced by power plants and 10,394,200 Gcal were produced by boiler houses, from which 76% use natural gas as a primary fuel [39]. The average capacity factor of boiler houses in the region is 68%. The average level of boiler house wear is 70%, and in some municipalities it reaches more than 80%. The vast majority of boiler houses (boiler room equipment) and heating grids were built more than 20–25 years ago, which means not only a low level of their technical condition but also high backwardness of existing capacities from modern analogues designed for the production and transmission of thermal energy [40,41]. The technical backwardness of the regional heat supply system has a significant impact on the energy intensity of the economy. The energy intensity of the Krasnodar Territory is 1.5 times higher than the global average and 2-2.5 times higher than in developed countries. Maintaining high energy intensity indicators compared with the world average is a limiting factor for the economic growth [41].

Considering that the total volume of thermal energy produced in the region in recent years is about average 15,711,700 Gcal, the introduction industrial solar collectors only in food industry enterprises can bring energy saving up to 16–17% of the total amount of thermal energy produced in the region annually and have a positive impact on improving the energy efficiency of the region's economy.

Another positive effect for the economy of the region may be the creation of new jobs in the industry of manufacturing and maintenance of solar collectors. Given the estimates of the average productivity of solar collectors in the region obtained in the previous paragraph (573.24 kWh/m^2), it is possible to calculate the area of solar collectors needed to convert the industrial processes listed in Table 5 to solar energy. According to our calculations, the required number of solar collectors is estimated at 5,300,000 m^2. The study [23] claims that in countries with low labor costs and advanced automated production of solar collectors, on average systems with a total of 87 m^2 solar collector area have to be installed per full-time job. With this assumption, we can estimate that the potential for creating new jobs in production, installation, and maintenance of solar thermal systems is around 61,000.

5. Discussion and Policy Applications

As our analysis demonstrated, the market for industrial solar collectors is a rapidly growing segment of the global market for renewable energy technologies, which is currently developing with minimal government support measures, mainly due to market mechanisms. This indicates the commercial attractiveness of the technology, especially in countries with a high level of solar insolation and well-developed industries characterized by seasonality and a significant proportion of low-temperature processes.

For the southern regions of Russia, the use of industrial solar collectors can be considered a commercially viable alternative to the construction of new boiler houses and heating grids in the case of the creation of new enterprises for processing agricultural products in areas where there are no central heating systems and poorly developed heating grids. Also, the installation of solar collectors can be considered an investment-attractive option for the modernization of worn-out and outdated equipment of traditional boiler houses of industrial enterprises, which allows partially replacing hydrocarbon sources. In both cases, the development of industrial solar collectors will help to reduce the energy and carbon intensity of the Russian economy [42,43] and increase the share of renewable energy in the country's energy balance. This will help the country to fulfill its obligations under the Paris Climate Agreement [35,44].

It is also crucial that the use of solar collectors instead of hydrocarbon alternatives for the heat supply of industrial processes has a positive impact on the environment not only at the stage of direct operation of the solar collector but throughout the entire life cycle, including the stages of extraction and processing of raw materials for production of solar collectors, the stage of their production, the stages of transportation and installation, as well as disposal after use [45]. Thus, the use of solar collectors not only helps to approach the achievement of the goals of decarbonization of the economy, but also is fully integrated into the concept of transition to a circular economy.

Nevertheless, despite the relatively high level of commercial attractiveness and environmental efficiency, the development of solar collectors in Russia, in our opinion, needs certain measures of state support. Given the fact that the general concept of government incentives for development of renewable energy in Russia is aimed primarily at developing national production of equipment for renewable energy sources [46], it is advisable to direct measures of state support for the development of solar collectors not only to create effective demand (for example, through the system of purchase and installation subsidies for the solar collector), but, first of all, aiming to support existing Russian manufacturers of solar collectors with their technologies and competencies in this field. The creation and expansion of effective demand for solar collectors through customer subsidies can lead to the occupation of the domestic market with products of world leaders that can compete in price due to large-scale production [47]. Based on international practices, several types of government incentives can be proposed: (i) the government co-financing of demonstration projects of Russian manufacturers; (ii) the introduction of property tax benefits for enterprises using solar collectors in industrial processes; and (iii) the introduction of accelerated depreciation on solar collectors for industrial enterprises [48,49]. Also, the Russian practice of stimulating innovative production is rich in positive examples of the development of new enterprises through obtaining the status of a resident of the Skolkovo innovation cluster.

An analysis of the official websites of leading Russian manufacturers of solar collectors (JSC VPK NPO Mashinostroeniya (Moscow), LLC Novy Pole (Moscow), LLC Altenergiya (Krasnodar Region), ANDI Group (Moscow), GreenSun Technologies (Vladivostok)) shows that all of them (except the Novy Pole company, which has created its own brand and a whole line of products) need serious adjustments to their marketing strategy, development of after-sales services, and diversification of product sales channels. Given this situation, the co-financing of demonstration projects can be proposed as the most likely effective form of state support.

6. Conclusions

In our study, we investigated the prospects for the development of solar collectors in the industry of the southern regions of Russia. As a basis for calculating the demand for industrial solar collectors, we considered the structure of heat supplies and the industrial production of a specific territory. It was shown using the example of the Krasnodar Region that the potential for energy savings in the region's industry due to the introduction of solar collectors is at least 16–17% of the total volume of thermal energy produced. Converting such a volume of thermal generation to solar energy will require the installation of 5,300,000 m^2 of solar collectors, which will create more than 60,000 new jobs in the region.

The economic efficiency of solar collectors is still insufficient to compete with conventional boiler houses operating on cheap hydrocarbon fuels (expected average LCOE 3.8–6.6 rubles/kWh comparing current tariffs 1.5–2 rubles/kWh in a district heating area). However, the installation of solar collectors may well be considered as an investment-attractive option for the modernization of worn-out and outdated equipment of traditional boiler houses of industrial enterprises, which allows partially replacing hydrocarbon sources and lowering the carbon intensity of the Russian economy. As a measure of state incentives for the development of industrial solar collectors in Russia, we offer state co-financing of demonstration projects of Russian manufacturers. This will increase the level of awareness of the population and businesses about the capabilities of this technology, as well as increase the technical competencies and innovation potential of companies involved in the production and installation of solar collectors.

The results of our study can be useful in developing and improving federal and regional programs for the development of renewable energy in Russia and improving the energy efficiency of the Russian economy, as well as the industrial policy of the Russian Federation. They allow policymakers to more clearly classify the industrial enterprises that are most suitable for the introduction of solar collectors and to introduce more effective incentives for the development of this type of renewable energy. Also, these results can be used to calculate the required size of subsidies (or carbon taxes) which allow balancing the economic efficiency of solar collectors with the current hydrocarbon thermal generation technologies in the region.

The results of our study may be applicable outside of Russia. The proposed "demand-side" approach and algorithm for developing a regional strategy for increasing energy efficiency and the reduction of carbon intensity in the industrial sector can be used in other countries and regions.

Author Contributions: Conceptualization and methodology, S.R. (Svetlana Ratner); software, S.R. (Svetlana Ratner); validation, S.R. (Svetlana Ratner), K.G. and S.R. (Svetlana Revinova); formal analysis, I.L.; investigation, S.R. (Svetlana Ratner); resources, K.G. data curation, S.R. (Svetlana Revinova) writing—original draft preparation, S.R. (Svetlana Ratner) writing—review and editing, K.G. visualization, K.G. supervision, S.R. (Svetlana Ratner); project administration, S.R. (Svetlana Ratner); funding acquisition, I.L. All authors have read and agreed to the published version of the manuscript.

Funding: The publication has been prepared with the support of the «RUDN University Program 5-100».

Conflicts of Interest: The authors declare no conflict of interest.

Abbreviations

RES	renewable energy sources
LCOE	levelized cost of energy
FPC water	flat plate solar water collectors
ETC water	evacuated tube collector
SHIP	solar heat for industrial processes
MWth	megawatts thermal

References

1. Olabi, A. Circular economy and renewable energy. *Energy* **2019**, *181*, 450–454. [CrossRef]
2. Donia, E.; Mineo, A.; Sgroi, F. A methodological approach for assessing business investments in renewable resources from a circular economy perspective. *Land Use Policy* **2018**, *76*, 823–827. [CrossRef]
3. Renewable Energy Statistics 2019. Available online: https://www.irena.org/publications/2019/Jul/Renewable-energy-statistics-2019 (accessed on 13 December 2019).
4. REN21. Renewables 2019 Global Status Report. Paris: REN21 Secretariat. Available online: https://www.ren21.net/wp-content/uploads/2019/05/gsr_2019_full_report_en.pdf (accessed on 13 December 2019).
5. Renewable Energy Policies in a Time of Transition. IRENA, OCED/IEA and REN21. 2018. 112 P. Available online: https://www.irena.org/-/media/Files/IRENA/Agency/Publication/2018/Apr/IRENA_IEA_REN21_Policies_2018.pdf (accessed on 13 December 2019).
6. Nicolli, F.; Vona, F. Energy market liberalization and renewable energy policies in OECD countries. *Energy Policy* **2019**, *128*, 853–867. [CrossRef]
7. Erdiwansyah Mamat, R.; Sani, M.; Sudhakar, K. Renewable energy in Southeast Asia: Policies and recommendations. *Sci. Total Env.* **2019**, *670*, 1095–1102. [CrossRef] [PubMed]
8. Ratner, S.; Nizhegorodtsev, R. Analysis of renewable energy projects' implementation in Russia. *Therm. Eng.* **2017**, *64*, 429–436. [CrossRef]
9. Energy Efficiency Market Report 2017. OECD/IEA, Paris. 2017. 142 P. Available online: https://www.nrcan.gc.ca/sites/www.nrcan.gc.ca/files/energy/energy-resources/Energy_Efficiency_Marketing_Report_2017.pdf (accessed on 13 December 2019).
10. Schmitt, B. Classification of Industrial Heat Consumers for Integration of Solar Heat. *Energy Procedia* **2016**, *91*, 650–660. [CrossRef]
11. Butuzov, V. Solar Heat Supply: World Statistics and Peculiarities of the Russian Experience. *Therm. Eng.* **2018**, *65*, 741–750. [CrossRef]
12. Kumar, L.; Hasanuzzaman, M.; Rahim, N. Global advancement of solar thermal energy technologies for industrial process heat and its future prospects: A review. *Energy Convers. Manag.* **2019**, *195*, 885–908. [CrossRef]
13. Schramm, S.; Adam, M. Storage in Solar Process Heat Applications. *Energy Procedia* **2014**, *48*, 1202–1209. [CrossRef]
14. Farjana, S.; Huda, N.; Mahmud, M.; Saidur, R. Solar process heat in industrial systems – A global review. *Renew. Sustain. Energy Rev.* **2018**, *82*, 2270–2286. [CrossRef]

15. Weiss, W.; Spörk-Dür, M. Solar Heat Worldwide. Global Market Development and Trends in 2017. Available online: https://www.iea-shc.org/Data/Sites/1/publications/Solar-Heat-Worldwide-2018.pdf (accessed on 13 December 2019).
16. Chang, K.; Lin, W.; Leu, T.; Chung, K. Perspectives for solar thermal applications in Taiwan. *Energy Policy* **2016**, *94*, 25–28. [CrossRef]
17. Suresh, N.; Rao, B. Solar energy for process heating: A case study of select Indian industries. *J. Clean. Prod.* **2017**, *151*, 439–451. [CrossRef]
18. Statistical Digest Industrial "Production in Russia – 2016", Part 5.4 "Consumption of certain types of fuel and energy resources by type of economic activity. Available online: https://gks.ru/bgd/regl/b16_48/Main.htm (accessed on 14 December 2019).
19. Frid, S.; Kolomiets, Y.; Sushnikova, E.; Yamuder, V. Effectiveness and prospects of using different solar water heating systems under the climatic conditions of the Russian Federation. *Therm. Eng.* **2011**, *58*, 910–916. [CrossRef]
20. Gridasov, M.; Kiseleva, S.; Nefedova, L.; Popel', O.; Frid, S. Development of the geoinformation system Renewable sources of Russia: Statement of the problem and choice of solution methods. *Therm. Eng.* **2011**, *58*, 924–931. [CrossRef]
21. Ueckerdt, F.; Hirth, L.; Luderer, G.; Edenhofer, O. System LCOE: What are the costs of variable renewables? *Energy* **2013**, *63*, 61–75. [CrossRef]
22. Reichenberg, L.; Hedenus, F.; Odenberger, M.; Johnsson, F. The marginal system LCOE of variable renewables–Evaluating high penetration levels of wind and solar in Europe. *Energy* **2018**, *152*, 914–924. [CrossRef]
23. Weiss, W.; Spörk-Dür, M. *Solar Heat Worldwide. Global Market Development and Trends in 2018*; AEE-Institute for Sustainable Technologies: Gleisdorf, Austria, 2019.
24. Wang, H.; Chen, W. Modelling deep decarbonization of industrial energy consumption under 2-degree target: Comparing China, India and Western Europe. *Appl. Energy* **2019**, *238*, 1563–1572. [CrossRef]
25. Sahoo, N.; Mohapatra, P.; Sahoo, B.; Mahanty, B. Rationality of energy efficiency improvement targets under the PAT scheme in India–A case of thermal power plants. *Energy Econ.* **2017**, *66*, 279–289. [CrossRef]
26. Huang, J.; Tian, Z.; Fan, J. A comprehensive analysis on development and transition of the solar thermal market in China with more than 70% market share worldwide. *Energy* **2019**, *174*, 611–624. [CrossRef]
27. Imtiaz Hussain, M.; Ménézo, C.; Kim, J. Advances in solar thermal harvesting technology based on surface solar absorption collectors: A review. *Sol. Energy Mater. Sol. Cells* **2018**, *187*, 123–139. [CrossRef]
28. Shafieian, A.; Khiadani, M.; Nosrati, A. A review of latest developments, progress, and applications of heat pipe solar collectors. *Renew. Sustain. Energy Rev.* **2018**, *95*, 273–304. [CrossRef]
29. Oosthuizen, D.; Goosen, N.; Hess, S. Solar thermal process heat in fishmeal production: Prospects for two South African fishmeal factories. *J. Clean. Prod.* **2020**, *253*, 119818. [CrossRef]
30. Svetlana, R.; Yuri, C.; Larisa, D.; Anna, P. Management of Energy Enterprises: Energy-efficiency Approach in Solar Collectors Industry: The Case of Russia. *Int. J. Energy Econ. Policy* **2018**, *8*, 237–243.
31. Loewen, J. LCOE is an undiscounted metric that distorts comparative analyses of energy costs. *Electr. J.* **2019**, *32*, 40–42. [CrossRef]
32. Fan, J.; Wei, S.; Yang, L.; Wang, H.; Zhong, P.; Zhang, X. Comparison of the LCOE between coal-fired power plants with CCS and main low-carbon generation technologies: Evidence from China. *Energy* **2019**, *176*, 143–155. [CrossRef]
33. Bruck, M.; Sandborn, P.; Goudarzi, N. A Levelized Cost of Energy (LCOE) model for wind farms that include Power Purchase Agreements (PPAs). *Renew. Energy* **2018**, *122*, 131–139. [CrossRef]
34. Administration and City Council of Krasnodar. Tariffs for Housing and Communal Services. Available online: https://krd.ru/upravlenie-tsen-i-tarifov/elektronnyy-sbornik-1-07-2013/analytic_info/informatsiya-ob-urovne-tsen-i-tarifov-na-tovary-i-uslugi-v-gorode-krasnodare-za-2019-god/informatsiya-ob-urovnyakh-tsen-i-tarifov-na-tovary-i-uslugi-po-munitsipalnomu-obrazovaniyu-gorod-krasnodar-za-1-kvartal-2019-goda/tarify-na-uslugi-zhilischno-kommunalnogo-khozyaystva/ (accessed on 14 December 2019).
35. Klimenko, V.; Klimenko, A.; Mikushina, O.; Tereshin, A. To avoid global warming by 2 °C—mission impossible. *Therm. Eng.* **2016**, *63*, 605–610. [CrossRef]
36. Orlov, A.; Grethe, H.; McDonald, S. Carbon taxation in Russia: Prospects for a double dividend and improved energy efficiency. *Energy Econ.* **2013**, *37*, 128–140. [CrossRef]

37. Liobikienė, G.; Butkus, M. The European Union possibilities to achieve targets of Europe 2020 and Paris agreement climate policy. *Renew. Energy* **2017**, *106*, 298–309. [CrossRef]
38. Saheb, Y.; Shnapp, S.; Johnson, C. The Zero Energy concept: Making the whole greater than the sum of the parts to meet the Paris Climate Agreement's objectives. *Curr. Opin. Environ. Sustain.* **2018**, *30*, 138–150. [CrossRef]
39. Ratner, S.; Iosifov, V.; Ratner, M. Optimization of the regional energy system with high potential of use of bio-waste and bioresources as energy sources with respect to ecological and economic parameters: The Krasnodar Krai case. *Reg. Econ. Theory Pract.* **2018**, *16*, 2383–2398. [CrossRef]
40. Ratner, S.V.; Ratner, P.D. Regional Energy Efficiency Programs in Russia: The Factors of Success. *Region* **2016**, *3*, 68–85. [CrossRef]
41. Ministry of Fuel and Energy Complex and Housing and Communal Services of the Krasnodar Territory. State Program of the Krasnodar Territory Development of the Fuel and Energy Complex. Available online: http://www.gkh-kuban.ru/kcpvcp19.html (accessed on 14 December 2019).
42. Matraeva, L.; Solodukha, P.; Erokhin, S.; Babenko, M. Improvement of Russian energy efficiency strategy within the framework of green economy concept (based on the analysis of experience of foreign countries). *Energy Policy* **2019**, *125*, 478–486. [CrossRef]
43. Matraeva, L.V.; Goryunova, N.A.; Smirnova, S.N.; Babenko, M.I.; Erokhin, S.G.; Solodukha, P.V. Methodological approaches to the assessment of energy efficiency within the framework of the concept of green economy and sustainable development. *Int. J. Energy Econ. Policy* **2017**, *7*, 231–239.
44. Berezin, A.; Ratner, S. Policy transition to low-carbon economy in Russia: State support measures. In Proceedings of the 13th International Days Of Statistics And Economics Conference Proceedings, Prague, Czech Republic, 5–7 September 2019; Libuše Macáková, MELANDRIUM, 2019 Fügnerova 691 274 01 Slaný IČO: 48709395. ISBN 978-80-87990-18-6.
45. Kylili, A.; Fokaides, P.; Ioannides, A.; Kalogirou, S. Environmental assessment of solar thermal systems for the industrial sector. *J. Clean. Prod.* **2018**, *176*, 99–109. [CrossRef]
46. Boute, A.; Zhikharev, A. Vested interests as driver of the clean energy transition: Evidence from Russia's solar energy policy. *Energy Policy* **2019**, *133*, 110910. [CrossRef]
47. Ratner, S.V.; Klochkov, V.V. Scenario Forecast for Wind Turbine Manufacturing in Russia. *Int. J. Energy Econ. Policy* **2017**, *7*, 144–151.
48. Gawel, E.; Lehmann, P.; Purkus, A.; Söderholm, P.; Witte, K. Rationales for technology-specific RES support and their relevance for German policy. *Energy Policy* **2017**, *102*, 16–26. [CrossRef]
49. Huntington, S.; Rodilla, P.; Herrero, I.; Batlle, C. Revisiting support policies for RES-E adulthood: Towards market compatible schemes. *Energy Policy* **2017**, *104*, 474–483. [CrossRef]

© 2020 by the authors. Licensee MDPI, Basel, Switzerland. This article is an open access article distributed under the terms and conditions of the Creative Commons Attribution (CC BY) license (http://creativecommons.org/licenses/by/4.0/).

Article

Granger Causality Network Methods for Analyzing Cross-Border Electricity Trading between Greece, Italy, and Bulgaria

George P. Papaioannou [1,2,†], Christos Dikaiakos [1,3,*,†], Christos Kaskouras [1,†], George Evangelidis [4,†] and Fotios Georgakis [5,†]

1. Research, Technology & Development Department, Independent Power Transmission Operator (IPTO) S.A., 104 43 Athens, Greece; g.papaioannou@admie.gr (G.P.P.); c.kaskouras@admie.gr (C.K.)
2. Center for Research and Applications in Nonlinear Systems (CRANS), Department of Mathematics, University of Patras, 26 500 Patras, Greece
3. Department of Electrical and Computer Engineering, University of Patras, 26 500 Patras, Greece
4. Department of Transmission System Operations & Control, Independent Power Transmission Operator (IPTO) S.A., 104 43 Athens, Greece; gevangelidis@admie.gr
5. Market Management Department, Independent Power Transmission Operator (IPTO) S.A., 104 43 Athens, Greece; fgeorgakis@admie.gr
* Correspondence: c.dikeakos@admie.gr; Tel.: +30-210-9466873; Fax: +30-210-5192263
† These authors contributed equally to this work.

Received: 15 January 2020; Accepted: 11 February 2020; Published: 18 February 2020

Abstract: Italy, Greece, and, to a lesser degree, Bulgaria have experienced fast growth in their renewable generation capacity (RESc) over the last several years. The consequences of this fact include a decrease in spot wholesale prices in electricity markets and a significant effect on cross border trading (CBT) among neighboring interconnected countries. In this work, we empirically analyzed historical data on fundamental market variables (i.e., spot prices, load, RES generation) as well as CBT data (imports, exports, commercial schedules, net transfer capacities, etc.) on the Greek, Italian, and Bulgarian electricity markets by applying the Granger causality connectivity analysis (GCCA) approach. The aim of this analysis was to detect all possible interactions among the abovementioned variables, focusing in particular on the effects of growing shares of RES generation on the commercial electricity trading among the abovementioned countries for the period 2015–2018. The key findings of this paper are summarized as the following: The RES generation in Italy, for the period examined, drives the spot prices in Greece via commercial schedules. In addition, on average, spot price fluctuations do not affect the commercial schedules of energy trading between Greece and Bulgaria.

Keywords: cross border trading; Granger causality; electricity trading; spot prices

1. Introduction

Over the several past years, European electricity markets have gone through a process of significant developmental changes which contributed to the further liberalization of the energy sector. In addition to this, European countries are committed to reach specific targets for 2030 and 2050 regarding the percentage of renewable energy sources (RES) participation in the energy generation mix, alongside the targets set by the European Green Deal [1]. Variable energy sources (VRES), especially solar and wind, cannot produce constant power since they are highly correlated with weather conditions.

European electricity markets are becoming increasingly integrated under the target of a single European electricity market. Greece has not yet market-coupled with any of the interconnected countries, but it is expected to do so under the "European Target Model" (ETM) mechanism by June 2020. The traded energy volume appears to have a severe degree of seasonality with human activities slowing down towards summer and then increasing again, based on European electricity market reports [2]. In general, energy trading follows demand. The implemented policies led the traded energy volume within European countries to reach the 33% of the total consumption in the first quarter of 2019 which indicates a 4% increase compared to the corresponding period of the previous year [3]. Nevertheless, in the second quarter of 2019, a reduction of 14% compared to the same period of the previous year in trading volume appeared [4]. The aforementioned policy of increasing the number of market-coupled countries within Europe and hence enforcing the pan-European energy market was initially designed in order to achieve energy price convergence and thus ensuring the highest level of safety and security of supply [5].

On the other hand, European countries following the 2030 climate and energy framework have pledged to reduce the use of fossil fuels and increase RES to improve their share in the energy generation mix [6]. This has led to a remarkable increase of RES, particularly solar and wind power. The European Union has committed to obtaining 20% of the consumed energy until 2020 from RES with this percentage increasing to the 32% in 2030 [7]. These targets will be revaluated for an upward revision in 2023 [6]. By 2017, the respective percentage was 17.5% with some countries having already achieved their individual targets [8].

The increased share of RES has a direct effect on spot prices since, as investigated, it is highly correlated with solar and wind intermittent availability. In general, deployment of renewable (mainly wind and solar) generation capacity affects the dynamics of electricity spot prices in a negative way, while the opposite happens with its volatility [9–11]. This should be anticipated since RES are more efficient than conventional generators in terms of marginal cost, but on the other hand, they cannot guarantee a constant and secure supply, since they are highly dependent on exogenous parameters.

This combination of uncertainty and lower spot prices discourage stakeholders from investing in new capacities, either from renewables or conventional sources. However, market designs can, to some extent, smoothen the volatility of electricity prices and enhance investments [12].

In this paper, we studied the abovementioned issues by considering the interconnection of Greece with Bulgaria and Italy and the effect of one country's intermittent generation to the other's market clearing price and, more importantly, on their cross border trading (CBT). In addition to this, we tried to leverage the available data and detect any hidden (causal as we will mention in the results section) connections among CBT fundamentals. All the examined countries invest in VRES deployment while taking advantage of the interconnection. The level of the system marginal price (SMP) in Greece, which is consistently one of the highest in Europe, discourages new investments in energy capacity. However, the interest is high, since the Greek energy market has room for improvement, especially under the implementation of the ETM. Therefore, we believe that our work will highlight the effect of RES growth in these countries and how this contributes to the changes in Greek SMP and commercial schedules.

Currently, Greece has active electricity interconnections with Turkey, Albania, North Macedonia, Bulgaria, and Italy with the latter being the most active regarding total exchanges. In 2017, 22.72% of energy imports in Greece were from Italy, while the respective percentage of energy exports was 26.82% of the total. The total imports in Greece for 2018 were 11,223.913 MWh, and exports were 4983.061 MWh [11].

However, it should be mentioned that the above countries are connected to each other via physical interconnection and not by market coupling, hence comparison with other studies that investigate countries that are connected via market coupling might lack common ground. However, misfunctions of simple interconnection compared to market coupling can be detected.

The rest of this paper is organized as follows: A thorough literature review is presented in Section 2. Previous works on CBT using multiple models are illustrated in this section. In addition to this, detailed information on the methodology is provided. Section 3 briefly presents the market structure of the examined countries and the interconnection between policy and procedure. Sections 4 and 5 illustrate the data (time series) we used for the completion of the current work on the preprocessing procedure, their summary statistics, and the correlations among them. The Granger causality connectivity analysis is briefly presented in Section 6. The methodology and the results exploitation as well as the data preprocessing and the model's validation are shown in this section. Finally, the outcomes of the simulations along with a discussion are provided in Sections 7 and 8, respectively.

2. Literature Review

Various studies have been conducted examining how the increasing penetration of renewables combined with interconnections among neighboring countries affect the dynamics of electricity spot prices domestically and cross border as well as their inter-trading commercial schedules. Since Germany is the leading country in Europe implementing renewable generation in its energy mix and one of the countries with the most cross border interconnections (CBI), it has been the center of research by the scientific community [10,13,14].

The first layer of the investigation consists of the consequences the increasing RES capacity has on the domestic SMP. Ketterer (2014) [13] used daily data on a GARCH model to examine the effect of domestic wind electricity generation on the volatility of electricity price. Specifically, the study showed that increasing wind generation reduces the price levels while at the same time increases its volatility.

In addition to the previous study, Wozabal et al. (2014) [10] using hourly data and ordinary least squares regression (OLS) showed that Intermittent Energy Sources (IES)affects SMP in a complex way. For a small to moderate quantity of IES, the price variance tends to decrease, while large quantities of IES have the opposite effect. Furthermore, in their study they highlighted some policy measures that might support variance absorbing technologies. Grid interconnections were one of the suggested measures, in the sense that they might operate as a means of ensuring the development of sufficient capacities over time.

Extensive investigation has been conducted regarding the Germany–France interconnection. German intermittent electricity generation has been proven to affect the dynamics of French spot prices. Specifically, increasing German renewable generation leads to a decrease in the level of French spot prices but has a positive effect on its volatility [12]. A step forward has been made by Haximusa (2018) [14], who showed that wind and solar generation in Germany have an ambiguous effect on the volatility of French spot prices. He considered three levels of French demand (i.e., low, medium, and high) and applied a GARCH analysis to show the effects on the spot price regarding the different demand levels. He proved that during medium and high French demands, importing energy from Germany decreases the spot price volatility, while the opposite happens when the demand is low.

Another example of strong interconnection coupled with increasing competition is the Nordic electricity market. Bask et al. (2008) [15] applied a stochastic model to generate electricity prices at Nord Pool and then used a GARCH analysis to show that prices became less sensitive to external shocks while the Nordic power market was being expanded and the degree of competition was being increased.

Denny et al. (2010) [16] investigated how the increasing penetration of wind generation affects the spot price dynamics of countries coupled to the UK and Ireland. The results indicate that increasing interconnections will reduce the level as well as the volatility of spot prices in both countries.

A paper that is very related to our work here, as far as its main topic is concerned, is the work of Zugno et al. 2013 [17], dealing with the influence of wind power generation on European cross border power flows. Wind power generation and spot price have been found to have a non-linear effect on cross border power exchange across Europe. Using Principal Component (PC) analysis (to reduce the problem's dimensionality), cross border power exchange hourly data were used as dependent variables in local polynomial regression using the PC (extracted from the matrix of cross border flow variables) as exogenous factors. The main findings in this work were that an increase in forecasted wind power generation causes a fall in the German import of power (or rise in the export), while rising spot prices show the opposite direction. Another very significant finding was that, from a global perspective, variations in wind power generation in Germany had significant effects on power flows in Europe. More specifically, import and export patterns were largely altered and loop flows were originated. The abovementioned paper was a data-driven research study made possible due to the availability of data on wind power generation, consumption, and power flows provided by the European Network of Transmission System Operators for Electricity (ENTSO-E). In general, the higher the wind power penetration in Germany, the more this country exports to its direct neighbors to the South. Also, it appears that there is a loop flow in the power transit from Germany to Switzerland via Austria, since the flow from Austria to Switzerland is positively correlated with German wind power generation. In our work, presented in this paper, we tried to detect similar interactions using, however, Granger causality analysis, on the CBT between Italy–Greece–Bulgaria.

In the present work, we have applied Granger causality (GC) on hourly data extracted by ENTSO-E database for four years (2015–2018). To the best of our knowledge, the cross-border electricity exchange among the abovementioned countries has not yet been investigated and, hence, our work appears to provide a valuable contribution to the scientific community. Another aspect of our work is that we examined how a country's spot price (e.g., Greece) and commercial (trading) schedule was affected by VRES generation in a neighboring country (e.g., Italy) without the existence of a market coupling policy.

Granger causality, an established time series technique, was used to analyze the interactions ("cause and effects") of a representative sample of 13 European electricity spot prices for the period 2007–2012 [18]. The study applied GC via network theory and provided inferences regarding the European electricity network's state and dynamic evolution over time, therefore assessing spot price convergence and "quality" of market coupling. The causal interactions among European electricity spot prices were modeled as a connectivity network on which the spot prices constituted the nodes of the network, while the links corresponded to the significant influences among relative pair-wise price changes.

One of the early applications of causal flow modeling in electricity markets is the work by Park et al. (2006) [19], in which the authors used advanced techniques in causal analysis (vector autoregressive (VAR), vector error correction model (VECM), and directed acyclic graphs (DAGs)) to find the dynamic relationships between electricity spot prices and the prices of major electricity-generation fuel sources (i.e., oil, gas, and coal prices) in US electricity spot markets.

A multivariate version of GC was used in a paper by Narayan et al. (2009) [20] to examine the causal relationships between electricity consumption, exports and gross domestic product for a panel of Middle Eastern countries.

In their causal modeling and inference for electricity markets, Ferkingstad et al. (2011) [21] used a combination of VAR, VECM, and a linear non-Gaussian acyclic model (LiNGAM) which they call time-lagged causal flow, a concept very close to the GC. They applied their hybrid model to weekly Nordic and German electricity prices, using oil, gas, and coal prices with German wind power and Nordic water reservoir levels as exogenous. They showed that, in contemporaneous time, Nordic and German spot prices were interlinked through gas prices.

A combination of out-of-sample Granger causality tests and DAGs were also used by Yang and Zhao (2014) [22] to investigate the temporal linkages among economic growth, energy consumption, and carbon emissions in India.

The methodology applied in this work, referred onwards as Granger causality connectivity analysis (GCCA), is based on modern network theory, an efficient approach to characterizing connecting systems [23]. Our purpose was to study the dynamic evolution of the network and the nodes which included the following criteria of three countries: the spot prices; imports–exports, the demand (load), the VRES generation; and the commercial schedules of CBT. This study will inform how the spot prices of the above countries are influenced by their own interactions as well as by the interactions of the other variables. Based on the results of this study, we will be able to draw conclusions regarding the cross-border trading development of this "peripheral" sub-system of the European electricity system.

The identification of directed operation connectivity among the components of a system is a challenging work. The GCCA is a powerful tool in detecting such connectivities in complex systems such as an electricity market. Granger Causality is a way to investigate "causality" between two variables, a probabilistic account of causality, and a concept closely related to the idea of cause and effect but not the same. GC allows one to know which particular variable comes before another in a time series but does not describe a causal link in the true sense. In Econometrics, "cause" is actually realized as "Granger cause".

The most recent approach to complex systems is the contemporary network theory, originating from the small world theory of Watt and Strogatz (1988) [24], as well as its scale-free "version" of Barabasi and Albert (1999) [25]. Recently, a large number of papers has focused on various applications of complex networks [26]. We applied this fascinating approach to the CBET among three neighboring countries to extract valuable information regarding their interactions. The results are expected to be useful in policy design among other areas. All calculations in this paper were based on the theoretical approach to GCCA presented in the work of Seth (2008, 2010 2014) [27–29] and implemented by using his toolbox run on MATLAB (ver.2019a, The MathWorks Inc., Natick, MA, USA).

A review paper presenting the application of GC in energy economics research is given by Narayan and Sangth (2014) [20]. The examination of causal relationships between the electricity consumption and economic growth using linear and non-linear GC tests in the case of Turkey is presented in the paper by Nazlioglou et al. (2014) [30]. The tests reveal a bi-directional GC in both the short and long run between the electricity consumption and economic growth in Turkey. The relationship between economic output and energy use is also the focus of a work by Brun et al. (2014) [31]. Woo et al. (2006) [32] applied a Granger instantaneous causality test in order to examine the potential causal relationships between wholesale electricity and natural gas prices in California, revealing bi-directional relationships between these two markets [32]. An in-depth analysis of the causal relationships in oil markets globally is given in a thesis work by Antoniadou (2015) [33]. In this work, the GC and the Toda–Yamamoto approaches were applied, revealing that the causality between crude oil prices and natural gas prices is not bi-directional, since only crude oil prices Granger causes natural gas prices. The relationship between investor attention and crude oil prices is examined by the paper of Li et al. (2019) [34]. Using the Google search volume index (GSVI) that "captures" investor attention, the author used linear and non-linear GC tests. The results show that a bi-directional GC exists only between WTI future crude oil returns and investor attention.

The latent volatility GC for four renewables energy exchanged traded funds (ETFs) and crude oil ETF (USO) were examined in a work by Chang et al. [35]. The empirical results showed that there are significant positive latent volatility GC relationships between solar, wind, nuclear, and crude oil ETFs as well as significant volatility spillovers shocks for renewable energy ETFs. Using a GC in quantiles analysis (evaluating causal relations in each quantile of the distribution), they succeeded in discriminating between causality affecting the median and the tails of the conditional distribution and provided evidence for the existence of a bi-directional causality between changes in RES consumption and economic growth using RES consumption, oil prices, and economic activity data in the US (July 1989 to July 2016).

The GC test was also used in analyzing spillover effect of oil and natural gas prices between emerging and developed countries in a paper by Zhong et al. (2019) [36]. The main findings were that oil and natural gas markets have significant GC and that emerging markets have a strong impact on many developed markets regarding returns and volatility spillover systems.

A combination of GC test and the (recently developed) cross-quantilogram was used on crude oil, natural gas, heating oil, electricity, and gasoline market data to evaluate their directional predictability in a work by Scarcioffolo et al. (2019) [37]. They found no strong evidence to support the decoupling between crude oil and natural gas markets. Natural gas and heating oil were found to be strongly linked across all quantiles.

3. Wholesale Electricity Markets: The Case of Greek, Italian, Iberian, French, and Bulgarian Markets

3.1. The Greek Market

Greece's liberalized electricity market was established according to European Directive 96/92/EC. The Greek Wholesale Electricity Market (GEM) currently operates as a day ahead mandatory pool which is based on an optimization algorithm for both energy and ancillary services [38] taking into account the following:

- Predictions of demand (on different time scales);
- Offers from generators and bids from suppliers;
- Units' availability and technical constraints (min, max, ramp up, ramp down);
- Must-run production (e.g., hydro power plants);
- Commercial schedules of the interconnections.

Greek Wholesale Energy Market GEM is in a transitional period towards its final design, namely, ETM where both an intra-day market and a balancing market are expected to operate in order to enhance GEM's liquidity and efficiency. More specifically, the new regulatory framework for the ETM makes provisions for a day-ahead and intra-day market managed by Hellenic Energy Exchange S.A. (Athens, Greece) and a balancing market managed by Greece's Transmission System Operator (IPTO) (Athens, Greece). The new day-ahead market will be a semi-compulsory market, where orders should cover the availability and should be compatible with Price Coupling of Regions algorithm (PCR EUPHEMIA) standards, and the biddings will be on a physical asset basis except RES which could be on a portfolio basis. There are also provisions for exchange-based futures and over-the-counter (OTC) contract limits on the volumes.

3.2. The Italian Market

In Italy, Gestore del Mercato Elettrico (GME) is responsible for the operation of the power market as well as gas and environmental ones. Gestore del Mercato Elettrico runs under the framework of the Italian Regulatory Authority (ARERA) which sets the rules and activities of the abovementioned sectors. The Italian Wholesale Electricity Market essentially runs on the Italian Power Exchange (IPEX), where producers and suppliers trade blocks of energy. More specifically, GME operates a day-ahead market (MGP) based on auctions, an intra-day auction market (MI), a forward market (MTE), and a market (MPEG) for continuous trading of energy daily products. Moreover, it operates on behalf of TERNA which is the Italian Transmission System Operator the Ancillary Services Market (MSD) while at the same time issues the OTC transactions (PCE).

Italy is divided into six zonal markets based on geographical criteria. These are Northern Italy (NORD), Central Northern Italy (CNOR), Central Southern Italy (CSUD), Southern Italy (SUD), the Sicily (SICI), and Sardinia (SARD). Greece is connected to SUD, which is the South part of the Italian electricity system. The zonal price, in the case of no congestion, is unique and is the interception of the aggregated demand and supply curves which are calculated after an algorithmic procedure. In the case of congestion, the market is split into the four aforementioned regions, and the system calculates different price equilibriums. The national unique price (PUN), which is the consumer price, is the weighted average of the individual zonal prices [38].

3.3. The Iberian and French Electricity Markets

Italy has already established a market coupling relationship with France and hence with the Iberian wholesale electricity market. The Iberian wholesale electricity market MIBEL, is a joint wholesale electricity market which comprises Spain and Portugal [39]. OMI-Polo Español S.A. (OMIE) belongs to the Iberian Market Operator business group, and it is subject to the rules and regulatory framework governing Spain's electricity sector [40]. The Iberian Wholesale Electricity Market is composed of an intra-day market and a day-ahead market producing a common spot price for both Spain and Portugal, unless there is a violation in the interconnection capacity between them; in such a case, a market splitting mechanism is activated which results in different prices for both countries. The Spanish Transmission System Operator (REE) and the Portuguese one (REN) are responsible for the technical implementation of the daily schedules. Following the gate closure, six intra-day market sessions are held four hours ahead of the physical delivery [41].

The French wholesale electricity market plays a key role in the French power system by allowing the balance between supply and demand. Most of the electricity (95%) injected into the French power system comes from nuclear and other sources (hydro, gas, coal, renewable energy sources) with a small percentage (5%) covered from imports [42]. Electricity products of the French Wholesale Electricity Market are traded on power exchanges or over the counter (OTC) via brokers or through bilateral agreements between the involved parties. The spot products are daily (day-ahead) or weekend products delivered at base load hours or during peak hours (from 8 a.m. to 8 p.m. for the working days) [43]. The abovementioned products are available on an hour or half-hour basis or as complex blocks covering several hours. The spot price of the French Wholesale Electricity Market is the price of the day-ahead market which is run on the European Power Exchange (EPEX SPOT) [44].

France is connected to the Central European System through its six interconnection lines to England, Belgium, Germany, Switzerland, Italy, and Spain. Currently the interconnection capacity between Spain and France is approximately 5% of its installed capacity which is far below the EU targets for interconnections; this limit has been set at 15% for 2030 at the EU level [45]. At the moment, the planned interconnections between France and Spain through the Gulf of Bizkaia is under consultation; this project has been characterized as a project of common interest with an interconnection capacity of 5000 MW. The latter is expected to be in operation between 2024–2025 [46,47].

3.4. The Bulgarian Market

The Bulgarian electricity market operates under the command of the Independent Bulgarian Energy Exchange (IBEX) which was established in 2014 as a 100% subsidiary of Bulgarian Energy Holding and is among the last countries introducing such an exchange market. A major change that was introduced by IBEX was the establishment of an organized day-ahead market, in 2016, to replace the pre-existing model.

Currently, IBEX operates three trading platforms: a day-ahead market, and intra-day market, and a centralized market for bilateral contracts; it is mandatory for generators with an installed capacity of 1 MW or more to sell their electricity through IBEX.

The electricity market consists of two segments.
- Regulated Market

Prices are set by the regulator (Energy and Water Regulatory Commission) and consumers are supplied based on territory. This segment includes households and small businesses connected to the low voltage distribution network.
- Free (Liberalized) Market

Electricity is freely negotiated and bought by suppliers and consumers directly from the electricity generators or via IBEX. According to the last amendment (May 2019), all RES and co-generation power plants with an installed capacity of 1 MW or more have the obligation to sell their electricity only via the power exchange.

3.5. Markets Comparison

One of the outmost criteria to measure a market's concentration and hence define its competitiveness is the Herfindahl–Hirschman index (HHI). Even though in various cases it fails to consider the complexities of various markets, it remains a reliable index. Specifically, regarding the EU energy markets, the European Commission has set the limits for the HHI as shown in Table 1 [46].

Table 1. Herfindahl–Hirschman index (HHI) in the power generation "market" of the analyzed electricity markets for 2015-2018 as depicted in the national reports of regulators sent to the European Commission (EC).

Year	HHI (Greek Market)	HHI (Italian Market)	HHI (Bulgarian Market)
2015	6804 (vhc *)	882 (d *)	
2016	6423 (vhc *)	881 (d *)	
2017	6357 (vhc *)	755 (d *)	
2018	5627 (vhc *)	730 (d *)	>4700 (hc *)
Mean (2015–2018)	6302 (vhc *)	812 (d *)	>4700 (hc *)

* Legend (according to the European Commission 2014 [46]): vhc = very high concentration (HHI > 5000), hc = high concentration (1800 < HHI < 5000), mc = moderate concentration (1000 < HHI < 1800), d = deconcentration (HHI < 1000).

Figure 1 illustrates the energy generation mix in 2018 for the examined countries.

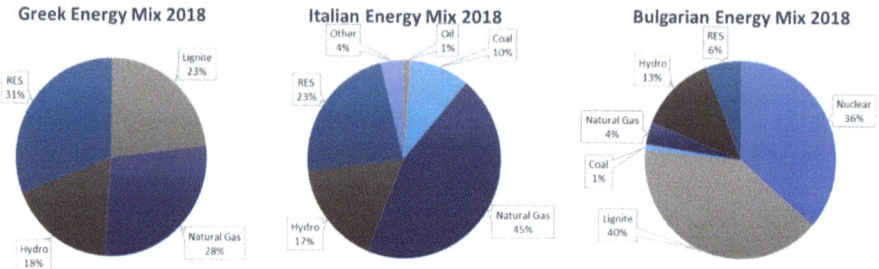

Figure 1. Energy generation mix for Greece, Italy, and Bulgaria (left to right) for 2018 [47–49] (outcome of IPTO's elaboration).

From the energy generation mix above, we can observe that Bulgaria was primarily dependent on nuclear power (36%) as a source of energy, which does not appear in either the Greek generation mix nor in the Italian, and on lignite (40%). On the other hand, Italy devoted 45% of its generation mix to natural gas, and the corresponding amounts for Greece and Bulgaria were 28% and 4%, respectively.

The common element that is being illustrated is the share of hydro in the examined countries' energy mix. Specifically, hydro in Greece accounts for 18%, in Italy for 17%, and Bulgaria for 13% of the total energy generation. Finally, Greece appears to have the highest percentage of RES generation in the energy mix (31%), followed by Italy (23%) and Bulgaria (6%).

3.6. Cross Border Trade in Electricity

The future "architecture" of the European CBT will be based on capacity allocation and congestion management (CACM). According to CACM, all TSOs are required to develop deliverables towards implementing the markets coupling, the so-called single intra-day (SIDC) and single day-ahead coupling (SDAC) [50]. The regulation also outlines the methods for calculating how much capacity the participants in the market can use on cross border lines without putting the system at risk. Also, the document harmonizes cross border market operations in Europe in order to enhance the competition and the integration of renewables.

The SIDC creates a single EU cross-zonal intra-day electricity market, in which buyers and sellers of energy cooperate to trade electricity continuously on the day the energy is needed. The SIDC enhances the efficiency of intra-day trading across Europe by competition promotion, liquidity increase (facilitating the mechanism of buying and selling without affecting the price), making the sharing of energy generation resources easier [51].

The SIDC is implemented in three phases or waves. In the first wave, 14 countries went live in June 2018, in the second wave 7 countries, including Bulgaria, went live in November 2019. Italy and Greece will be included in the third wave, foreseen for Quarter 4 of 2020 [52].

The SDAC is the pan-European single day-ahead coupling serving 27 countries (at the time of the writing of this paper). Under this agreement, 33 TSOs and 16 nominated market electricity operators (NEMOs) will cooperate. The SDAC uses (PCR EUPHEMIA), that calculates, across Europe, electricity spot prices and does not allocate implicitly the auction-based cross border capacity [53]. According to the Agency for the Cooperation of Energy Regulators (ACER), SDAC has improved the level of efficiency regarding the usage of the interconnections from 60% in 2010 to 87% in 2018 [54].

On the 1st of October 2018 (first delivery date of the projects related to SDAC), the Germany–Austria bidding zone split was successfully implemented. In South East Europe (SEE), the target times for the bidding zone borders adhering to CACM are as follows [52]:

- Bulgaria–Greece: to be defined;
- Greece–Italy: Quarter 4, 2020.

The electricity trade can be unidirectional or bidirectional over long time horizons. Electricity flows in either direction due to the demand changes in both countries, for example, when the latter is not perfectly correlated due to the different seasonal or diurnal patterns.

The correlation coefficient between Greece's and Italy's forecasted load was 0.524 and statistically significant (so the loads were not perfectly correlated, indicating the existence of strong bidirectionality of CBT).

According to Antweiler [55], higher correlations diminish trade, a very essential finding, since if demand is strongly correlated between two countries, they will both have high and low demand simultaneously, allowing a limited space for additional trade. Another very significant finding in Reference [55] is that the intensity of cross-border electricity trading increases along with the coefficient of variation (cost/total demand) and that the difference or dissimilarity in the size of the countries (e.g., Italy versus Greece) encourages more trade.

A top priority for the European Commission is the harmonization and integration of national electricity markets to a single pan-European market [5]. Energy policy, however, remains strongly a tool of national sovereignty which means that a greater level of integration corresponds to the fact that unilateral national policies can impact interconnected markets.

In this work, we investigated the impact of interconnections on the expansion of renewables promoted by fixed-in tariffs and (unlimited) priority on DA-prices between Italy, Greece, and Bulgaria, as they are revealed through CBT. The contribution of the specific hub-selection lies on the lack of market coupling among the examined countries, and hence the CBT operates as a simple interconnection. It would be of high importance to identify any differences between market-coupled countries' CBT and the specific work.

3.7. Greece's Imports and Exports

Greece relies heavily on interconnections to meet the demand. This is due to the fact of high wholesale electricity prices which remain high despite the penetration of RES into the system. Moreover, due to the economic recession investment in RES (which, in general, lowers spot prices) had been discouraged.

The average relative margin available for cross zonal trade (MACZT) between 2016 and 2018, as calculated by ACER [54], for Greece and Italy, relative to maximum admissible active power flow (Fmax), set to at least 70% by the clean energy package (CEP), is almost 100%, and the percentage of hours when MACZT is at least 70% for the GR–IT border is also almost 100%.

According to the same report [54], for the border between Bulgaria and Greece, the average DA price differential (€/MWh) and average of absolute DA price differential (€/MWh) during the period 2016–2018 is as follows (Table 2).

Table 2. Average and average absolute DA price differential for the Bulgaria–Greece border.

Year	Average DA Price Differential	Average Absolute DA Price Differential
2016	−6.0	14.6
2017	14.6	19.8
218	−20.5	24.2
2016–2018	−4.0	19.5

In 2018, the annual average DA prices in Greece (60.4 €/MWh) and Italy (62.04 €/MWh) were among the highest in European bidding zones, whereas Bulgaria's DA prices were among the lowest (39.89 €/MWh). This difference in DA prices justifies the fact that during this specific period, Greece was a major exporter, as shown in Figure 2, in all neighboring countries except Italy.

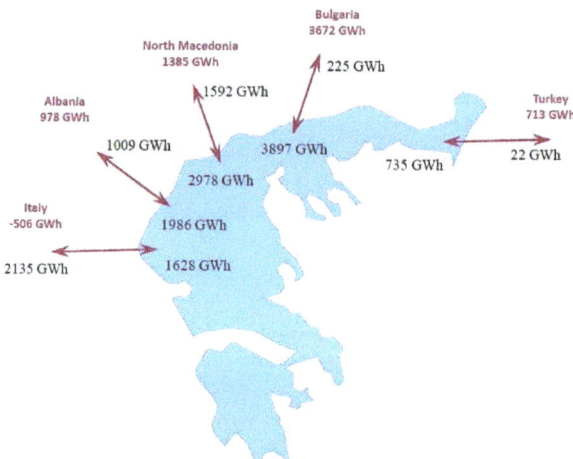

Figure 2. Greece's energy imports and exports for 2018. The net position is marked with red (outcome of IPTO's elaboration).

Figure 3 illustrates the monthly amount of energy trading between Greece and all the interconnected countries for the examined period (2015–2018).

From Figure 3 we observed a reduction in the difference between imports and exports (interconnection balance). It was continuously positive, which means that Greece exported less energy than it imported (it is a net importer) during the whole period between 2015 and 2018.

As far as the rate of implementation of the ETM for DA markets is concerned, significant progress has been made in recent years. The ETM foresees a single DA coupling, enabling the efficient use of cross-zonal capacity in the "right economic flow (direction)" (from low to high price), when there is a price differential across a bidding zone border. The level of efficient use of interconnectors in the DA market timeframe and the estimated social welfare gains still to be obtained, from extending DA market coupling per border, are the two indicators that illustrate the progress toward ETM.

The estimated social welfare gains, still to be obtained, from further extending DA market coupling in the borders of Greece–Italy and Greece–Bulgaria, during the period between 2017 and 2018, as calculated by ACER [56] are illustrated in Table 3.

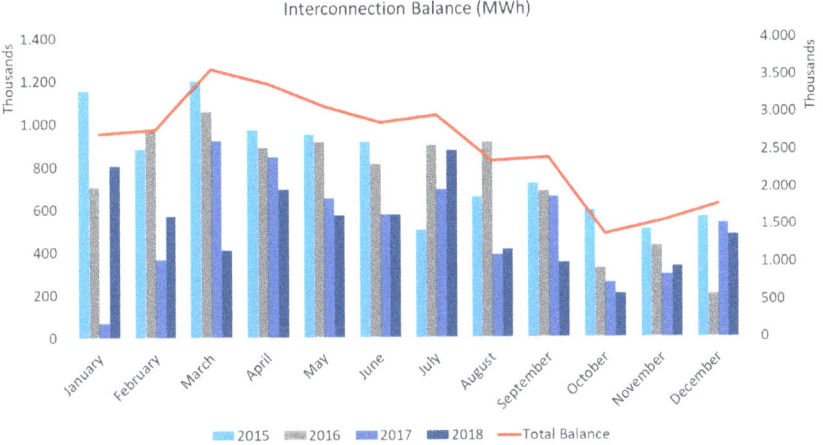

Figure 3. Energy balance between Greece and interconnected countries.

Table 3. The estimated social welfare gains from further extending Day Ahead (DA) market coupling in Greece and Italian/Bulgarian borders.

Border	2017	2018
GR–IT	6	10
BG–GR	22.5	18.5

Historically, Greek Wholesale Electricity Prices (SMPs) were driven by the low prices of fossil fuels and more specifically from the lignite ones. The Green Deal alongside climate change awareness made the EU impose stricter rules for CO_2 emissions. The 2030 and 2050 emission targets (90% reduction compared to 1990 EU levels) led to higher prices for CO_2 emission rights which, in turn, had a dramatic impact on lignite prices. The latter seriously affected the GEM, since gas prices are now compared to lignite prices which, in turn, incurs higher wholesale electricity prices according to the merit order mechanism. Also, even though Greece has already achieved the EU targets for RES penetration levels by 2020, this is not mirrored in market prices due mostly to the financial crisis of 2011 but also to the low volatility and liquidity of GEM.

3.7.1. Joint Allocation Office (JAO)

The Joint Allocation Office (JAO) is a joint service company that consists of 22 TSOs in 19 countries and facilitates the electricity market by organizing auctions for cross border transmission capacity. It was created in June 2015 and was the result of the merge between the Central Allocation Office (CAO) and ASC.EU S.A. located in Germany and Luxembourg respectively, the two former regional allocation offices for cross border transmission capacities. On the 1st of October 2018, JAO became the Single Allocation Platform (SAP) of forward capacity for all European TSOs, operating in accordance with EU legal rules. Currently, JAO performs long- and short-term auctions of transmission capacity and is the duty of the national regulatory authorities to decide which auction should be performed.

3.7.2. Greece–Italy (GR–IT) Interconnection

Currently, Greece and Italy operate as two separate electricity markets, while no market coupling has been applied.

Before the annual and monthly capacity allocation process, IPTO (ADMIE) calculates the GR–IT Net Transmission Capacity (NTC) and agrees on it with TERNA. The final NTC values are stored in the system and then communicated via email to JAO. After receiving the NTC values, JAO is responsible for the annual and monthly transmission rights allocation on GR–IT border. Joint Allocation Office (JAO) performs the auction and informs IPTO and TERNA of the capacity auction outcome.

Two days before the scheduling day (until 12:30 CET), IPTO receives from JAO the annual and monthly final capacity right documents which are aggregated into the long-term capacity rights.

The values of daily ATC and relative data are submitted by ADMIE and TERNA to JAO every day (until 09:30 CET). Once the daily auctions are calculated, they are sent by JAO to ADMIE and TERNA.

After having received the capacity right document, IPTO interacts with the capacity holders or their counterparties who inform IPTO of the capacity they are going to use, based on their capacity rights documents (long term or short term).

The day-ahead scheduling process, performed by IPTO, includes two phases of nominations, submission and matching.

The long-term (LT) phase: until 08:30 CET, capacity holders submit their nominations based on long-term rights. Consequently, IPTO and TERNA exchange and match the LT nominations until 09:00 CET. In the case of a mismatch, the export value prevail rule is applied.

The remaining capacity (non-nominated LT capacity rights), until the NTC of the GR–IT interconnection, comprises the offered capacity for daily auctions (performed by JAO). The IPTO sends both LT and ST capacity rights to ENEX which is the NEMO for the Greek DAM. The SMP of the Greek DAM is the hourly solution of the algorithm of the daily co-optimization of the energy production bids (taking into account the techno-economic constraints of the plants) as well as the energy bids of importers and exporters and the Load declaration of load representatives. The total commercial schedules are determined up to 99% by the Greek DAM clearing, and they have a causal relationship with the price difference between Greek DAM and South Italy marginal prices.

The short-term phase: until 14:30 CET, capacity holders submit their nominations based on both LT and ST rights. Both IPTO and TERNA exchange and match the LT and ST nominations until 15:35 CET. In the case of a mismatch, the minimum value prevail rule is applied.

3.7.3. Greece–Bulgaria (GR–BG) Interconnection

The interconnection between Greece and Bulgaria operates more or less similarly to the GR–IT one, with no market coupling and through the JAO platform. The responsible party for the Bulgarian side is ESO (the Bulgarian system operator).

The auctions between Greece and Bulgaria are bidirectional and are separated to:

- Daily auctions;
- Monthly auctions;
- Yearly auctions.

The NTCs, which are the auctioning basis, are agreed to between IPTO and ESO. Afterwards, the NTCs and ATCs for each interconnection and direction are published on the auction websites. If there is a change to the ATCs regarding the monthly auctions, users can be informed timely regarding the relevant auction website.

4. Data and Preprocessing

We focused on the Italian, Greek, and Bulgarian electricity markets, their actual and forecasted solar and wind power productions, demands (loads), DA-prices (spot or wholesale prices), and finally their commercial schedules and transfer capacities of cross-border trading with their interconnected countries. Below, we may refer to them as cross-border trading fundamentals (CBTFs). Table 4 below contains the description of the data which consists of hourly observations, converted to daily values (average of 24 hourly values), covering the period 2015 to 2018 (1460 data points).

Table 4. Data Overview.

Data (Variable)	Description	Unit of Measurement	Resolution
D1	Values of forecasted solar production in the South Italian region that is connected to Greece	MWh	Daily
D2	Values of forecasted wind production in the South Italian region that is connected to Greece	MWh	Daily
D3	Values of forecasted Solar production in Greece	MWh	Daily
D4	Values of forecasted Wind production in Greece	MWh	Daily
D7	Total forecasted demand (load) in Italy	MWh	Daily
D9	Total forecasted demand (load) in Greece	MWh	Daily
D11	Wholesale electricity price in the South Italy region that is connected to Greece	€/MWh	Daily
D12	Wholesale electricity price in Greece	€/MWh	Daily
D13	Total commercial schedule from Italy to Greece [1]	MW	Daily
D14	Total commercial schedule from Greece to Italy	MW	Daily
D15	NTC from Greece to Italy [2]	MW	Daily
D16	NTC from Italy to Greece	MW	Daily
D17	Values of forecasted solar production in Bulgaria	MWh	Daily
D18	Values of forecasted wind production in Bulgaria	MWh	Daily
D21	Wholesale electricity price in Bulgaria	lev/MWh	Daily
D22	Total commercial schedule from Bulgaria to Greece	MW	Daily
D23	Total commercial schedule from Greece to Bulgaria	MW	Daily
D24	Total forecasted demand (load) in Bulgaria	MWh	Daily

[1] Total commercial schedule is the summation of all the agreed transactions for the delivery and receipt of power and energy among the traders in the interconnection. That schedules are also agreed to by the TSO of the two control areas. [2] Transfer capacity is the allocated through auctions transfer ability to traders that allows them to exchange electricity among geographic areas for each market time unit and for a given direction.

Regarding Bulgaria, there were no available data for the DA price (D21) before 2017 which meant that in the examined cases (study A and C) of this work, which included the specific variable, we limited the data range to the years 2017 to 2018 (725 data points).

In general, we used data for the South Italy region (SUD) which connects to Greece. However, when it came to Load Forecasted, we decided that it was more accurate to use the Total Load Forecasted and not the South one. This decision was based on the procedure IPEX follows to determine the PUN, in which the Load is considered as aggregated for the whole Italy and is not split into zonal regions. Moreover, the total load forecasted in Italy is highly correlated to the load forecasted in South Italy (0.70), so we can safely use the total demand.

Hence, in the forecasted load time series for both Greece and Italy (D7 and D9), we considered in all our calculations the total forecasted load in the Italian and Greek markets.

Testing Examined Time Series for Stationarity

Since the primary precondition for Granger causality analysis is that the variables must be covariance stationary (CS), also known as weak or wide-sense stationarity, we tested our data for stationarity. Augmented Dickey–Fuller (ADF) and Kwiatkowski–Phillips–Schmidt–Shin (KPSS) tests were performed for testing the null hypothesis that the examined series had unit roots which indicates stationarity or trend stationarity.

Since we used two sets of data (2015 to 2018 and 2017 to 2018 to incorporate the Bulgarian data), the tests were performed for each one of the time series and for both data sets. Table 5 illustrates the tests results for the acceptance or not of the null hypothesis. Rejection of the null hypothesis indicates data stationarity, while the inability to reject it indicates non-stationarity.

The ADF unit root test did not reject the null hypothesis for D24 for the whole series and for D9, D15, D16, and D24 for the reduced data series scenario. On the other hand, KPSS test indicates that D11 and D12 were not trend stationary when the whole series scenario was examined and D21 when the reduced series scenario was examined.

Therefore, for the abovementioned time series, we took their first difference to make them stationary. Then, we performed the tests again and all series appeared to be stationary and trend stationary.

Table 5. Augmented Dickey-Fuller (ADF) and Kwiatkowski–Phillips–Schmidt–Shin (KPSS) stationarity test results.

Data (Variable)	ADF		KPSS	
	2015 to 2018 (1460 Data Points)	2017 to 2018 (725 Data Points)	2015 to 2018 (1460 Data Points)	2017 to 2018 (725 Data Points)
D1	Stationary	Stationary	Stationary	Stationary
D2	Stationary	Stationary	Stationary	Stationary
D3	Stationary	Stationary	Stationary	Stationary
D4	Stationary	Stationary	Stationary	Stationary
D7	Stationary	Stationary	Stationary	Stationary
D9	Stationary	Non-Stationary	Stationary	Stationary
D11	Stationary	Stationary	Non-Stationary	Stationary
D12	Stationary	Stationary	Non-Stationary	Stationary
D13	Stationary	Stationary	Stationary	Stationary
D14	Stationary	Stationary	Stationary	Stationary
D15	Stationary	Non-Stationary	Stationary	Stationary
D16	Stationary	Non-Stationary	Stationary	Stationary
D17	Stationary	Stationary	Stationary	Stationary
D18	Stationary	Stationary	Stationary	Stationary
D21	N/A	Stationary	N/A	Non-Stationary
D22	Stationary	Stationary	Stationary	Stationary
D23	Stationary	Stationary	Stationary	Stationary
D24	Non-Stationary	Non-Stationary	Stationary	Stationary

To continue further with our research, we defined three studies (models) in order to find the best models to test the Granger causality of cross-border trading among the examined countries. Our aim was to conduct an analysis for all the possible combinations, hence we separated the model structure as follows:

- Study A contained variables of the three countries, treating them as consisting of a "whole" system (although Bulgaria is not directly connected to Italy), in an effort to identify any connection and interaction as a system, including "pass-through" or "hidden" causalities (transit flows);
- Study B contained variables of Greece and Italy;
- Study C contained variables of Greece and Bulgaria.

Table 6 shows all the variables (or nodes in the network) and the corresponding studies (models) they were included as an input.

Table 6. The case studies and the variables that were included [1].

Model	D1	D2	D3	D4	D7	D9	D11	D12	D13	D14	D15	D16	D17	D18	D21	D22	D23	D24
A	1	2	3	4	5	6	7	8	9	10	11	12	13	14	15	16	17	18
B	1	2	3	4	5	6	7	8	9	10	11	12						
C			1	2		3		4					5	6	7	8	9	10

[1] The numbers below each time series indicate the corresponding node of the specific variable to the Granger causality connectivity analysis model.

5. Summary Statistics

5.1. Summary Statistics

Since Granger causality Analysis is a multivariate (MVAR) method and since all MVAR methods rely heavily on the assumption of normality or near-normality which is often hard to achieve in practice, we proceed in presenting summary statistics of all variables, focusing on normality aspects. The results are illustrated in Table 7.

Table 7. Summary statistics.

Time Series	N	Minimum	Maximum	Mean	SD	Skewness	Kurtosis	JB Test (p-Value)
D1	1460	36.44	875.84	470.62	192.15	−0.233	1.938	81.86 (0.00)
D2	1460	20.68	3165.36	800.64	582.93	1.077	3.794	320.61 (0.00)
D3	1460	47.36	838.60	391.42	145.73	−0.336	1.988	89.74 (0.00)
D4	1460	76.40	1671.20	516.04	320.10	0.839	3.056	171.66 (0.00)
D7	1460	20,687.76	46,082.68	33,032.04	4647.46	−0.370	2.496	48.82 (0.00)
D9	1460	4060.80	8210.96	5865.18	713.22	0.482	2.627	64.87 (0.00)
D11	1460	20.07	127.30	49.72	11.47	0.779	4.978	385.58 (0.00)
D12	1460	26.91	111.40	52.41	10.63	1.121	5.365	645.96 (0.00)
D13	1460	0.00	530.40	223.52	181.02	−0.005	1.448	146.44 (0.00)
D14	1460	0.00	531.72	122.40	148.53	1.243	3.275	380.84 (0.00)
D15	1460	0.00	500.00	350.78	223.41	−0.871	1.809	271.02 (0.00)
D16	1460	0.00	500.00	351.34	223.25	−0.878	1.819	272.18 (0.00)
D17	1460	0.00	336.04	151.95	78.47	−0.016	1.913	71.98 (0.00)
D18	1460	1.92	645.04	184.43	139.82	0.983	3.251	239.15 (0.00)
D21 [1]	728	6.96	157.2	39.8	13.8	2.107	13.49	3864 (0.00)
D22	1460	0.00	700.00	450.60	144.32	−0.617	3.783	130.04 (0.00)
D23	1460	0.00	529.84	23.26	51.14	3.598	19.971	20,669.56 (0.00)
D24	1460	3160.84	6756.96	4321.93	694.95	0.916	3.022	204.23 (0.00)

[1] We converted Bulgarian lev to euros taking an exchange rate of 1 euro = 0.511 lev.

As illustrated in Table 7, none of the examined time series follow normal distribution. Some variables that worth highlighting regarding their skewness values are D2, D12, D14, D21, and D23 whose value appears to be higher than 1, which indicates strong positive skewness and hence data concentration to the right tail (maximum) of the distribution. This was anticipated for the wind generation (D2), since the specific variable was highly stochastic. However, regarding DA prices in Greece and Bulgaria (D12 and D21, respectively), the high value of skewness revealed a difficulty in the forecasting model construction. The same happens with the commercial schedules from Greece to Italy and to Bulgaria (D14 and D23). The contribution of this work is to provide a way of enhancing the forecasting by revealing and indicating any causal connections among CBTF.

Furthermore, from Table 7 we observe that all the variables appeared to have values of kurtosis close to three which was our benchmark of normality, except for D11, D12, D21, and D23. This indicates that the degree of uncertainty in the specific time series (DA prices) was high which contributes more to the difficulty of prediction.

The turnover ("liquidity") of the South Italy electricity market was much smaller than the turnover of the Greek Market. Thus, prices of South Italy zone are more volatile than those at Greece. As illustrated in Table 7 the maximum price at South Italy was 127.30 €/MWh, and the standard deviation was 11.5 €/MWh (mean value 49.72 €/MWh). In Greece, the maximum price was 111.4 €/MWh, and standard deviation was 10.63 €/MWh (mean value 52.41 €/MWh).

Finally, observing the Jarque–Bera (JB) test results, we see that none of the examined series were normally distributed, which means that we need to further analyze the data and apply decomposing methods. It is worth mentioning the corresponding value of the JB test for the Bulgarian DA price and the commercial schedule from Greece to Bulgaria (D21 and D23, respectively) which were extremely high, revealing a deviation from the normal distribution.

5.2. Correlation Coefficients

In order to have a "rough insight" into how the variables were interrelated or interacted with each other, we observed their correlation coefficients matrix. However, GC is actually a much more powerful tool to reveal any hidden causality, both in magnitude and direction.

The following tables (Table 8, Table 9) depict the correlation coefficient among the examined time series. We separated the results according to the models we tested to highlight each interconnection and hence make clearer their connection with the results.

Table 8. Correlation Matrix: Greece–Italy interconnection (2015–2018).

Time Series	D1	D2	D3	D4	D7	D9	D11	D12	D13	D14	D15	D16
D1	1											
D2	−0.241	1										
D3	0.824	−0.301	1									
D4	−0.195	0.271	−0.276	1								
D7	−0.030	−0.029	−0.040	0.056	1							
D9	−0.050	0.014	−0.022	0.078	0.526	1						
D11	0.029	−0.210	0.018	0.033	0.208	0.061	1					
D12	0.018	−0.113	0.013	−0.151	0.181	0.061	0.302	1				
D13	0.046	0.206	0.050	−0.176	−0.080	0.148	0.023	−0.001	1			
D14	0.153	−0.048	0.175	0.112	0.120	0.149	0.004	−0.008	0.010	1		
D15	0.198	0.061	0.190	−0.065	−0.006	0.195	−0.009	−0.022	0.772	0.477	1	
D16	0.194	0.061	0.186	−0.063	−0.005	0.199	−0.006	−0.013	0.776	0.475	0.998	1

Table 9. Correlation matrix: Greece–Bulgaria interconnection (2017–2018).

Time Series	D3	D4	D9	D12	D17	D18	D21	D22	D23	D24
D3	1									
D4	−0.280	1								
D9	−0.007	−0.036	1							
D12	−0.271	−0.108	0.155	1						
D17	−0.029	0.005	0.000	−0.030	1					
D18	−0.064	0.026	−0.079	0.096	−0.396	1				
D21	0.007	−0.012	0.355	0.047	−0.018	−0.032	1			
D22	−0.045	0.040	0.019	−0.009	0.426	−0.184	−0.067	1		
D23	0.003	0.002	−0.033	−0.034	−0.129	0.043	−0.007	−0.269	1	
D24	−0.068	0.051	0.626	0.118	−0.014	−0.049	0.304	−0.021	−0.030	1

Table 8 shows the correlation coefficients of the variables regarding the Greece and Italy interconnection for the period 2015–2018. As illustrated from the coefficients among D1, D2, D3, and D4, there was a relatively high correlation (either positive or negative) that reveals a similarity (similar diurnal effects) in weather conditions which was anticipated between Greece and South Italy. Specifically, there was a high positive correlation between solar (D1 and D3) and wind (D2 and D4) generation in two countries. On the other hand, correlation among solar and wind generation appeared to be negative which reveals that high solar was not followed by high wind generation and vice versa. We did not expect to observe this relation to the results of the current work, since solar and wind generation are exogenous and highly stochastic to be interpreted in the results.

The Pearson's correlation coefficient between PV generation (D1) and spot price (D11) in South Italy was not significant (0.029), while for Greece it was even smaller (0.013). This result is in compliance with results in a similar study conducted by Mayer and Luther [57] on the correlation between PV generation and spot prices in the European Power Exchange (EEX) and Amsterdam Power Exchange (APX) electricity markets. However, these weak correlational linkages between PV generation and spot prices do not "constitute" Granger Causal interactions as shown below.

Another insight from Table 8 is the strong positive correlation (0.526) between total load forecasted of the two countries (D7 and D9). This should be anticipated since the weather conditions, which are the main electricity load driver, were similar. The DA prices (D11 and D12) were also positively correlated with a coefficient of 0.302, something that can also be explained by the weather conditions similarities, since when RES generation is high, the wholesale price is or shows a tendency to be reduced.

Furthermore, a relatively moderate positive correlation (0.206) appears between the commercial schedule from Italy to Greece (D13) and wind generation in Italy (D2). However, we did not observe a similar correlation between commercial schedule from Greece to Italy (D14) and wind generation in Greece (D4). We anticipated an outcome that would partially explain this difference in the specific coefficients.

Table 9 depicts the correlation coefficients of the examined variables for the interconnection of Greece and Bulgaria. Now the examined period was 2017–2018, since as mentioned, we could not obtain previous reliable data for D21.

In contrast to the case of Greece and Italy, Table 9 did not reveal any correlation on weather conditions between Bulgaria and Greece. However, the wind and solar generation in the same country appeared to have a relatively strong negative correlation, −0.280 and −0.396 for Greece and Bulgaria, respectively.

Despite the fact of non-similar weather conditions, the total load forecasted for the two countries had a strong positive correlation of 0.626. A high positive correlation also appeared between D9 and D21, hence it would be interesting to see if a relation between Greek load and Bulgarian DA price does really exist and if the one Granger causes the other. When the demand in Greece was high, the Greek spot price also became high with a consequence to examine import opportunities of cheaper energy from Bulgaria.

A similarity with the case of Greece–Italy interconnection is shown in Table 9 with respect to RES generation and commercial schedules. Specifically, solar generation in Bulgaria (D17) was strongly correlated with commercial schedules from Bulgaria to Greece (D22) (correlation coefficient 0.426). Again, we did not observe a bidirectional relation which means that high RES generation in Greece does not imply high commercial schedules to the opposite direction.

Finally, Table 9 shows a strong relation between the Bulgarian load forecasted (D24) and the domestic DA price (D21), as expected, with a correlation coefficient of 0.304.

6. Granger Causality Connectivity Analysis (GCCA)

6.1. Granger Causality Test

The purpose of this paper was to investigate the relationships among the variables related to CBT between two countries, in the same multivariate framework, in which the well-known GC test (based on the MVAR approach) is the most widely used technique.

However, the traditional (in sample) GC test was not adequate to explore the important contemporaneous causal pattern among the variables which is used to conduct a data-determined structural decomposition of the VAR model. As it well known, the conventional VAR analysis relies heavily on the Cholesky decomposition in order to achieve a just-identified system in contemporaneous time. This decomposition is severely criticized for imposing a somehow not so realistic assumption of a recursive contemporaneous causal structure [58]. In order, therefore, to explore the important contemporaneous causal pattern, it is necessary to apply also the DAG tool [59,60].

The advantage of GCCA adopted in this paper, however, is that it combines the above two separate tools into one, "producing" a "new tool", providing similar as well as enhanced results. For information about the typical Granger causal connectivity computational technique as well as for a more theoretical approach to causal networks, the reader is referred to References [27–29].

In the current study, we applied the GC test to examine whether there is an overflow relation among the examined variables. According to GC, if a time series x "Granger-causes" time series y, then past values of x should contain information that helps predict y above and beyond the information contained in past values of y alone [61]. In one binary p-order VAR model.

$$\begin{pmatrix} y_t \\ x_t \end{pmatrix} = \begin{pmatrix} \varphi_{10} \\ \varphi_{20} \end{pmatrix} + \begin{pmatrix} \varphi_{11}^{(1)} & \varphi_{12}^{(1)} \\ \varphi_{21}^{(1)} & \varphi_{22}^{(1)} \end{pmatrix} \begin{pmatrix} y_{t-1} \\ x_{t-1} \end{pmatrix} + \begin{pmatrix} \varphi_{11}^{(2)} & \varphi_{12}^{(2)} \\ \varphi_{21}^{(2)} & \varphi_{22}^{(2)} \end{pmatrix} \begin{pmatrix} y_{t-2} \\ x_{t-2} \end{pmatrix} + \ldots + \begin{pmatrix} \varphi_{11}^{(p)} & \varphi_{12}^{(p)} \\ \varphi_{21}^{(p)} & \varphi_{22}^{(p)} \end{pmatrix} \begin{pmatrix} y_{t-p} \\ x_{t-p} \end{pmatrix} + \begin{pmatrix} \varepsilon_{1t} \\ \varepsilon_{2t} \end{pmatrix} \quad (1)$$

Considering the equation above, if all the coefficients $\varphi_{12}^{(q)}$ ($q = 1, 2, 3, \ldots, p$) are zero, variable x is not Granger cause of y, which means that x cannot change y. GC test looks at whether lagged variable of one variable can be brought into alternative variable equations. In case that variable x can help explain variable y, then variable x is the Granger cause of variable y. A tool that helps judging the Granger cause is the F-test:

$$H_0 : \varphi_{12}^{(q)} = 0 \qquad (2)$$

$$q = 1, 2, \ldots, p \qquad (3)$$

Hypothesis 1. *Existence of at least one q such that* $\varphi_{12}^{(q)} \neq 0$:

$$S_1 = \frac{(RSS_0 - RSS_1)/p}{RSS_1/(T - 2p - 1)} \sim F(p, T - 2p - 1) \qquad (4)$$

where: RSS_1 is the residual sum of squares in Equation (4) and RSS_0 is the residual sum of squares in the y equation when $\varphi_{12}^{(q)} = 0$.

$$RSS_1 = \sum_{t=1}^{T} \varepsilon_{1t}^2 \qquad (5)$$

The above statists fit in the F distribution.

- If S_1 is higher than the critical value of F, then the null hypothesis is rejected and, hence, variable x is the Granger cause of variable y;
- In any other case, the null hypothesis is accepted.

The selection of p in VAR model is crucial while applying the GC test, since the results are strongly related to the number of lag orders. P is expected to be sufficiently large to reflect dynamic features but at the same time not excessively large since it brings several parameters for estimation and thus reducing the degrees of freedom (DOFs) of the model.

Akaike Information Criterion (AIC) and Schwarz Criterion (SC) are commonly used methods of selecting the number of lag orders. This selection should be conducted considering the quantity of lag items and DOFs. The calculations of those two methods are as follows:

$$AIC = \frac{-2l}{T} + \frac{2n}{T} \qquad (6)$$

$$SC = \frac{-2l}{T} + n\frac{\ln T}{T} \qquad (7)$$

where n is the total number of estimated parameters, k is the number of endogenous variables, T is the sample length, d is the number of lag orders, l is the logarithmic likelihood calculated as follows, by hypothesizing that the multivariate normal distribution is obeyed:

$$l = -\frac{Tk}{2}(1 + \ln 2\pi) - \frac{T}{2}\ln\left(det\left(\frac{1}{T-m}\sum_{t}\varepsilon_t\varepsilon'_t\right)\right) \qquad (8)$$

Ideally, values of AIC and SC should be as low as possible.

6.2. Granger Causality Network and Visualization

Visualization of the connectivity among different time series is especially powerful when it comes to understanding the relation between different examined variables and how the one "Granger causes" the other.

In the network depiction of the corresponding G-causalities, nodes represent the different variable or system elements, in our case different time series and directed edges (arrows) represent causal interactions. Green lines depict unidirectional connections, with the direction of the arrow stating the relation between the nodes, meaning which one G-causes the other, while red lines depict bidirectional connections between two nodes. The thicker the line, the stronger the connection between the two corresponding nodes.

The same depiction, but in a matrix format, is also useful for the representation of causal interactions. In this setup, instead of lines there are colored boxes which correspond to two different nodes. The direction of the lines are replaced by the x- and y-axis with nodes of on x-axis being the nodes of the G-cause of the nodes on the y-axis. The darker the color of the box, the higher the G-causality effect.

6.2.1. Causal Density

The causal density (cd) reflects the portion of interactions among nodes that are causally significant and provides a useful measure of system dynamical complexity. Causal density cd is defined as:

$$cd = \frac{gc}{2N(N-1)} \tag{9}$$

where gc is the number of the total number of non-zero interactions observed, and N is the total number of nodes. The higher the value of cd the more the coordination of the system elements (nodes).

However, the term unit causal density (cdu) is often being used as a derivative of the cd. This is defined as the total number of significant interactions involving a specific node i. Nodes with relatively high values of cdu are considered to be hubs within the system.

6.2.2. Causal Flow

Causal flow (CF) is a measure that helps defining if a node is a causal source or causal sink. It is defined as the difference between a node's in-degree and out-degree. It is another mean of network's representation which considers all the lines and their thickness in order to define if a node exerts causal influence or not. Nodes with positive CF are causal sources while those with negative CF are causal sinks.

6.3. Data Preconditioning and Preprocessing in GCCA

In order to perform a GCCA, a primary precondition must hold, that the variables must be CS (also known as wide-sense stationary). Covariance Stationarity is equivalent to saying that the first and second statistical moments (mean and variance) of each variable are constant, not varying with time. Deviations from CS can be tested by detecting "unit roots" in the data. We used the ADF test to assess the presence of a unit root in the variables (Section 4). A variable is that CS exhibits a tendency to return to a constant mean (or to a deterministic trend line). According to this mean-reversion tendency, large values will tend to be followed by smaller values, and on the opposite, small by larger ones. The ADF test detects the absence of this behavior. For non-CS variables, the following steps are: (a) linear (deterministic) trends are removed and (b) unit roots are removed by differencing [28].

6.4. Model Validation in GCCA

The inferences in GCCA are valid only in the case that a MVAR model captures well the "hidden" correlation in the data.

The amount of the variance "explained" by the model (in terms of the adjusted sum-square error, or adjusted R^2), is the simplest indicator of the model's adequacy.

Typically, a value $R^2 < 0.30$ indicates that the model cannot explain an adequate amount of the variance in the data.

Another, similar to the abovementioned indicator is the model consistency, defined in Reference [28]. Within the frame of GCCA, this indicator is estimated as:

$$C = 1 - \frac{|R_S - R_r|}{|R_r|} \times 100 \tag{10}$$

where R_r is the correlation vector of the real data and R_s the correlation vector of the data generated by the MVAR model. Generally, if $C < 80\%$, a cause for concern regarding model's ability arises.

If the model captures effectively the data, the residuals of the MVAR model must resemble a white noise process, i.e., they must be serially uncorrelated.

The Durbin–Watson statistic d [62], tests this behavior, and is given by:

$$d = \frac{\sum_{t=2}^{T}(\varepsilon_t - \varepsilon_{t-1}^2)}{\sum_{t=2}^{T}\varepsilon_t^2} \qquad (11)$$

If $d < 1.0$ there may be, also, cause for concern.

6.5. Research Assumptions

Correlation coefficients (Tables 8 and 9) reveal some connections among CBTF. We tried to apply a more powerful tool (GC) to detect if there was any hidden relation and to identify the direction of such a relation.

7. Results

We tried a number of MVAR models (or case studies, see Section 4, Table 6) to study the interaction of the CBT variables between Greece–Italy–Bulgaria, although there was no physical interconnection between Italy and Bulgaria. The first complete model, A, consisted of all three countries in order to detect all possible interactions, direct and indirect (transit flows). The other two models, B and C, consisted of Greece–Italy and Greece–Bulgaria, respectively, in order to emphasize the direct interactions among interconnected countries.

From Table 10, we observe that the MVAR models can explain the largest amount of the variance of the dependent variable, based on the shown validity measures, for cases A and B, i.e., the interconnections that involve the variables as shown in Table 6(as described Table 4). Variables D9, D15, D16, D21, and D24 for model A and D11 and D12 for model B were entered in the respective MVAR models as first differenced variables to guarantee their covariance stationarity (Table 5). We focused on models A and B having the largest consistencies.

Table 10. Models' consistency.

Model Studies	Model's Consistency (%)
A	85.5815
B	83.2861
C	56.592

7.1. Study A: Cross-border Trading between Greece, Italy, and Bulgaria

In case study A (model A), nodes 1, 5, 7, 11, 12, and 16 were identified as causal sources (positive bars in the causal flow bar-chart), while nodes 3, 6, 8, 9, 17, and 18 as causal sinks (negative bars) as can be seen in Figure 4.

In general, we observed that node 5 (Italy's total load forecasted) was the most active variable (strongest driver), while node 8 (DA-Greece) was the most passive (biggest sink).

A connection was observed between nodes 1 and 3 (Italy's solar forecasted generation and Greece's solar forecasted generation, respectively) which is fairly explained due to the similar weather conditions of the two countries (diurnal effects).

Italy's load forecasting (node 5) seems to be strongly correlated with the first differences (changes) in Greece's and Bulgaria's load forecasting (nodes 6 and 18, respectively) which is explained mostly due to the similar weather conditions as well.

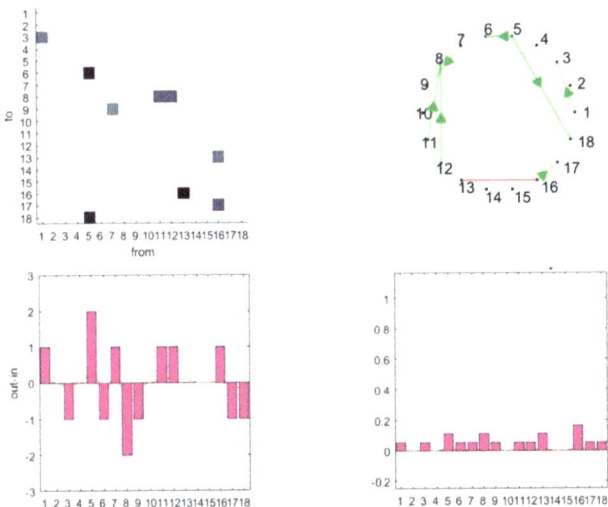

Figure 4. Study A: Demonstration of the GCCA. Matrix and network representations of the corresponding G-causalities are shown in the top panels. Causal flow and unit causal density are shown in the bottom panels.

Furthermore, we observed that DA Italy (node 7) affected the commercial schedule from Italy to Greece (node 9), since the Italian market was highly volatile and liquid, so it faced large excursions (spikes) in its wholesale electricity prices which, in turn, affected the power exchanged between the two countries.

Regarding DA Greece (node 8), which as mentioned was the most passive variable in the specific model, we observed that it was moderately correlated and affected by transfer capacity from Greece to Italy (node 11) and from Italy to Greece (node 12) which as shown in Table 8 were significantly highly correlated (0.998). A rational explanation seems that when the Greece–Italy interconnection was operational, NTC (500MW) comprised a significant value relative to the total demand in Greece. Consequently, DA Greece was highly affected whether Greece–Italy interconnection was operational or not.

A strong bi-directional relation was observed between solar generation in Bulgaria (node 13) and commercial schedules from Bulgaria to Greece which is also explained by the fact that Greece is a major importer from Bulgaria, since the latter has one of the lowest market clearing prices in EU driven by the low marginal costs of its nuclear plants and the low demand needs. So, it is rational to assume that a change in the solar generation of Bulgaria represents an excessive energy production which normally is exported by Bulgarian companies, taking into account that domestic demand for energy is covered by the conventional plants.

Finally, no relation was illustrated between the commercial schedule of Bulgaria and Greece (nodes 16 and 17) and DA prices in Bulgaria (node 15) which firstly seems rather strange. A possible explanation could be the wide margin between DA prices in Greece (higher prices) and Bulgaria. This fact leads electricity traders to commit to long-term contracts exporting energy from Bulgaria to Greece, as there is no commercial risk in that.

7.2. Study B: Cross-Border Trading between Greece and Italy

In case study B (model B), nodes 1, 2, 5, 7, and 11 were identified as casual sources (positive bars in the casual flow bar-chart), while nodes 3, 8, 9, and 10 as casual sinks (negative bars) as can be seen in Figure 5.

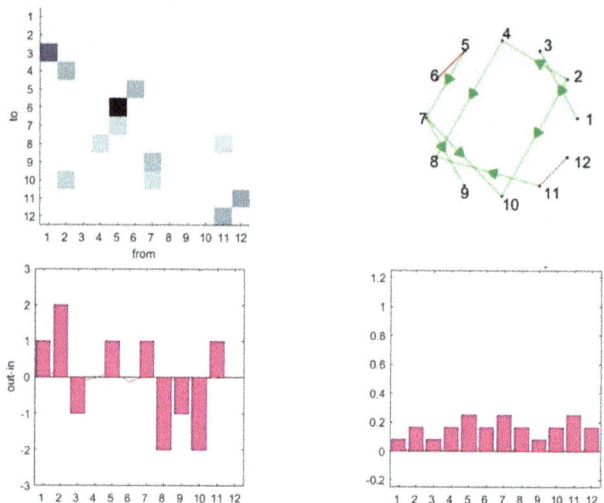

Figure 5. Study B: Demonstration of the GCCA. Matrix and network representations of the corresponding G-causalities are shown in the top panels. Causal flow and unit causal density are shown in the bottom panels.

In general, we observed that node 2 (wind generation in South Italy) was the strongest driver of the specific network with nodes 8 and 11 (DA Greece and commercial schedules from Greece to Italy, respectively) being the most remarkable sinks. This reflection may hide a relation of active and passive roles in the specific interconnection.

The solar forecasted generation, in South Italy (node 1), was strongly correlated to the Greek forecasted solar generation (node 3). A moderate correlation was also observed between South Italy's and Greece's wind generation (nodes 2 and 4). This seems reasonable, taking in consideration the geographical locations and the meteorological conditions of the two regions, South Italy and Greece, having a very similar climate, and it is something we observed from the correlation coefficient in Table 8. However, as mentioned earlier, in the case of wind and solar generation there was no point to discuss the Granger causality, since they are highly stochastic time series. The forecasted South Italy wind generation (node 2) also Granger causes strongly the commercial schedule from Greece to Italy (node 10).

We observed that node 7 (the change in DA price in South Italy), Granger causes nodes 9 (commercial schedule IT-GR) and 10 (commercial schedule GR–IT). But node 7 was influenced or Granger caused by node 5 (Italy's total load forecast). Also, node 5 had a bi-directional connection with node 6 (Greek total load forecast). This was something expected, since when the wholesale price in Italy was high, traders tried to buy cheap electricity for neighboring interconnected countries, hence the commercial schedules were directly affected. So, load forecasts in both countries Granger cause the DA price in South Italy (which means that the error in forecasting Italy's load were minimized when considering Greek load forecasts) which, in turn, Granger causes the total commercial schedules between IT–GR and vice versa. This is a very interesting result, indicating that the spot price in South Italy drives the commercial programs in the cross-border trading of both countries.

Node 4 (Greek forecast wind generation) Granger causes node 8 the changes in DA Greek price, as expected (wind generation reduces, in general, Spot prices). The changes in the DA Greek price were driven (Granger caused) by node 11 (the transfer capacity from Greece to Italy) which had a bi-directional connection to the node 12 (transfer capacity from Italy to Greece).

Regarding the casual density of variables under consideration, nodes 5 (Italian forecast load), 7 (changes in DA-price of South Italy) and 11 (transfer capacity from Greece to Italy) showcase the largest values, reflecting their total amount of casual interactivity, which is also a form of dynamical complexity in the network. These three variables were globally coordinated in their activity, within the system produced by model B. This means that they are useful in predicting each other's activity, as we have seen, but also useful in predicting other variables with different power.

The South Italian forecasted wind generation is the largest casual source, i.e., the variable (node) that exerts the strongest casual influence on the system as a whole, since Granger causes variables 10 and is strongly correlated to 4 (total commercial schedule GR–IT and Greek forecasted wind generation).

8. Conclusions

In the current study, we tried to identify and explain any causalities among energy fundamentals between Greece, Italy, and Bulgaria, using Granger causality theory. Causal connectivity implies a connection that lies on the enhancement of forecasting between two time series. That being said, and according to our knowledge, the lack of similar research on non-market coupled countries makes the current work unique and the outcomes that arise should be taken under consideration especially in the era of the implementation of the ETM.

We designed two models of which the one was reliable (highly consistent) enough to be analyzed. The model that was presented is the one that examines the interconnection between Greece and Italy. The authors could not design a reliable model for the Greece–Bulgaria interconnection. That might be due to the lack of data for the Bulgarian DA price (D21) which forced us to use half of the available data for the rest of the time series as well (in order to have the same number of observations) when examining the specific interconnection.

The results show that there were some causal connections among some of the electricity trading fundamentals which indicate that in order to facilitate the forecasting and minimize errors, we need to include them in the same model. There were also some weaknesses revealed which bring to the surface the need for market coupling among those countries.

It seems that the Italian market has a more active role in the specific interconnection followed by Greece. This could be explained by the fact that Italy is already part of the Central Western European (CWE) region which includes 19 market coupled countries, while Greece is not coupled yet with any of the interconnected countries and thus not so volatile and liquid like the Italian one

The lack of designing a reliable model that explains the causalities between Greece and Bulgaria offers opportunities for further research. Also, it would be interesting to see how the imminent market coupling between Greece and Italy will affect the relations described above.

Author Contributions: G.P.P. with C.D. and C.K. contributed significantly to the conceptualization, development, methodology and formal analysis of the paper. Moreover the former authors have contributed to software development, supervision and project administration as well. G.E. and F.G. contributed to Writing-Review & Editing, Data Curation and supervision. All authors have read and agreed to the published version of the manuscript.

Funding: This research received no external funding

Acknowledgments: Special thanks to ADMIE's Market Management Department for sharing their experience and knowledge with regards to the cross border trading.

Conflicts of Interest: The authors declare no conflict of interest.

References

1. The European Commission. *The European Green Deal*; EU Commission: Brussels, Belgium, 2019.
2. The European Commission. *Market Analysis*; EU Commission: Brussels, Belgium, 2019; Available online: https://ec.europa.eu/energy/en/data-analysis/market-analysis (accessed on 4 February 2020).
3. The European Commission. *Quarterly Report on European Electricity Markets (First Quarter of 2019)*; EU Commission: Brussels, Belgium, 2019; Volume 12.

4. The European Commission. *Quarterly Report on European Electricity Markets (Second Quarter of 2019)*; EU Commission: Brussels, Belgium, 2019; Volume 12.
5. The European Commission. *Energy 2020-A strategy for Competitive, Sustainable and Secure Energy*; EU Commission: Brussels, Belgium, 2011.
6. The European Commission. *A policy Framework for Climate and Energy in the Period from 2020 to 2030*; EU Commission: Brussels, Belgium, 2014.
7. The European Commission. *Renewable Energy Progress Report*; EU Commission: Brussels, Belgium, 2019.
8. Afionis, S.; Dupont, C.; Sokolowski, M.M.; Kalantzakos, S.; Sedlacko, M. *EU Environmental Policy*; Routledge: Abingdon, UK, 2015.
9. Jónsson, T.; Pinson, P.; Madsen, H. On the market impact of wind energy forecasts. *Energy Econ.* **2010**, *32*, 313–320. [CrossRef]
10. David, W.; Christoph, G.; David, H. *The Effect of Intermittent Renewables on Electricity Prices in Germany*; Technical University of Munich: Munich, Germany, 2014.
11. Regulatory Authority for Energy (RAE). *Regulation and Performance of the Electricity Market and the Natural Gas Market in Greece, in 2017*; RAE: Athens, Greece, 2018.
12. Phan, S.; Roques, F. *Is the Depressive Effect of Renewables on Power Prices Contagious? A Cross Border Econometric Analysis*; Cambridge Working Papers in Economics 1527; University of Cambridge: Cambridge, UK, 2015.
13. Ketterer, J.C. The impact of wind power generation on the electricity price in Germany. *Energy Econ.* **2014**, *44*, 270–280. [CrossRef]
14. Haximusa, A. *The Effects of German Wind and Solar Electricity on French Spot Price Volatility: An Empirical Investigation*; Vienna University of Economics and Business: Vienna, Austria, 2018.
15. Bask, M.; Widerberg, A. *Market Structure and the Stability and Volatility of Electricity Prices*; University of Gothenburg: Gothenburg, Sweden, 2008.
16. Denny, E.; Tuohy, A.; Meibom, P.; Keane, A.; Flynn, D.; Mullane, A.; O'Malley, M. The impact of increased interconnection on electricity systems with large penetrations of wind generation: A case study of Ireland and Great Britain. *Energy Policy* **2010**, *38*, 6946–6954. [CrossRef]
17. Zugno, M.; Pinson, P.; Madsen, H. Impact of Wind Power Generation on European Cross-Border Power Flows. *IEEE Trans. Power Syst.* **2013**, *28*, 3566–3575. [CrossRef]
18. Castagneto Gissey, G. *Electricity and Energy Price Interactions in Modern EU Markets*; Imperial College: London, UK, 2015.
19. Park, H.; Mjelde, J.W.; Bessler, D.A. Price dynamics among U.S. electricity spot markets. *Energy Econ.* **2006**, *28*, 81–101. [CrossRef]
20. Narayan, P.K.; Smyth, R. Multivariate granger causality between electricity consumption, exports and GDP: Evidence from a panel of Middle Eastern countries. *Energy Policy* **2009**, *37*, 229–236. [CrossRef]
21. Ferkingstad, E.; Løland, A.; Wilhelmsen, M. Causal modeling and inference for electricity markets. *Energy Econ.* **2011**, *33*, 404–412. [CrossRef]
22. Yang, Z.; Zhao, Y. Energy consumption, carbon emissions, and economic growth in India: Evidence from directed acyclic graphs. *Econ. Model.* **2014**, *38*, 533–540. [CrossRef]
23. Costa, L.D.F.; Rodrigues, F.A.; Travieso, G.; Villas Boas, P.R. Characterization of complex networks: A survey of measurements. *Adv. Phys.* **2007**, *56*, 167–242. [CrossRef]
24. Watts, D.J.; Strogatz, S.H. Collective dynamics of "small world" networks. *Lett. Nat.* **1998**, *393*, 440–442. [CrossRef]
25. Barabasi, A.-L.; Albert, R. Emergence of scaling in random networks. *Science* **1999**, *286*, 509–512. [CrossRef]
26. Boccaletti, S.; Latora, V.; Moreno, Y.; Chavez, M.; Hwang, D.U. Complex networks: Structure and dynamics. *Phys. Rep.* **2006**, *424*, 175–308. [CrossRef]
27. Seth, A.K. Causal networks in simulated neural systems. *Cogn. Neurodyn.* **2008**, *2*, 49–64. [CrossRef] [PubMed]
28. Seth, A.K. A MATLAB toolbox for Granger causal connectivity analysis. *J. Neurosci. Methods* **2010**, *186*, 262–273. [CrossRef] [PubMed]
29. Barnett, L.; Seth, A.K. *The MVGC Multivariate Granger Causality Toolbox: A New Approach to Granger-causal Inference*; Sackler Centre for Consciousness Science and School of Informatics, University of Sussex: Brighton, UK, 2015.

30. Nazlioglu, S.; Kayhan, S.; Adiguzel, U. Electricity Consumption and Economic Growth in Turkey: Cointegration, Linear and Nonlinear Granger Causality. *Energy Sources Part B Econ. Plan. Policy* **2014**, *9*, 315–324. [CrossRef]
31. Bruns, S.B.; Gross, C.; Stern, D.I. *Is There Really Granger Causality Between Energy Use and Output?* RWTH Aachen University: Aachen, Germany, 2013.
32. Woo, C.-K.; Olson, A.; Horowitz, I.; Luk, S. Bi-Directional causality in California's electricity and natural-gas markets. *Energy Policy* **2006**, *34*, 2060–2070. [CrossRef]
33. Antoniadou, S. *Causal Relationship in Oil Markets*; Interantional Hellenic University: Thessaloniki, Greece, 2016.
34. Li, S.; Zhang, H.; Yuan, D. Investor attention and crude oil prices: Evidence from nonlinear Granger causality tests. *Energy Econ.* **2019**, *84*, 104494. [CrossRef]
35. Chang, C.-L.; McAleer, M.; Wang, Y.-A. *Latent Volatility Granger Causality and Spillovers in Renewable Energy and Crude Oil ETFs*; Tinbergen Institute: Amsterdam, The Netherlands, 2018.
36. Zhong, J.; Wang, M.; Drakeford, B.; Li, T. Spillover effects between oil and natural gas prices: Evidence from emerging and developed markets. *Green Financ.* **2019**, *1*, 30–45. [CrossRef]
37. Scarcioffolo, A.R.; Etienne, X.L. Testing directional predictability between energy prices: A cross-Quantilogram analysis. In Proceedings of the AAEA Annual Meeting, Atlanta, GA, USA, 21–23 July 2019.
38. Papaioannou, G.P.; Dikaiakos, C.; Stratigakos, A.C.; Papageorgiou, P.C.; Krommydas, K.F. Testing the efficiency of electricity markets using a new composite measure based on nonlinear TS tools. *Energies* **2019**, *12*. [CrossRef]
39. Ardian, F. *Empirical Analysis of Italian Electricity Market*; Universite Paris-Saclay: Paris, France, 2016.
40. Iberian Electricity Market Iberian Electricity Market -MIBEL. Available online: https://www.mibel.com/en/home_en/ (accessed on 4 February 2020).
41. Ferreira, M.; Sebastiao, H. *The Iberian Electricity Market: Price Dynamics and Risk Premium in An Illiquid Market*; Centre for Business and Economics Research, University of Coimbra: Lisbon, Portugal, 2018.
42. Herráiz, Á.C.; Monroy, C.R. Analysis of the Iberian electricity forward market hedging efficiency. *Int. J. Financ. Eng. Risk Manag.* **2013**, *1*, 20–34. [CrossRef]
43. Réseau de Transport d'Électricité (RTE). *Electricity Report 2017*; RTE: Paris, France, 2017.
44. Frontier Economics. *Overview of European Electricity Markets*; European Commssion: Brussels, Belgium, 2016.
45. Commission de Regulation d' Energie Central to the operation of the French Power System. Available online: https://www.cre.fr/en/Electricity/Wholesale-electricity-market/wholesale-electricity-market (accessed on 4 February 2020).
46. *The European Commission Achieving the 10% Electricity Interconnection Target Making Europe's Electricity Grid Fit for 2020*; European Commission: Brussels, Belgium, 2015.
47. The European Commission. *The European Commission European solidarity on Energy: Better integration of the Iberian Peninsula into the EU Energy Market 2018*; EU Commission: Brussels, Belgium, 2018; Available online: https://ec.europa.eu/commission/presscorner/detail/en/IP (accessed on 4 February 2020).
48. The European Commission. *EU Energy Markets in 2014*; EU Commission: Brussels, Belgium, 2014.
49. Italian Regulatory Authority for Energy Networks and Environment (ARERA). *Annual Report on Electricity and Gas Markets*; ARERA: Milan, Italy, 2019.
50. Regulatory Authority for Energy (RAE). *National Report 2019*; RAE: Athens, Greece, 2019.
51. Energy and Water Regulatory Commission (EWRC) Bulgaria. *Annual Report to the European Commission*; EWRC: Sofia, Bulgaria, 2019.
52. The European Commission Comission Regulation (EU) 2015/1222 establishing a guideline on capacity allocation and congestion management. *Off. J. EU Comm.* **2015**, *197*, 24–72.
53. ENTSO-E Single Intraday Coupling. Available online: https://www.entsoe.eu/network_codes/cacm/implementation/sidc/ (accessed on 24 January 2020).
54. ENTSO-E. *Market Report 2019*; ENTSO-E: Brussels, Belgium, 2019.
55. ENTSO-E Single Day-ahead Coupling. Available online: https://www.entsoe.eu/network_codes/cacm/implementation/sadc/ (accessed on 24 January 2020).
56. ACER; CEER. *ACER Market Monitoring Report 2018-Electricity Wholesale Markets Volume*; ACER: Ljubljana, Slovenia, 2019.
57. Antweiler, W. Cross-Border trade in electricity. *J. Int. Econ.* **2016**, *101*, 42–51. [CrossRef]

58. Swanson, N.R.; Granger, W.J.C. Impulse response functions based on a casual approach to residual orthogonalization in vector autoregressions. *J. Am. Stat. Assoc.* **2012**, *92*, 357–367. [CrossRef]
59. Bernanke, B.S. Alternative explanations of the money-Income correlation. *Carnegie-Rochester Conf. Ser. Public Policy* **1986**, *25*, 49–99. [CrossRef]
60. Pearl Judea Causality: Models, reasoning and inference. *Econom. Theory* **2003**, *19*, 675–685.
61. Zheng, Q.; Sone, L. *Dynamic Contagion of Systemic Risks on Global Main Equity Markets based on Granger Causality Networks*; Hindawi, Discrete Dynamics in Nature and Society: London, UK, 2018.
62. Durbin, J.; Watson, G.S. Testing for serial correlation in least squares regression. *Biometrika* **1950**, *37*, 409–428.

© 2020 by the authors. Licensee MDPI, Basel, Switzerland. This article is an open access article distributed under the terms and conditions of the Creative Commons Attribution (CC BY) license (http://creativecommons.org/licenses/by/4.0/).

Article

An Integrated Energy Simulation Model for Buildings

Nikolaos Kampelis [1,*], Georgios I. Papayiannis [2,3], Dionysia Kolokotsa [1], Georgios N. Galanis [2], Daniela Isidori [4], Cristina Cristalli [4] and Athanasios N. Yannacopoulos [3]

1. Energy Management in the Built Environment Research Lab, Environmental Engineering School, Technical University of Crete, 73100 Chania, Greece; dkolokotsa@isc.tuc.gr
2. Mathematical Modeling and Applications Laboratory, Section of Mathematics, Hellenic Naval Academy, 18539 Piraeus, Greece; g.papagiannis@hna.gr (G.I.P.); ggalanis@hna.gr (G.N.G.)
3. Stochastic Modeling and Applications Laboratory, Department of Statistics, Athens University of Economics & Business, 10434 Athens, Greece; ayannaco@aueb.gr
4. Research for Innovation, AEA srl, Angeli di Rosora, 60030 Marche, Italy; d.isidori@loccioni.com (D.I.); c.cristalli@loccioni.com (C.C.)
* Correspondence: nkampelis@isc.tuc.gr

Received: 14 January 2020; Accepted: 25 February 2020; Published: 4 March 2020

Abstract: The operation of buildings is linked to approximately 36% of the global energy consumption, 40% of greenhouse gas emissions, and climate change. Assessing the energy consumption and efficiency of buildings is a complex task addressed by a variety of methods. Building energy modeling is among the dominant methodologies in evaluating the energy efficiency of buildings commonly applied for evaluating design and renovation energy efficiency measures. Although building energy modeling is a valuable tool, it is rarely the case that simulation results are assessed against the building's actual energy performance. In this context, the simulation results of the HVAC energy consumption in the case of a smart industrial near-zero energy building are used to explore areas of uncertainty and deviation of the building energy model against measured data. Initial model results are improved based on a trial and error approach to minimize deviation based on key identified parameters. In addition, a novel approach based on functional shape modeling and Kalman filtering is developed and applied to further minimize systematic discrepancies. Results indicate a significant initial performance gap between the initial model and the actual energy consumption. The efficiency and the effectiveness of the developed integrated model is highlighted.

Keywords: deformable models; electric energy demand; functional statistics; Kalman filtering; shape-invariant model

---

1. Introduction

Energy consumption in the building sector (combined with buildings construction) is associated with 36% of global final energy consumption and approximately 40% of total direct and indirect CO_2 emissions. Associated figures rise every year mainly due to (a) improved energy access levels in developing countries, (b) increased ownership levels of energy consuming devices, and (c) the growth of global buildings floor area [1]. Measures to reduce energy consumption at building level include passive and active energy efficiency measures, storage, energy management, and building integrated renewable energy systems (e.g., solar, geothermal, and wind). Design and modeling of integrated energy systems in net zero energy buildings is discussed by Athienitis and O'Brien [2].

Assessing energy efficiency in buildings is a complex task which varies according to the aim of the analysis and the specificity of each case. In any case, expert knowledge and a set of technical and non-technical information is required. Technical information usually concerns the geometry and thermal characteristics of the building envelope, the design and operation of HVAC system, as well as data from measurements regarding indoor/outdoor climate conditions and energy end usage.

Non-technical data in some cases are associated with occupancy levels, clothing levels, users behavior, perceptions, personal economics, and preferences.

Several decades now, significant research efforts have been directed towards evaluating energy efficiency in buildings. Research results have been applied in advancing state-of-the-art, knowledge and understanding, fostering new policies, better regulation, and innovations. In this context, several methods have been developed and evolved along with the creation of custom software applications built to match the needs of certain research and development fields. Various new techniques have been developed and applied to include the physical or "white box" approach, the statistical or machine learning "black box" and the hybrid "gray box" approach. A comparison of the models and applications for predicting building energy consumption in terms of complexity, easiness to use, running speed, input requirements, and accuracy is conducted by Zhao and Magoulès [3].

Physical models are developed to evaluate energy consumption in buildings by using equations of heat and mass transfer between the building and the surrounding environment as well as energy conservation among the building spaces and system components. Such physical models are classified based on the deployed approach and whether it is defined as single (well-mixed) zone [4], multi-zone, or zonal as provided in the state-of-the-art building modeling and energy prediction review by Foucquier et al. [5]. The zonal method is a simplification of CFD in the sense that the thermal zone is divided into several cells and in some cases representation in 2D is feasible. The advantage of the zonal approach is linked to the capability of dealing with large volumes in moderate computation time. SPARK [6] is such an example of a zonal approach software application. According to the multi-zone approach, each one zone or building element (e.g., wall and window) or system (e.g., HVAC system) or specific load (e.g., due to occupancy) is considered as one node for which the heat transfer equations are calculated. Each thermal zone is considered as homogeneous and represented by uniform state variables such as temperature, pressure, relative humidity, etc. The multi-zone or nodal approach is particularly effective when evaluating the energy performance of a building with many thermal zones since computational time for a year round simulation is relatively small. Therefore, the multi-zone approach is suitable for testing the impact of alternative energy efficiency measures provided that a reliable validated model has been created. The main disadvantage of this approach is related to the difficulty in creating a valid building model especially in the absence of holistic information of the building as built, systems installed, operational aspects, and data from measurements. EnergyPlus, ESP-r, and TrnSys are among the most robust and frequently used multi-zone software programs. Benchmarking between building energy simulation software programs reported in the literature is available by Harish et al. [7].

Data-driven methods on the other hand require no physical information, i.e., thermal or geometrical parameters, as they do not deploy heat transfer equations. Regression, Artificial Neural Network (ANN), Genetic Algorithm (GA), and Support Vector Machine (SVM) are some of the techniques used in building energy forecasting based solely on measurements of parameters such as temperature, relative humidity, solar radiation, wind velocity, and energy consumption/production. Machine learning techniques for estimating building energy consumption are exploited by Naji et al. [8] and by Robinson et al. [9]. The main drawback of data-driven methods concerns the interpretation of results in physical terms. Data-driven methods for building energy prediction and classification studies are reviewed by Amasyali and El-Gohary [10] and by Wei et al [11]. Hybrid or "gray box" methods combine physical modeling with data-driven techniques to counterbalance the weaknesses of the "white" and "black" box approaches. Machine learning techniques can be used for parameter estimation, e.g., by coupling a multi-zone model with GA. Another hybrid approach is to use statistical models to improve the performance of physical models with respect to end-uses, which are often unknown and hard to be modeled in a deterministic way.

In contrast with data-driven methods, detailed simulation models do not require conditions monitoring to predict the building performance. However, various sources of uncertainty can lead to significant discrepancies between model predictions and metered energy use. Such sources of uncertainty can be distinguished to specification-related, modeling-related, and scenario-related as discussed by De Wit and Augenbroe [12]. Furthermore, embedding realistic occupant behavior on building modeling and its impact on energy predictions is investigated and addressed as a significant and complex problem by Ryan and Sanquist [13]. Overall, performance gaps can be attributed to a variety of discrepancies in building modeling. A methodology to identify building energy performance problems related to operational, measurements, or simulation aspects is proposed by Maile et al. [14].

Essentially, validation is a critical step to assess the simulation model's plausibility and reliability. This is especially important for the investigation of energy efficiency measures and the assessment of the actual (not the relative) potential impact in terms of energy and cost savings, environmental performance, thermal comfort, etc. Methods for model validation (otherwise referred to as calibration) are reviewed by Coakley et al. [15], and indicators used in relevant standards (ASHRAE Guideline 14, IPMVP, FEMP) to address acceptance thresholds are discussed [16] by Royapoor and Roskilly [17]. Calibration of a building energy model based on actual measurements is extensively investigated by O'Neil et al. [18]. The building under study hosts offices and conferences rooms, and the authors perform an extensive sensitivity analysis including more than 2000 parameters automatically refined using analytic meta-model-based optimization. A similar approach is followed by O'Neil et al. [19] incorporating EnergyPlus and TRNSYS in their investigation and over 1000 parameters for model calibration.

This paper investigates (i) the validation of a Near-Zero Energy Building (NZEB) simulation model through trial and error approach of important model parameters and (ii) a novel postprocessing approach is proposed which improves the simulation model's accuracy, combining techniques of functional statistics through appropriate energy shape modeling (employing the concept of deformable models) and Kalman filtering to reduce the remaining biases. Initially, a physical model of the Leaf Lab industrial NZEB in Italy is used to conduct a simulation of the building's energy consumption for two consecutive years using weather data recorded from onsite meteorological stations. HVAC electric consumption obtained from the simulation is compared to actual measured values to establish the initial (baseline) performance gap. Subsequently, through a trial and error approach, important model parameters are identified and fine-tuned until an improved acceptable correlation between simulated and recorded HVAC electric energy consumption is reached. At a second stage, inconsistencies in the shape of the optimized energy prediction are corrected through an appropriate functional shape/reshape modeling task by comparing recent measured energy demand outputs to predictions and appropriately estimating expected deviations which are used to improve the shape model's output. At the last step, Kalman filtering is incorporated to remove remaining systematic biases where is needed. Therefore, the original physical model's results are passed through this integrated model where meaningful interventions are performed to better predict the true situation without neglecting the physical interpretation of the model output. The benefits from the proposed approach are illustrated in the later section (Section 3) with substantial reduction in the error magnitude.

2. Methodology and Modeling Approaches

In this section, we describe the modeling approaches that are used throughout the paper for the prediction of the energy demand. In particular we present (a) the initial energy simulation model and its optimized version by appropriate parameter tuning, (b) the shape model approach in which the optimized output of the energy simulation model is reshaped to further reduce systematic inefficiencies related to discrepancies from the actual shape of the energy demand, and (c) postprocessing using Kalman filtering which reduces remaining systematic errors not captured by the shape model approach. A flowchart of the methodology is presented in Figure 1.

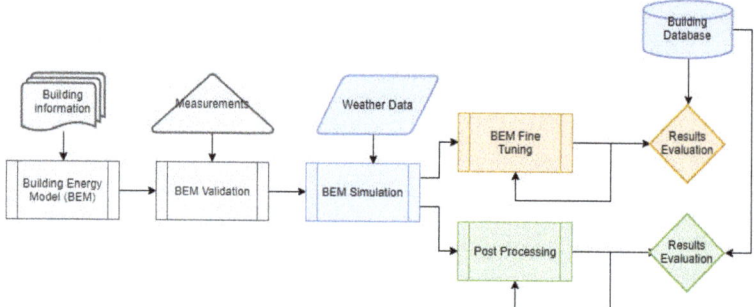

Figure 1. Flowchart of the high level methodology followed.

2.1. The Energy Simulation Model: General Presentation of the Model, Areas of Application, Advantages, and Shortcomings

EnergyPlus is the simulation engine software used to conduct an integrated simulation of the building, system, and plant whereby supply and demand are matched based on successive iteration substitution following Gauss–Seidel updating [20]. Open Studio is used as the API software for developing and parameterizing the model following the principles outlined by Brackney et al. [21]. Ambient temperature, relative humidity, solar radiation, and wind speed data for 2017 and 2018 was obtained from local meteorological equipment and converted to two yearly weather files. Data of total HVAC energy consumption, for the years 2017 and 2018, are exploited for providing the baseline against which model based results are evaluated.

The simulation model contains, on the one hand, the geometry, construction components and materials of the building under study. For opaque material thickness (m), thermal conductivity (W/mK), density (kg/m^3), and thermal absorptance (*dimensionless*) properties are edited. For transparent materials, such as glass in windows and sky windows thickness (m), thermal conductivity (W/mK) and optical properties, such as solar, visible, and infrared transmittance, are inserted. On the other hand, a model of the HVAC system is designed based on the installed technologies and adjusted accordingly to the actual key performance heat pump technical parameters such as Coefficients of Performance (COP), fan maximum flow power (m^3/s), pressure rise, and efficiency. Other parameters such as rated total heating/cooling capacity, and rated and maximum air flow rated are automatically sized based on the software's calculations. Furthermore, with respect to the operation of the major installed systems, the simulation model takes into account the temperature set points of the HVAC system, ventilation, and infiltration rates (ACH^{-1}) and a number of schedules to determine artificial lighting, electric equipment, and occupancy. Moreover, humidity control is exercised using appropriate schedules controlling the operation of the HVAC to ensure that the relative humidity during working hours in the various thermal zones varies between 40% and 60%. Subsequently, an intensive search of the parameters that affected the daily, monthly, and annual power distribution profiles is followed to improve the initial results of the model based by minimizing deviation from HVAC power consumption data. Through the trial and error various combinations and fine-tuning of the all of the above parameters is carried out to reach the optimum results when assessing intra-day, monthly, and annual deviation levels.

EnergyPlus simulation is based on heat balance calculations solved simultaneously with the aid of on an integration solution manager, which includes surface heat balance, air heat balance, and building systems simulation blocks. The heat balance of outside surfaces is calculated based on the equation

$$q''_{asol} + q''_{LWR} + q''_{conv} - q''_{ko} = 0 \qquad (1)$$

where

q''_{asol} is the absorbed direct and diffuse solar (short wavelength) radiation and heat flux
q''_{LWR} is the net long wavelength (thermal) radiation flux exchange with the air and surroundings
q''_{conv} is the convective flux exchange with the outside air
q''_{ko} is the conduction heat flux (q/A) into the wall

Clearly, q''_{asol} is influenced by parameters such as location, surface angle and tilt, surface material, and weather conditions. q''_{LWR} is determined by radiation exchange between the surface and the ground, sky and air. It is dependent on the absorptivity and emissivity of the surface; the temperature of the surface, sky, ground, and air; and corresponding view factors. Assumptions such that each surface is at uniform temperature and energy flux leaving a surface is evenly distributed are considered reasonable for building energy simulation. Using the Stefan–Boltzmann Law in the above equation yields

$$q''_{LWR} = \epsilon \sigma F_{gnd}(T^4_{gnd} - T^4_{surf}) + \epsilon \sigma F_{sky}(T^4_{sky} - T^4_{surf}) + \epsilon \sigma F_{air}(T^4_{air} - T^4_{surf}) \qquad (2)$$

where

ϵ is the long-wave emittance of the surface
σ is the Stefan–Boltzmann constant
F_{gnd} is the view factor of wall surface to ground surface temperature
F_{sky} is the view factor of wall surface to sky temperature
F_{air} is the view factor of wall surface to air temperature
T_{surf} is the outside surface temperature
T_{gnd} is the ground surface temperature
T_{sky} is the sky temperature
T_{air} is the air temperature

The above equation is converted by introducing linear radiative heat transfer coefficients such that

$$q''_{LWR} = h_{r,gnd}(T_{gnd} - T_{surf}) + h_{r,sky}(T_{sky} - T_{surf}) + h_{r,air}(T_{air} - T_{surf}) \qquad (3)$$

where

$$h_{r,gnd} = \epsilon \sigma F_{gnd}(T^4_{surf} - T^4_{gnd})/(T_{surf} - T_{gnd}) \qquad (4)$$
$$h_{r,gnd} = \epsilon \sigma F_{gnd}(T^4_{surf} - T^4_{sky})/(T_{surf} - T_{sky}) \qquad (5)$$
$$h_{r,gnd} = \epsilon \sigma F_{gnd}(T^4_{surf} - T^4_{air})/(T_{surf} - T_{air}) \qquad (6)$$

Exterior convection is modeled using equation

$$q''_{conv} = h_{c,ext} A (T_{surf} - T_{air}) \qquad (7)$$

where

q''_{conv} is the rate of exterior convective heat transfer
$h_{c,ext}$ is the exterior convection coefficient
A is the surface area
T_{surf} is the surface temperature
T_{air} is the outdoor air temperature

Conduction heat fluxes are modeled based on equation

$$q''_{ko}(t) = \sum_{j=0}^{\infty} X_j T_{o,t-j\delta} - \sum_{j=0}^{\infty} Y_j T_{i,t-j\delta} \qquad (8)$$

where

$q''_{ko}(t)$ is the conductive heat flux for the current time step

T is temperature
i indicates the internal element of the building
o indicates the external element of the building
X, Y are the response factors

In more detail, Conduction Transfer Functions (CTFs) as shown in (9) and (10) below are used to estimate the heat fluxes on either side of the building elements based on previous temperature values of interior and exterior surfaces as well as previous interior flux values.

$$q''_{ki}(t) = -Z_o T_{i,t} - \sum_{j=1}^{nz} Z_j T_{i,t-j\delta} + Y_o T_{o,t} + \sum_{j=1}^{nz} Y_j T_{o,t-j\delta} + \sum_{j=1}^{nq} \Phi_j q''_{ko,t-j\delta} \quad (9)$$

$$q''_{ko}(t) = -Y_o T_{i,t} - \sum_{j=1}^{nz} Y_j T_{i,t-j\delta} + X_o T_{o,t} + \sum_{j=1}^{nz} X_j T_{o,t-j\delta} + \sum_{j=1}^{nq} \Phi_j q''_{ko,t-j\delta} \quad (10)$$

where

X_j is the outside CTF coefficient, j = 0,1,...nz
Y_j is the cross CTF coefficient, j = 0,1,...nz
Z_j is the inside CTF coefficient, j = 0,1,...nz
ϕ_j is the flux CTF coefficient, j = 0,1,...nq
T_i is the inside surface temperature
T_o is the outside surface temperature
q''_{ko} is the conduction heat flux on the outside face
q''_{ki} is the conduction heat flux on the inside face

In addition, for each thermal zone EnergyPlus simulation is based on an integration of energy and moisture balance as shown in the Equation (11) below.

$$C_z \frac{dT_z}{dt} = \sum_{i=1}^{N_{sl}} \dot{Q}_i + \sum_{i=1}^{N_{surfaces}} h_i A_i (T_{si} - T_z) + \sum_{i=1}^{N_{surfaces}} \dot{m}_i C_p (T_{zi} - T_z) + \dot{m}_{inf} C_p (T_{zi} - T_z) + \dot{Q}_{sys} \quad (11)$$

where

$\sum_{i=1}^{N_{sl}} \dot{Q}_i$ is the sum of convective heat transfer from the zone surfaces
$\sum_{i=1}^{N_{surfaces}} h_i A_i (T_{si} - T_z)$ is the convective heat transfer from the zone surfaces
$\dot{m}_{inf} C_p (T_{zi} - T_z)$ is the heat transfer due to infiltration of outside air
$\sum_{i=1}^{N_{surfaces}} \dot{m}_i C_p (T_{zi} - T_z)$ is the heat transfer due to interzone air mixing
$\dot{Q}_{sys}$ is the air systems output
$C_z \frac{dT_z}{dt}$ is the energy stored in zone air, and
$C_z = \rho_{air} C_p C_T$

Infiltration is outdoor air unintentionally entering the building due to the opening of doors as well as air leakage through windows and other openings. Infiltrated air is mixed with air in the various thermal zones of the building. Determining infiltration (or air tightness) values contains significant uncertainty, as it requires a complex and elaborate procedure often referred to as blower door test. Infiltrated air is commonly modeled as the number of air changes per hour (ACH) and taken into account in the air heat balance at temperature equal to that of ambient air. In EnergyPlus, infiltration is modeled based on Equation (12).

$$Infiltration = (I_{design})(F_{schedule})[A + B|(T_{zone} - T_{odb})| + C(Windspeed) + D(Windspeed)^2] \quad (12)$$

where

I_{design} is the user defined infiltration value (ACH^{-1})
T_{zone} is the zone air temperature at current conditions (deg C)
T_{odb} is the outdoor air dry-bulb temperature (deg C)
$F_{schedule}$ is a user defined schedule value between 0 and 1
A is the constant term coefficient
B is the temperature term coefficient
C is the velocity term coefficient
D is the velocity squared coefficient

Similarly, ventilation can be modeled using a schedule, maximum and minimum values, as well as delta temperature values, and is determined by the equation

$$Ventilation = (V_{design})(F_{schedule})[A + B|(T_{zone} - T_{odb})| + C(Windspeed) + D(Windspeed)^2] \quad (13)$$

where

V_{design} is the user defined ventilation value (ACH^{-1})
T_{zone} is the zone air temperature at current conditions (deg C)
T_{odb} is the outdoor air dry-bulb temperature (deg C)
$F_{schedule}$ is a user defined schedule value between 0 and 1
A is the constant term coefficient
B is the temperature term coefficient
C is the velocity term coefficient
D is the velocity squared coefficient

Furthermore, the energy provided to each thermal zone by the HVAC system, $\dot{Q}_{sys}$, is given by Equation (14).

$$\dot{Q}_{sys} = \dot{m}_{sys} C_p (T_{sys} - T_z) \quad (14)$$

Equations (11) and (14) can be transformed to yield zone air temperature as shown in Equation (15) below.

$$T_z^t = \frac{\sum_{i=1}^{N_{sl}} Q_i^t + \dot{m}_{sys} C_p T_{supply}^t + (C_z \frac{T_z}{t} + \sum_{i=1}^{N_{surfaces}} h_i A_i T_{si} + \sum_{i=1}^{N_{zones}} \dot{m}_i C_p T_{zi} + \dot{m}_{inf} C_p T_\infty)^{t-\delta t}}{\frac{C_z}{\delta t} + (\sum_{i=1}^{N_{surfaces}} h_i A_i + \sum_{i=1}^{N_{zones}} \dot{m}_i C_p + \dot{m}_{inf} C_p + \dot{m}_{sys} C_p)} \quad (15)$$

2.2. Shape Modeling and Systematic Inefficiencies Correction of the Prediction Model: Presentation of the Properties and Capabilities of the Shape Invariant Model and Implementation in the Current Study

At the first stage, the energy simulation model was carefully tuned to reduce prediction errors. However, there are some sources of error that cannot be effectively treated only through parameter tuning. Therefore, advanced statistical modeling approaches are considered to further reduce the prediction model's deviance from the true situation. At this postprocess stage, the actual phenomena of energy and mass transfer between the environment and the building as well as within the building itself are not explicitly modeled. Instead, this is a functional modeling approach, wherein the shape of the prediction for the energy demand is appropriately modeled and rearranged to better approximate the actual (measured) energy shape. As a result, the proposed shape modeling approach obsoletes deficiencies caused on the difference of the shape between prediction–reality while the remaining biases are further treated through appropriate Kalman filtering procedures discussed in Section 2.3.

2.2.1. The Shape Model Approach

Let us denote by $(t, X_j(t))$ the observations representing the energy demand at a specified day j where $t \in [0, 24)$ represents the hour of the day and $X_j(t)$ the observed energy demand at time instant t. In general, these measurements are available only on certain time segments (e.g., per hour); therefore, the accurate daily shape is not known. That means that although the true shape of the energy demand is a continuous function with respect to time parameter t, in practice only some points of this shape are known at specified time segments t_i; therefore, the available data are of the form $\{t_i, X_j(t_i)\}_{i=1}^n$, which can be considered as landmarks. As we are interested in working with the shapes of the daily energy demands, we need to consider the shape of the day j as a function with respect to time, i.e., $f_j : [0, 24) \to \mathbb{R}$. Then, using the available data $\{t_i, X_j(t_i)\}_{i=1}^n$ from day j we are able to estimate a smoothed version of the energy shape employing any typical interpolation method or nonparametric filters (e.g., spline smoothers, kernel-based smoothing methods, etc. [22–24]) by choosing appropriately the mollification parameters in order not to lose important aspects of the information. In this manner, the shape function of the intra-day power demand is sufficiently recovered with the advantage that we can get estimates for the demand even in time instants that no data are available and allowing to treat data with functional statistics techniques.

The daily shape of the energy demand is not expected to change dramatically between two days given that we consider typical working days (i.e., not weekends or public holidays). As a result, a standard energy-demand picture is expected to be observed with small fluctuations from one day to another, where these fluctuations can be efficiently calibrated by appropriate shape model considerations. Consider, for example, that for an arbitrary day j, $f_j(t)$ denotes the observed energy demand and $\tilde{f}_j(t)$ denotes the prediction obtained by the simulation model discussed in Section 2.1. Clearly, as the simulation model prediction is not expected to coincide with the true state of the energy demand, if systematic inefficiencies are presented between the prediction and reality, then a shape correction procedure could remarkably reduce errors caused to daily energy shape deviances. We discuss an approach under which we expect to provide corrections to the simulation model prediction by properly "reshaping" the simulation output in order to better match the observed energy shapes based on previous data. Such an approach is possible under the framework of deformation models (see, e.g., [25–28]) where the observed function f_j (i.e., observed energy shape) is considered as a deformation of the prediction model $\tilde{f}_j$ (predicted energy shape), and this relation is mathematically expressed through the model

$$f_j(t) = R_j(\tilde{f}_j(t)) + \epsilon_j(t) \qquad (16)$$

where $R_j : \mathbb{R} \to \mathbb{R}$ is called a deformation function and $\epsilon_j(t)$ is considered as a white noise process. Although several models can be proposed to parameterize the deformation function (e.g., shape-invariant model [25–27]), for the particular nature of the data we consider in this paper, we may propose a simpler model which consists of modeling the reshape function α_j defined by

$$\alpha_j(t) := [\tilde{f}_j(t)]^{-1} f_j(t). \qquad (17)$$

If the modeler had knowledge of the reshape function, then he/she could perfectly adjust the initial prediction $\tilde{f}_j(t)$, provided by the simulation model discussed in Section 2.1, to obtain exactly the true energy demand $f_j(t)$. In particular, knowledge of the exact functional form for $\alpha(\cdot)$ would allow predictions even at intermediate time instants for which observations are not available. In this section, we propose a functional statistical model to estimate the reshape function from past data of initial simulation model discrepancies from the actual measured energy shapes, so that it can be used for future predictions.

As the exact knowledge of the reshape function is not an option, we can estimate it using data from the previous days (we should choose a small time window ~5–10 days) to model the "typical" reshape function that is observed in the near past. Clearly, using the information provided from the last N days, let us define the set of reshape functions $\mathcal{A} := \{\alpha_1, \alpha_2, ..., \alpha_N\}$. For the case under consideration, it is expected that there are certain aspects of the daily energy shape which the prediction model cannot efficiently calibrate and systematically does not capture, and as a result, the reshape functions must be very similar objects as shown for example for a typical set of observations in Figure 2. In such a case, the approach we propose here is valid.

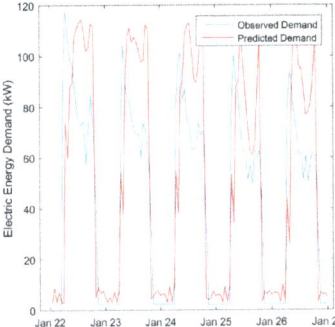
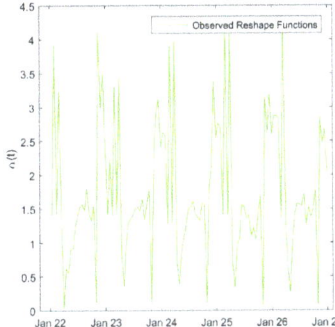

Figure 2. Example of the observed and simulated daily energy demand (per hour) and the corresponding reshape functions for the time period 22/01/2018 to 26/01/2018.

Under this consideration, for an appropriately chosen period of time each function, α_j should not present significant fluctuations from the reshape function that is considered as the "typical" one. Therefore, an appropriate notion of mean regarding the set of reshape functions $\mathcal{A}$ is needed to properly define/represent the mean element in the set. The latter task requires the calculation of the mean element among a number of functions living in a space which does not necessarily has linear structure, therefore the notion of Fréchet mean needs to be exploited [29,30] for this purpose. In the context we discuss here, each reshape function $\alpha_j(\cdot)$ is considered as a deformation to the mean element (i.e., Fréchet mean) $\bar{\alpha}(\cdot)$ with respect to which an appropriate notion of deviance is defined and its minimization will provide the mean element (please see [31] for technical details on this subject). A general model like the *shape-invariant model* [26] can be used for the parameterization of the functional characteristics of the reshape functions in order to define a consistent parametric form for the Fréchet mean. According to the shape-invariant model, shape deformations like vertical (scale) and horizontal (time) shifts can be efficiently captured. The standard shape-invariant model applied to the energy reshape functions can be written as

$$\alpha_j(t) = \beta_j + \kappa_j \bar{\alpha}(t - \zeta_j) + \epsilon_j(t), \quad \epsilon_j(t) \sim WN(0, \sigma^2) \tag{18}$$

where $\bar{\alpha}(t)$ denotes the mean pattern of the energy reshape function, β_j and κ_j introduce vertical shifts parameterization, while ζ_j introduces time-shift parameterization. Recall that any α_j is considered as a deformation of the observed mean pattern (Fréchet mean) of the set $\mathcal{A}$. As a result, the incurred mean energy reshape function estimated by the data from the previous N days can be calculated through the equation

$$\bar{\alpha}(t; \theta^*) = \frac{1}{N} \sum_{j=1}^{N} \left(\frac{1}{\kappa_j} \alpha_j(t + \zeta_j) - \beta_j \right) \tag{19}$$

where vector $\theta^* = (\kappa, \zeta, \beta)' = (\kappa_1, ..., \kappa_N, \zeta_1, ..., \zeta_N, \beta_1, ..., \beta_N)'$ contains all the deformation parameters. Clearly, as these parameters uniquely define the mean reshape function must be selected to minimize the mean model variance from the set $\mathcal{A}$, i.e., the vector θ^* is chosen as

$$\theta^* := \arg\min_{\theta \in \Theta} \Lambda(\theta) = \arg\min_{\theta \in \Theta} \sum_{j=1}^{N} \int_{\mathcal{T}} (\alpha_j(t) - \bar{\alpha}(t; \theta))^2 \, dt \tag{20}$$

where $\mathcal{T}$ represents the time period within a day and Θ represents the space of the deformation parameters which are subjected to some normalization constraints, i.e.,

$$\Theta := \left\{ \theta : \sum_{j=1}^{N} \beta_j = 0, \ \sum_{j=1}^{N} \zeta_j = 0, \ \prod_{j=1}^{N} \kappa_j = 1, \ \kappa_j \geq 0, \text{ for all } j = 1, 2, ..., N \right\}. \tag{21}$$

Clearly, the estimation of the reshape function for the day $j = N + 1$ will be used to improve the initial prediction, provided by the simulation model, for the energy demand of this day through the model

$$f_j(t) = \hat{f}_j(t) + \eta_j(t) := \tilde{\alpha}(t; \theta)\tilde{f}_j(t) + \eta_j(t) \tag{22}$$

where the error term $\eta_j(t)$ is considered as a white noise process. After the initial mean reshape function estimation from the first batch of data corresponding to the initial N days has been obtained, an exponentially weighted scheme can be used to update appropriately the reshape functions taking into account both the effect of the initial reshape function and more recent observations on the reshape function. In particular, the proposed scheme can be described through the following steps.

Initial Step Set $k = 0$ and provide a choice for $\lambda \in (0, 1)$. Given the simulation model predictions $\{\tilde{f}_j\}$ and the corresponding measurements $\{f_j\}$ for $j = 1, 2, ..., N$ estimate the reshape functions $\{\alpha_j\}$ from (17). Then, set as $\hat{\alpha}^{(k)}(t)$ the Fréchet mean of $\{\alpha_j\}$ (calculated by (19)–(20)) and provide the prediction $\hat{f}_j(t) = \hat{\alpha}^{(k)}(t)\tilde{f}_j(t)$ for $j = N + 1$. For every new prediction task, repeat steps 1–3.
Step 1 Given the new measurement $f_j(t)$ set $k = k + 1$, $\alpha_0(t) := \hat{\alpha}^{(k-1)}(t)$ and $\alpha_1(t) := \tilde{f}_j^{-1}(t)f_j(t)$.
Step 2 Set as $\hat{\alpha}^{(k)}(t)$ the Fréchet mean of $\alpha_0(t)$ and $\alpha_1(t)$ with weights $1 - \lambda$ and λ, respectively.
Step 3 Given the simulation model prediction $\tilde{f}_{j+1}(t)$ provide the improved (reshaped) prediction $\hat{f}_{j+1}(t) = \hat{\alpha}^{(k)}(t)\tilde{f}_{j+1}(t)$.

The exponential weighting is used to reduce the effect of older observations to the new predictions, i.e., for the prediction on the day $N + 2$ using the weight parameter $\lambda \in (0, 1)$ we could weight the most recent information (last observed reshape function $\alpha_{N+1}(\cdot)$) by λ and the older observations by $(1 - \lambda)$. In this manner, choosing λ close to 1 we allocate more weight to the most recent realizations and reduce the effect of the older observations. Otherwise, choosing λ close to zero, we forget the older observations with a very slow rate allowing the past information to contribute more to the prediction. Clearly, the choice of this parameter is critical to the quality of the prediction, and the final choice of this parameter depends strongly to the nature of the application that the prediction model is employed to. Note also that at each step the Fréchet mean of previous reshape functions is taken into account as an observation through the term $\alpha_0(t)$ although it is not. However, this modification allows to simultaneously condense and appropriately weight (according to our preferences provided by the choice of λ) the past information into a single term.

2.2.2. The Weighted Shape Model Approach

In practice, it has been proved that there are periods of time that the prediction models do not provide reliable predictions and they may significantly deviate from the true situation. An example of such a situation is the energy demand prediction of the building during the period of summer holidays discussed in Section 3. In such cases, it is very important to quickly perceive when this happens and rapidly adjust the prediction to an acceptable level of deviance from the reality. For such purposes, we present here a small variation of the scheme presented in Section 2.2.1 where a weighted version of the reshape model is used.

The main idea is to divide the prediction into two parts: (a) the prediction provided by the reshape model and weight it by a proportion $w \in (0, 1)$ and (b) the prediction provided by previous observed energy shapes (measurements only) weighted by the proportion $1 - w$. For the second part, the same procedure that used for the estimation of the mean reshape function is used, i.e., given the past measurements of the daily energy shapes $\{f_1, f_2, ..., f_N\}$ their Fréchet mean $\tilde{f}(t)$ is estimated by (19)–(20) substituting α_j with f_j. The Fréchet mean in this case is interpreted as the most typical energy shape observed in the previous days without taking into account any information provided by the energy simulation model (predictive model). Then, one could shape the prediction either by constantly setting the weighting parameter w to a specific value or by changing this value each time new data become available to adjust the weighted shape model as close as it is possible to the true situation as observed until that time instant. Such a weight allocation criterion could be constructed as follows. Define by $g_{j+1}(t;w) := w\hat{f}_j(t) + (1-w)\tilde{f}_j(t)$ the weighted prediction, where $\tilde{f}_j$ denotes the mean energy shape as estimated until day j and $\hat{f}_j$ denotes the shape model's prediction for the day $j+1$ derived from the approach discussed in Section 2.2.1. Then, given the measurement of the day $j+1$, $f_{j+1}(t)$, the weight w for the next prediction (i.e., the day $j+2$) is chosen as the minimizer to the criterion

$$\min_{w \in (0,1)} \int \left(g_{j+1}(t;w) - f_{j+1}(t)\right)^2 dt. \qquad (23)$$

Clearly, if the prediction provided by the simulation model is completely misplaced, then this criterion will act rapidly as a safety filter and will provide a prediction that is based more on the empirical data (previously observed energy shapes); otherwise, the prediction will be based on the simulation model. Therefore, criterion (23) will act as a detection mechanism of significant inconsistencies of the estimates provided by the simulation model and if such a significant shift occurs, will immediately properly adjust the prediction and improve its accuracy. Moreover, monitoring the value of w for a reasonable period of time, provides a measure of efficiency for the prediction model that is used. Ideally, we expect the value of w to remain in high levels (near to 1) except of some periods that major changes happen where the simulation model has not yet re-adjusted.

2.3. Postprocessing Using Kalman Filtering: The General Algorithm, Capabilities and Areas of Application, and the Filter Proposed for the Present Work

Numerical simulation models in several applications, including energy prediction systems, are exposed to systematic or non-systematic biases, an issue in which a wide number of internal or external parameters are involved: inability of simulating sub-scale phenomena, limitations in the parameterization of all the engaged parameters, and numerical schemes problems can be listed among them.

Towards the limitation of the problems above and in the framework of integrated forecasting systems, statistical postprocesses and bias removal techniques have a critical role. In particular, Kalman filters [32–34] provide a very popular approach for systematic bias removal, combining recursively available records with direct modeled outputs, by using weights that minimize the corresponding biases, requiring limited CPU resources and data backlog. Kalman filters have been successfully applied to a wide number of numerical models including atmospheric parameters (see, for example, [35–39]), sea state (see, e.g., [40–42]), extreme events (see, e.g., [43,44]), as well as renewable power resources (see, e.g., [45–47]). A general description of the basic Kalman filter algorithm is summarized below.

The main target is the simulation of two parameters that evolve in time in parallel. The state vector x and the observation corresponding y. The change of the two processes in time are described by the *system* and the *observation* equations, respectively:

$$x_t = F_t x_{t-1} + w_t \qquad (24)$$
$$y_t = H_t x_t + v_t. \qquad (25)$$

The coefficient matrices, F_t and H_t, are defined as the *system* and the *observation* matrix, respectively. The corresponding covariance matrices W_t, V_t of the random vectors w_t and v_t, respectively, should be determined before the application of the filter while w_t and v_t need to be independently distributed according to Gaussian probability laws. The Kalman filter theory provides a method for the recursive estimation of the unknown state x_t utilizing all the observation values y up to time t. The following equations describe a full integration step of the Kalman algorithm.

$$x_t = F_t x_{t-1} + K_t(y_t - H_t x_{t|t-1}), \qquad (26)$$

where

$$K_t = P_{t|t-1} H_t^T (H_t P_{t|t-1} H_t^T + V_t)^{-1} \qquad (27)$$

and the covariance matrix of the unknown state x is given by

$$P_{t|t-1} = F_t P_{t-1} F_t^T + W_t \qquad (28)$$

where

$$P_t = (I - K_t H_t) P_{t|t-1}. \qquad (29)$$

In the present work, a linear filter has been adopted for the reduction of possible systematic biases in previous simulations steps. In particular, the estimation of the bias y_t in time is estimated by means of the direct model output m_t (where m_t):

$$y_t = x_{0,t} + x_{1,t} m_t + v_t$$

The above provides the observation equation of the filter with observation matrix $H_t = [1 \ m_t]$ and state vector $x_t = [x_{0,t} \ x_{1,t}]$. The covariance matrices V_t and W_t of the state and observation errors are estimated by utilizing a training window for the observed and modeled data (see [38,41,42]). The estimated by the filter bias values are utilized for the improvement of model outputs in the next time steps. It is worth noticing that the optimum training windows as well as initial values are case sensitive and should be estimated separately for every application under study.

3. Test Cases and Results

In this section, the proposed integrated model presented in Section 2 is implemented to the prediction of the daily and hourly energy demand of Leaf Lab for the time period 2017–2018.

The Leaf Lab is an industrial Near-Zero Energy Building (NZEB) of 6000 m² total floor area, integrating energy efficiency measures, advanced automations, renewable energy generation, and storage. The building envelope is highly insulated and consists of walls and double glazed windows with U values of 0.226 W/m²K and 1.793–3.194W/m²K, respectively. The HVAC system of the Leaf Lab is composed of ground water source heat pumps with a nominal heating COP of 4.8 and a cooling EER of 6.2–7. The HVAC is coupled to a thermal storage water tank which is heated or cooled using excess PV power. The roof of the Leaf Lab is covered by a PV system of 236.5 kWp. Energy systems are integrated by MyLeaf platform, which allows monitoring of measurements and advanced control functions [48]. Leaf lab is part of the Leaf Community, a microgrid integrating industrial and office buildings with various renewable energy systems (biPVs, tracker PVs, and micro-hydro power system), storage (electrochemical and thermal), and electric vehicles [49]. A complete description of the Leaf Lab and systems installed along with details of the building modeling and validation procedure are available in [50]. The 3D representation of the Leaf Lab simulation model is presented in Figure 3.

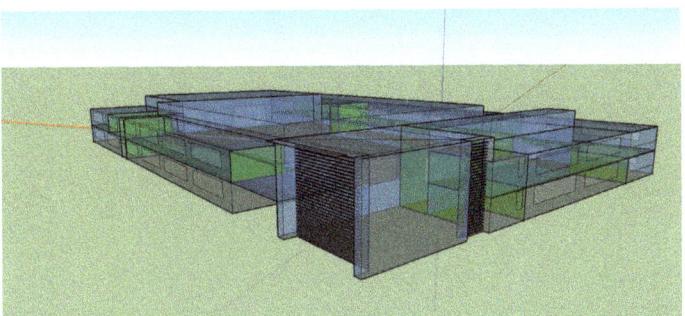

Figure 3. The Leaf Lab 3D simulation model.

First, an indicative analysis is performed to the initial simulation model outputs comparing to the energy demand measurements to reveal and discuss the weak points of the non-optimized simulation model (SM) and what improvements are obtained from its optimized version (OSM) discussed in Section 2.1. Next, we present the improvements in energy prediction that are obtained by adopting the shape modeling approach (Section 2.2) and Kalman filtering (Section 2.3) in the postprocessing stage. For the model performance assessment, standard statistical indices are incorporated that are widely used for measuring the performance of prediction models. Let us denote by $\{\hat{X}_t\}_{t=1}^T$ the under study model predictions for the energy demand at certain time instants $t = 1, 2, ..., T$ and as $\{X_t\}_{t=1}^T$ the true (measured/observed) energy demand stature at the same time instants. In particular, the following statistical indices are used.

- *Prediction Bias*, $Bias = \frac{1}{T}\sum_{t=1}^T (X_t - \hat{X}_t)$, indicating any systematic underestimation or overestimation of the quantity of interest.
- *Mean Absolute Error*, $MAE = \frac{1}{T}\sum_{t=1}^T |X_t - \hat{X}_t|$, indicating the mean absolute deviance of the model predictions from the true value.
- *Root Mean Squared Error*, $RMSE = \sqrt{\frac{1}{T}\sum_{t=1}^T (X_t - \hat{X}_t)^2}$, indicating the mean squared deviance of the model predictions from the true value.
- *Nash–Sutcliffe model efficiency coefficient*, which is used to assess the predictive power of the model:

$$NSE = 1 - \frac{\sum_{t=1}^T (X_t - \hat{X}_t)^2}{\sum_{t=1}^T (X_t - \tilde{X}_t)^2},$$

taking values on $(-\infty, 1]$ where an index equal to zero corresponds to a perfect prediction whereas values below zero corresponds to the case where the prediction model was outperformed by a reference model $\tilde{X}$.

3.1. Indicative Analysis of the Initial Simulation Results: Revealing the Weak Points

A snapshot of the simulated versus measured HVAC electric power for working days of the week from 7/8/17 to 11/8/17 is presented in Figure 4. Simulated versus actual measured total HVAC electric energy consumption for the year 2017 is presented in Figure 5a. It is noted that there are significant deviations on a monthly basis which become more evident for months of unstable environmental conditions such as March, May, and September, as well as in February. The total annual HVAC electrical energy consumption in 2017 for Leaf Lab was 285,719 kWh (47.61 kWh/m^2), whereas the respective simulated value is 197,305 kWh (32.88 kWh/m^2). The corresponding values excluding weekends is 232,669 kWh and 180,522 kWh, which correspond to an unacceptably high level of monthly CVRMSE equal to 40.7%.

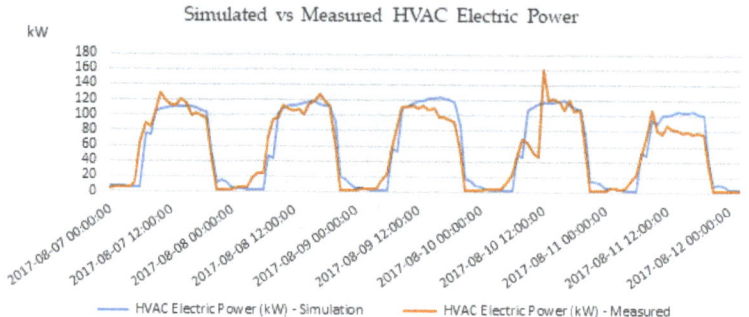

Figure 4. Simulated vs. measured HVAC electric power for the week from 7/8/17 to 11/8/17.

Following parameterization of the model monthly deviation in HVAC, electric energy consumption are minimized to a percentage difference ranging from 1.78% in February to 30.28% in September and an acceptable CVRMSE of 13.89% (Figure 5b). Total simulated HVAC electric energy in this optimized case is 252,689 kWh or 221,946 kWh excluding weekends associated to a percentage difference of 11.5% and 4.6%, respectively.

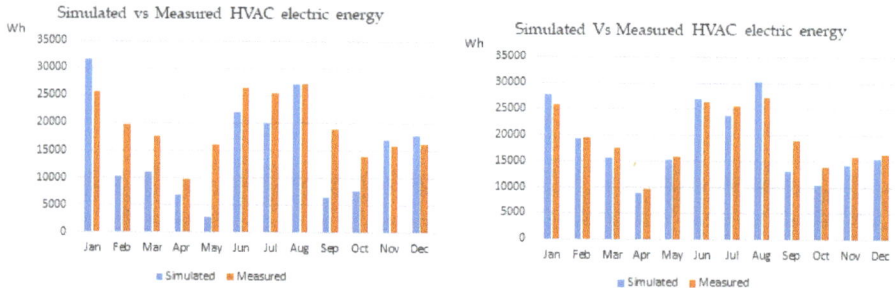

Figure 5. Measured total HVAC electrical energy consumption for the year 2017 versus: (**a**) initial model (baseline) (**left plot**) and (**b**) parameterized model (optimized) (**right plot**).

Table 1 contains monthly measured HVAC electrical energy consumption values and the corresponding simulated ones obtained from the baseline and optimized models for 2017.

Table 1. Monthly measured HVAC electrical energy consumption values and the corresponding simulated ones obtained from the baseline and optimized models for 2017.

2017	Simulated - Baseline Model (MWh)	Simulated - Optimized Model (MWh)	Measured (MWh)
Jan	31.55	27.79	25.72
Feb	10.26	19.27	19.61
Mar	11.09	15.76	17.55
Apr	6.79	8.92	9.64
May	2.90	15.40	16.02
Jun	21.97	27.10	26.34
Jul	20.02	23.69	25.55
Aug	27.19	30.31	27.27
Sep	6.39	13.15	18.87
Oct	7.62	10.56	14.00
Nov	17.05	14.38	15.86
Dec	17.70	15.62	16.24

Similarly, results from the initial simulation in 2018 are presented in Figure 6a. Notably, there are significant levels of underprediction especially in April, May, and September. The total annual HVAC electrical energy consumption in 2018 for Leaf Lab was 290,760 kWh (48.46 kWh/m^2), whereas the respective simulated value is 221,136 kWh (36.85 kWh/m^2), which equals a percentage difference of 23.9%. The corresponding values excluding weekends is 250,042 kWh and 200,461 kWh, which correspond to an unacceptably high level of monthly CVRMSE equal to 31.42%.

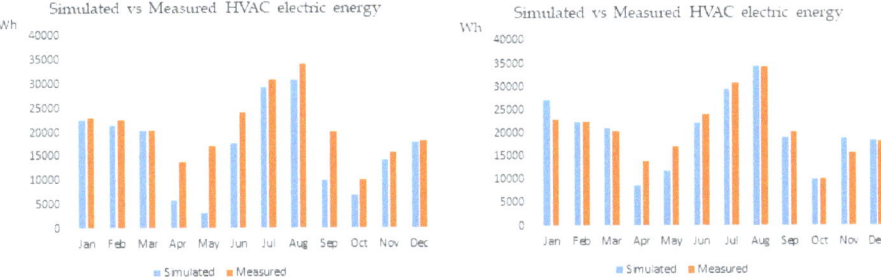

Figure 6. Measured total HVAC electrical energy consumption for the year 2018 versus: (**a**) initial model (baseline) (**left plot**) and (**b**) parameterized model (optimized) (**right plot**).

Table 2 contains monthly measured HVAC electrical energy consumption values and the corresponding simulated ones obtained from the baseline and optimized models for 2018.

Table 2. Monthly measured HVAC electrical energy consumption values and the corresponding simulated ones obtained from the baseline and optimized models for 2018.

2018	Simulated - Baseline Model (MWh)	Simulated - Optimized Model (MWh)	Measured (MWh)
Jan	22.60	27.29	23.02
Feb	21.42	22.52	22.49
Mar	20.29	21.07	20.40
Apr	5.92	8.76	13.85
May	3.27	11.76	17.01
Jun	17.65	22.30	24.09
Jul	29.32	29.45	30.90
Aug	30.94	34.56	34.16
Sep	9.92	18.99	20.13
Oct	6.88	10.06	10.11
Nov	14.29	18.80	15.75
Dec	17.96	18.46	18.14

An intensive search of the parameters that affect the daily, monthly, and annual power distribution profiles is presented in the following to improve the initial results of the model based by minimizing deviation from HVAC power consumption data. Through trial and error various combinations and fine-tuning of the all of key parameters are carried out to reach the optimum results both when assessing deviation at intra-day, monthly, and annual timescale. During this approach, the two key parameters identified as critical to decreasing the model's deviation from actual metered energy consumption are the COP and infiltration rates. COP nominal values are initially used but as they are representative of standard conditions, validation is required to better estimate their performance in dynamic conditions. Furthermore, the energy model uses performance curves to simulate the dynamic behavior of heat pump systems which is hard to define in the absence of very elaborate measuring equipment especially for systems customized to provide heating and cooling for large facilities. Besides, fine-tuning model parameters to match the actual performance in such systems is further justified by the fact that user behavior is not recorded and even if it was, modeling of such a complex activity is not feasible provided the current features of energy building software tools. Indicatively, the opening of external windows, the entry (or exit) of people or industrial equipment and the manual control of the temperature set point in the various thermal zones are some factors of uncertainty as to the actual thermal energy exchanges of a building such as the Leaf Lab.

Following careful fine-tuning of the model, simulation results are improved as shown in Figure 6b. Specifically, parameters related to the infiltration, ventilation, internal loads (electric equipment and lighting), and occupancy of the various zones were varied until the best possible correlation was established. Although some noticeable at monthly level differences remain especially in April and May, overall, the simulated results approach actual measured values as revealed by CVRMSE of 14.33%. In Figure 7, simulated versus measured HVAC electric power for a week in November 2018 is presented. In this case, the simulated pattern follows the actual measured values; however, there clear deviations especially with respect to the peaks early in the morning and late in the afternoon are observed.

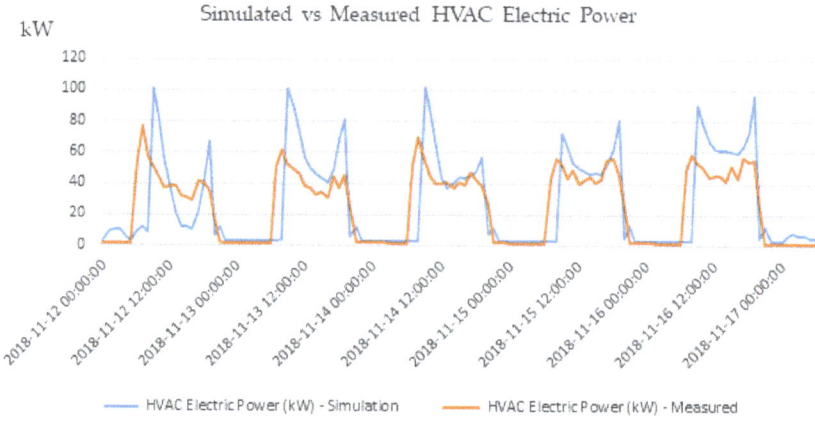

Figure 7. Simulated versus measured HVAC electric power for the week from 12/11/18 to 17/11/18.

In Figures 8 and 9, the evolution of statistical indices (Bias, MAE, and RMSE) regarding the deviance of the initial simulation model is illustrated, and its optimized parameterized version comparing to the measured energy demand for the years 2017–2018. With regards to Bias, the optimized model significantly outperforms the initial energy model of the building which underestimates energy consumption for most months in 2017. Concerning MAE and RMSE, it is demonstrated that the optimized model generally performs better except for months March, April, October, and November for which the initial model performs slightly better. Similarly, in 2018, the results of the optimized model are associated with a Bias error of lower magnitude for months from April to October. MAE and RMSE, in this case, are lower in the case of the optimized model compared to the baseline model for months from February to November. In contrast, the initial model performs marginally better for estimating energy consumption in January, October, and November.

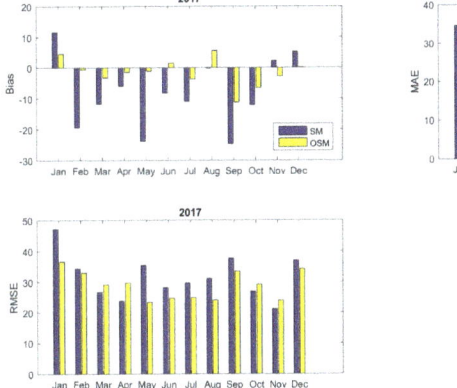

Figure 8. Statistical error indices for the energy prediction through the simulation model (SM) and the optimized simulation model (OSM) for the year 2017 (per month).

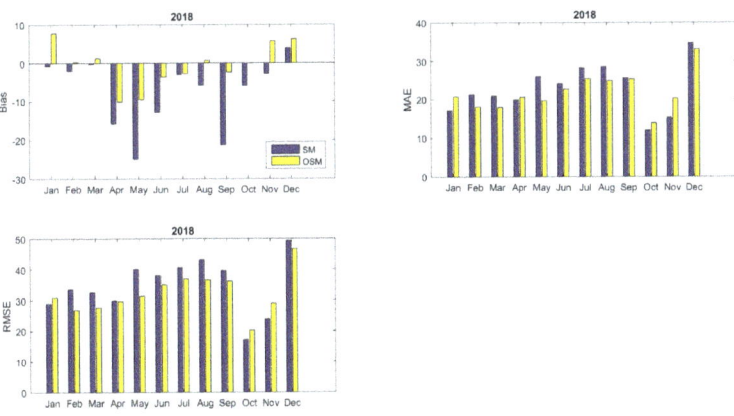

Figure 9. Statistical error indices for the energy prediction through the simulation model (SM) and the optimized simulation model (OSM) for the year 2018 (per month).

3.2. Diagnostic Results for the Complete Model Outputs

In this section, we discuss the improvements obtained to the energy consumption prediction for the Leaf Lab building for the time period 2017–2018 by implementing the postprocess part of the integrated model, i.e., the shape modeling approach and Kalman filtering. For convenience, let us introduce the abbreviations (SM) for the initial (benchmark) simulation model, (OSM) for the optimized/parameterized simulation model, (RS) for the reshaped simulation model, (w-RS) for the weighted reshaped model, and (KF-RS) for the integrated model by implementing Kalman filtering at the later stage of the reshape model output. For illustration purposes, we divide our analysis into two parts: (a) regarding the intra-day (hourly) energy demand prediction and (b) regarding the prediction of daily summary of the energy demand (total energy demand on the whole day).

3.2.1. Evaluating Intra-Day Energy Demand Predictions

In Table 3, the yearly diagnostic results of the intra-day energy demand predictions (hourly), provided by all incorporated prediction models for the years 2017 and 2018, are presented. Note that the NSE coefficient is calculated using as reference model the initial simulation model (SM). Based on the various indices, it is clearly demonstrated that the reshape modeling provides much higher levels of accuracy compared to the model's simulated outputs. Specifically, it is observed that RMSE in both years is reduced ~50%. The Kalman filtering procedure does not further improve the intra-day prediction performance of the reshaped model, indicating that the utilized shape model filters out efficiently most systematic sources of error affecting the intra-day (hourly) energy demand prediction.

Table 3. Model diagnostic results for intra-day energy demand predictions (hourly) for the years 2017 and 2018.

Model	Bias	MAE	RMSE	NSE	Bias	MAE	RMSE	NSE
		2017				2018		
SM	−8.67	21.32	31.66	0.00	−7.67	22.66	35.56	0.00
OSM	−1.79	19.39	28.80	0.17	−0.63	21.77	32.58	0.16
RS	−0.14	12.92	20.46	0.58	−0.25	12.91	22.01	0.62
w−RS	−0.51	9.91	17.03	0.71	−0.18	9.58	17.65	0.75
KF−RS	0.20	11.33	17.60	0.70	−0.36	12.67	18.73	0.73

Figures 10 and 11 illustrate the monthly Bias, MAE, and RMSE values for both reshaped models RS and w-RS computed for the intra-day performance of the models compared to the results obtained by the optimized simulation model (OSM). Both reshaped models perform better as all statistical indices in the majority of cases are significantly improved indicating the superior predictability of the reshaped approach.

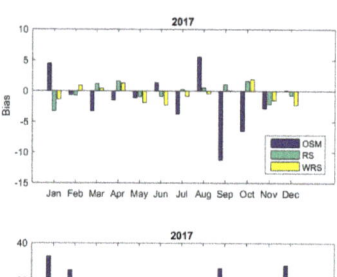
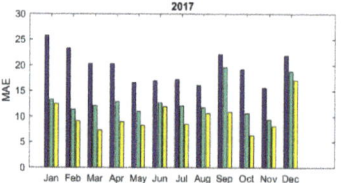
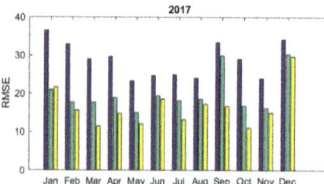

Figure 10. Statistical error indices for the Reshaped Simulation Model (RS) and the Weighted Reshaped Simulation Model (w-RS) for the year 2017 (per month).

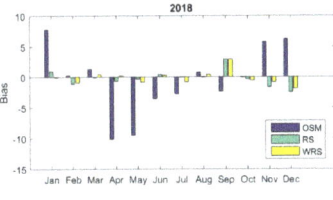
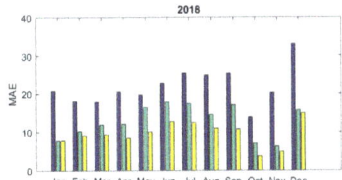
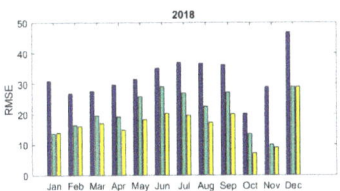

Figure 11. Statistical error indices for the Reshaped Simulation Model (RS) and the Weighted Reshaped Simulation Model (w-RS) for the year 2018 (per month).

3.2.2. Evaluating Daily Summaries Energy Demand Predictions

At a second stage, we are interested in the performance of the incorporated models in predicting the total energy demand per day. The same model approaches are tested, and their results are illustrated in Table 4. In this case, the Kalman filtering step further improves the prediction since the model biases are reduced significantly and the best NSE values are obtained through the KF-RS model. This happens because although the RS and w-RS models efficiently capture intra-day effects in the shape of the energy demand, when we are interested in the daily summary, these effects are mutually canceled and become obsolete. However, the KF-RS model, in general, improves the average biases, as it detects these inefficiencies as systematic ones from the perspective of daily summaries, and therefore it further improves the model outputs.

Table 4. Model diagnostic results for the predictions on the daily energy demand (summaries) for the years 2017 and 2018.

Model	Bias	MAE	RMSE	NSE	Bias	MAE	RMSE	NSE
		2017				2018		
SM	−195.00	344.65	423.71	0.00	−182.59	306.54	413.80	0.00
OSM	−37.54	237.38	304.32	0.30	−12.80	232.49	328.62	0.30
RS	−1.90	192.54	272.03	0.44	−6.09	194.10	295.32	0.32
w−RS	−11.18	144.97	226.93	0.61	−4.39	139.18	238.78	0.50
KF−RS	−0.73	143.46	207.51	0.66	−1.70	155.74	247.69	0.51

In Figures 12 and 13, the performance of the KF-RS model in predicting the daily energy demand (summaries) is illustrated, comparing again to the OSM model. It is evident that the later model is significantly improved by correcting major inconsistencies in certain time periods throughout the year and more accurately approximating the true levels of energy needed. In particular, for special cases where systematic biases are present (see, e.g., highlighted periods in Figures 12 and 13), the KF-RS model proves able to eliminate systematic over- or underestimations.

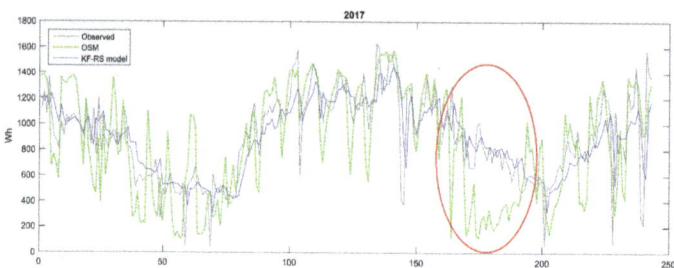

Figure 12. Prediction of the daily energy demand (summary) obtained by optimized simulation model (OSM) and Kalman filter-enhanced reshaped simulation model (KF-RS) for the year 2017.

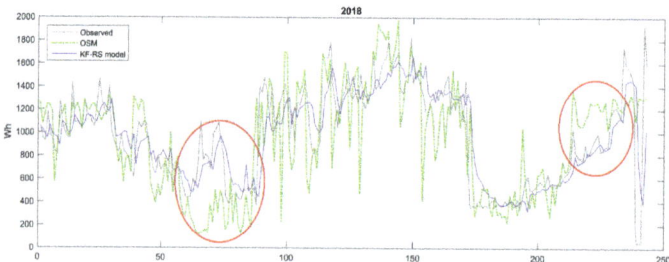

Figure 13. Prediction of the daily energy demand (summary) obtained by optimized simulation model (OSM) and Kalman filter-enhanced reshaped simulation model (KF-RS) for the year 2018.

4. Conclusions

In this paper, validation of the building energy model of an industrial near-zero energy building is investigated over a two-year period. It is demonstrated that manual extensive fine-tuning of key model parameters is valuable to improve initial overall error levels using trial and error of key parameters. However, this process requires a high level of expertise, deep knowledge of the facility under study, and it is very time-consuming. Furthermore, it is demonstrated that systematic deficiencies continue to occur even after careful fine-tuning of key model parameters. On the one hand, this is due to the complicated issue of modeling the behavior of users as well as of equipment use in a large industrial facility. On the other hand, the dynamic performance of custom-built HVAC systems requires specific measurements if it is not to be solely defined at a high level and modeled based on nominal performance coefficients. To address this problem two modeling approaches are integrated: (a) the shape and weighted shape model and (b) Kalman filtering postprocessing. The above methods are applied in a postprocessing stage, which is tested and evaluated. Initial, optimized simulation results and results from postprocessing are analyzed compared with the aid of bias error, mean absolute error, root mean squared error, and the Nash–Sutcliffe model efficiency coefficient. Results are explored to demonstrate the effectiveness of the method in capturing and reducing intra-day systematic deviations as well as the overall performance gap. Overall, the proposed integrated approach proves to be very effective in eliminating systematic over or under estimations and critically reduces the magnitude of deviations and as a result significantly reduces the performance gap compared to the optimized simulation model. In particular, the integrated prediction model succeeded in reducing the RMSE (in mean values) approximately 39% and 43% for the hourly predictions in years 2017 and 2018, respectively, and approximately 32% and 25% for the daily energy demand for the same years which are rather impressive improvements. Note also that in all cases, the integrated prediction model substantially increased model efficiency coefficient (NS) at least 70% comparing to the efficiency of the optimized simulation model (OSM). The proposed approach is applicable in several types of buildings (i.e., residential, commercial, public, etc.) provided that reliable data of energy consumption and of human activity are available over a significant period of time (ideally more than one year). As a future step, the presented approach could be deployed and tested across several types of buildings that fall into different energy efficiency categories to assess its performance over a wide spectrum of applications.

Author Contributions: Conceptualization, N.K., G.I.P., D.K., G.N.G. and A.N.Y.; Data curation, N.K., G.I.P. and D.I.; Formal analysis, N.K., G.I.P., G.N.G. and A.N.Y.; Funding acquisition, N.K., D.K. and C.C.; Investigation, N.K., G.I.P., D.K., G.N.G. and A.N.Y.; Methodology, N.K., G.I.P., D.K., G.N.G. and A.N.Y.; Project administration, C.C.; Resources, D.K., G.N.G., D.I., C.C. and A.N.Y.; Supervision, D.K. and C.C.; Validation, G.I.P.; Visualization, N.K. and G.I.P.; Writing – original draft, N.K., G.I.P., G.N.G. and A.N.Y.; Writing – review & editing, N.K., G.I.P., D.K., G.N.G., D.I. and A.N.Y. All authors have read and agreed to the published version of the manuscript.

Funding: This project has received funding from the European Union's Horizon 2020 research and innovation programme under the Marie Skeodowska-Curie grant agreement No 645677.

Conflicts of Interest: The authors declare no conflict of interest.

References

1. International Energy Agency. Available online: https://www.iea.org/topics/energy-efficiency/buildings/ (accessed on 18 September 2019).
2. Athienitis, A.K.; O'Brien, W. *Modeling, Design, and Optimization of Net-Zero Energy Buildings*; Wiley Online Library: Hoboken, NJ, USA, 2015.
3. Zhao, H.x.; Magoulès, F. A review on the prediction of building energy consumption. *Renew. Sustain. Energy Rev.* **2012**, *16*, 3586–3592. [CrossRef]
4. Santamouris, M. *Advances in Building Energy Research*; Taylor & Francis: Hoboken, NJ, USA, 2010; Volume 4.
5. Foucquier, A.; Robert, S.; Suard, F.; Stéphan, L.; Jay, A. State of the art in building modeling and energy performances prediction: A review. *Renew. Sustain. Energy Rev.* **2013**, *23*, 272–288. [CrossRef]
6. SPARK 2.0 REFERENCE MANUAL Simulation Problem Analysis and Research Kernel. Available online: https://simulationresearch.lbl.gov/VS201/doc/SPARKreferenceManual.pdf (accessed on 11 November 2019).
7. Harish, V.; Kumar, A. A review on modeling and simulation of building energy systems. *Renew. Sustain. Energy Rev.* **2016**, *56*, 1272–1292. [CrossRef]
8. Naji, S.; Keivani, A.; Shamshirband, S.; Alengaram, U.J.; Jumaat, M.Z.; Mansor, Z.; Lee, M. Estimating building energy consumption using extreme learning machine method. *Energy* **2016**, *97*, 506–516. [CrossRef]
9. Robinson, C.; Dilkina, B.; Hubbs, J.; Zhang, W.; Guhathakurta, S.; Brown, M.A.; Pendyala, R.M. Machine learning approaches for estimating commercial building energy consumption. *Appl. Energy* **2017**, *208*, 889–904. [CrossRef]
10. Amasyali, K.; El-Gohary, N.M. A review of data-driven building energy consumption prediction studies. *Renew. Sustain. Energy Rev.* **2018**, *81*, 1192–1205. [CrossRef]
11. Wei, Y.; Zhang, X.; Shi, Y.; Xia, L.; Pan, S.; Wu, J.; Han, M.; Zhao, X. A review of data-driven approaches for prediction and classification of building energy consumption. *Renew. Sustain. Energy Rev.* **2018**, *82*, 1027–1047. [CrossRef]
12. De Wit, S.; Augenbroe, G. Analysis of uncertainty in building design evaluations and its implications. *Energy Build.* **2002**, *34*, 951–958. [CrossRef]
13. Ryan, E.M.; Sanquist, T.F. Validation of building energy modeling tools under idealized and realistic conditions. *Energy Build.* **2012**, *47*, 375–382. [CrossRef]
14. Maile, T.; Bazjanac, V.; Fischer, M. A method to compare simulated and measured data to assess building energy performance. *Build. Environ.* **2012**, *56*, 241–251. [CrossRef]
15. Coakley, D.; Raftery, P.; Keane, M. A review of methods to match building energy simulation models to measured data. *Renew. Sustain. Energy Rev.* **2014**, *37*, 123–141. [CrossRef]
16. M & V Guidelines: Measurement and Verification for Performance-Based Contracts Version 4.0. Available online: https://www.energy.gov/sites/prod/files/2016/01/f28/mv_guide_4_0.pdf (accessed on 12 August 2019).
17. Royapoor, M.; Roskilly, T. Building model calibration using energy and environmental data. *Energy Build.* **2015**, *94*, 109–120. [CrossRef]

18. O'Neill, Z.D.; Bailey, T.G.; Eisenhower, B.A.; Fonoberov, V.A. Calibration of a building energy model considering parametric uncertainty. *ASHRAE Trans.* **2012**, *118*, 189–196.
19. O'Neill, Z.D.; Eisenhower, B.A.; Bailey, T.G.; Narayanan, S.; Fonoberov, V.A. Modeling and calibration of energy models for a DoD building. *ASHRAE Trans.* **2011**, *117*, 358–365.
20. EnergyPlus^{TM}, Engineering Reference Documentation Version 9.1.0, National Renewable Energy Laboratory, 2019. Available online: https://energyplus.net (accessed on 12 August 2019).
21. Brackney, L.; Parker, A.; Macumber, D.; Benne, K. *Building Energy Modeling with OpenStudio*; Springer: Berlin, Germany, 2018.
22. Wand, M.P.; Jones, M.C. *Kernel Smoothing*; Chapman and Hall/CRC: Boca Raton, FL, USA, 1994.
23. Ramsay, J.O.; Silverman, B.W. *Applied Functional Data Analysis: Methods and Case Studies*; Springer: New York, NY, USA, 2007.
24. Gu, C. *Smoothing Spline ANOVA Models*; Springer Science & Business Media: Berlin, Germany, 2013; Volume 297.
25. Lawton, W.; Sylvestre, E.; Maggio, M. Self modeling nonlinear regression. *Technometrics* **1972**, *14*, 513–532. [CrossRef]
26. Kneip, A.; Engel, J. Model estimation in nonlinear regression under shape invariance. *Annals Stat.* **1995**, *23*, 551–570. [CrossRef]
27. Gervini, D.; Gasser, T. Self-modeling warping functions. *J. R. Stat. Soc. Ser. B (Statistical Methodol.)* **2004**, *66*, 959–971. [CrossRef]
28. Papayiannis, G.I.; Giakoumakis, E.A.; Manios, E.D.; Moulopoulos, S.D.; Stamatelopoulos, K.S.; Toumanidis, S.T.; Zakopoulos, N.A.; Yannacopoulos, A.N. A functional supervised learning approach to the study of blood pressure data. *Stat. Med.* **2018**, *37*, 1359–1375. [CrossRef]
29. Fréchet, M. Les éléments aléatoires de nature quelconque dans un espace distancié. *Ann. De L'institut Henri Poincaré* **1948**, *10*, 215–310.
30. Izem, R.; Marron, J.S. Analysis of nonlinear modes of variation for functional data. *Electron. J. Stat.* **2007**, *1*, 641–676. [CrossRef]
31. Bigot, J.; Charlier, B. On the consistency of Fréchet means in deformable models for curve and image analysis. *Electron. J. Stat.* **2011**, *5*, 1054–1089. [CrossRef]
32. Kalman, R.E. A new approach to linear filtering and prediction problems. *J. Basic Eng.* **1960**, *82*, 34–45. [CrossRef]
33. Kalman, R.E.; Bucy, R.S. New results in linear filtering and prediction theory. *J. Basic Eng.* **1961**, *83*, 95–108. [CrossRef]
34. Kalnay, E. *Atmospheric Modeling, Data Assimilation and Predictability*; Cambridge University Press: Cambridge, UK, 2003.
35. Crochet, P. Adaptive Kalman filtering of 2-metre temperature and 10-metre wind-speed forecasts in Iceland. *Meteorol. Appl.* **2004**, *11*, 173–187. [CrossRef]
36. Galanis, G.; Anadranistakis, M. A one-dimensional Kalman filter for the correction of near surface temperature forecasts. *Meteorol. Appl.* **2002**, *9*, 437–441. [CrossRef]
37. Emmanouil, G.; Galanis, G.; Kallos, G. Statistical methods for the prediction of night-time cooling and minimum temperature. *Meteorol. Appl.* **2006**, *13*, 169–178. [CrossRef]
38. Galanis, G.; Louka, P.; Katsafados, P.; Pytharoulis, I.; Kallos, G. Applications of Kalman filters based on non-linear functions to numerical weather predictions. *Ann. Geophys.* **2006**, *24*, 2451–2460. [CrossRef]
39. Hua, S.; Wang, S.; Jin, S.; Feng, S.; Wang, B. Wind speed optimisation method of numerical prediction for wind farm based on Kalman filter method. *J. Eng.* **2017**, *2017*, 1146–1149. [CrossRef]
40. Emmanouil, G.; Galanis, G.; Kallos, G. Combination of statistical Kalman filters and data assimilation for improving ocean waves analysis and forecasting. *Ocean Model.* **2012**, *59*, 11–23. [CrossRef]
41. Galanis, G.; Emmanouil, G.; Chu, P.C.; Kallos, G. A new methodology for the extension of the impact of data assimilation on ocean wave prediction. *Ocean Dyn.* **2009**, *59*, 523–535. [CrossRef]
42. Galanis, G.; Chu, P.C.; Kallos, G. Statistical post processes for the improvement of the results of numerical wave prediction models. A combination of Kolmogorov-Zurbenko and Kalman filters. *J. Oper. Oceanogr.* **2011**, *4*, 23–31. [CrossRef]
43. Patlakas, P.; Drakaki, E.; Galanis, G.; Spyrou, C.; Kallos, G. Wind gust estimation by combining a numerical weather prediction model and statistical postprocessing. *Energy Procedia* **2017**, *125*, 190–198. [CrossRef]

44. Samalot, A.; Astitha, M.; Yang, J.; Galanis, G. Combined kalman filter and universal kriging to improve storm wind speed predictions for the northeastern United States. *Weather Forecast.* **2019**, *34*, 587–601. [CrossRef]
45. Louka, P.; Galanis, G.; Siebert, N.; Kariniotakis, G.; Katsafados, P.; Pytharoulis, I.; Kallos, G. Improvements in wind speed forecasts for wind power prediction purposes using Kalman filtering. *J. Wind Eng. Ind. Aerodyn.* **2008**, *96*, 2348–2362. [CrossRef]
46. Pelland, S.; Galanis, G.; Kallos, G. Solar and photovoltaic forecasting through postprocessing of the Global Environmental Multiscale numerical weather prediction model. *Prog. Photovoltaics: Res. Appl.* **2013**, *21*, 284–296. [CrossRef]
47. Stathopoulos, C.; Kaperoni, A.; Galanis, G.; Kallos, G. Wind power prediction based on numerical and statistical models. *J. Wind Eng. Ind. Aerodyn.* **2013**, *112*, 25–38. [CrossRef]
48. MyLeaf Platform - Loccioni Group. Available online: https://https://myleaf2.loccioni.com (accessed on 26 July 2018).
49. Provata, E.; Kolokotsa, D.; Papantoniou, S.; Pietrini, M.; Giovannelli, A.; Romiti, G. Development of optimization algorithms for the Leaf Community microgrid. *Renew. Energy* **2015**, *74*, 782–795. [CrossRef]
50. Kampelis, N.; Gobakis, K.; Vagias, V.; Kolokotsa, D.; Standardi, L.; Isidori, D.; Cristalli, C.; Montagnino, F.; Paredes, F.; Muratore, P.; et al. Evaluation of the performance gap in industrial, residential & tertiary near-Zero energy buildings. *Energy Build.* **2017**, *148*, 58–73.

© 2020 by the authors. Licensee MDPI, Basel, Switzerland. This article is an open access article distributed under the terms and conditions of the Creative Commons Attribution (CC BY) license (http://creativecommons.org/licenses/by/4.0/).

Article

Energy Calculator for Solar Processing of Biomass with Application to Uganda

Toby Green [1,*], Opio Innocent Miria [2], Rolf Crook [1] and Andrew Ross [1]

[1] School of Chemical and Process and Engineering, University of Leeds, Leeds LS2 9JT, UK; R.Crook@leeds.ac.uk (R.C.); a.b.ross@leeds.ac.uk (A.R.)
[2] Centre for Renewable Energy and Energy Conservation, Makerere University, Kampala, Uganda; opiomiria@gmail.com
* Correspondence: pmtag@leeds.ac.uk

Received: 24 February 2020; Accepted: 12 March 2020; Published: 21 March 2020

Abstract: Rural areas of developing countries often have poor energy infrastructure and so rely on a very local supply. A local energy supply in rural Uganda frequently has problems such as limited accessibility, unreliability, a high expense, harmful to health and deforestation. By carbonizing waste biomass streams, available to those in rural areas of developing countries through a solar resource, it would be possible to create stable, reliable fuels with more consistent calorific values. An energy demand calculator is reported to assess the different energy demands of various thermochemical processes that can be used to create biofuel. The energy demand calculator then relates the energy required to the area of solar collector required for an integrated system. Pyrolysis was shown to require the least amount of energy to process 1 kg of biomass when compared to steam treatment and hydrothermal carbonization (HTC). This was due to the large amount of water required for steam treatment and HTC. A resource assessment of Uganda is reported, to which the energy demand calculator has been applied. Quantitative data are presented for agricultural residues, forestry residues, animal manure and aquatic weeds found within Uganda. In application to rural areas of Uganda, a linear Fresnel HTC integration shows to be the most practical fit. Integration with a low temperature steam treatment would require more solar input for less carbonization due to the energy required to vaporize liquid water.

Keywords: biomass; energy resource assessment; developing countries; concentrated solar; thermochemical

1. Introduction

Those living in rural areas of developing nations face many daily problems, including but not limited to, having access to a clean cook fuel with an appropriate energy content. Women and children are often forced to spend multiple hours a day trekking for firewood and other materials to burn for cooking. Inefficient burning of inappropriate material leads to health issues, time spent collecting these materials is time away from education and earning a living. By carbonizing biomass to create a solid fuel that can be used for cooking, it would be possible to help in solving the above problems faced daily for those living in rural areas of developing nations. To aid the integration of concentrated solar technology with the thermochemical techniques used to create a biofuel, an energy demand calculator has been made and applied to Uganda. Uganda was selected as the case study nation due to the strong links between the University of Leeds and the Centre for Research in Energy and Energy Conservation (CREEC) based at the Makerere University, Kampala, Uganda.

The integration of biomass and concentrated solar technology (CST) has been proven successful multiple times shown within literature. With enough solar radiation to power the earth 4200 times, and multiple ways to thermochemically treat biomass into a fuel, integration of the two is only logical.

Solar pyrolysis is not a new nor a novel technology and is described many times within literature. It has been present in academia since the 1980s, using solar simulators (furnace images) and elliptic mirrors as a source of radiation [1]. Solar pyrolysis is defined as an endothermic process of converting biomass into an inert atmosphere in which the required heat for the reaction to occur is provided by concentrated solar energy [2]. Solar pyrolysis thus allows for solar energy to be stored within a chemical compound. Throughout literature, it has become apparent that solar pyrolysis can occur via two techniques, that being through either direct or indirect radiation. Either biomass is directly heated by concentrated solar radiation through either a borosilicate or quartz glass, or is indirectly heated via convection or a heat transferring fluid as shown by R. Adinberg et al. (2014) [1,3]. Indirect reactors have external walls heated by concentrated solar radiation. Conduction through the walls heats the reactants. Most indirect reactors are catalytic tubular reformers, whereby a gas flows across a heated catalyst [4,5]. Double cavity reactors have recently been developed for thermochemical purposes. In the double cavity reactor, the reaction chamber is physically separated from the reactor that receives the solar radiation [4,6].

There has been work shown in literature assessing various solar pyrolysis techniques, temperatures, hold times, and other variables, allowing a compilation of the key research challenges that lie ahead with the technology. Morales et al. (2014) studied the effect of solar pyrolysis on orange peel [2]. The study was conducted with the feedstock directly heated within a borosilicate glass tube with helium as the gas carrier as part of a parabolic trough array. The irradiance profile of the parabolic trough was plotted using the SolTrace software provided by the [US] National Renewable Energy Laboratory (NREL). The parabolic trough had an aperture width of 1.3 m and a reflectivity of 0.94. The borosilicate glass receiver had an external diameter of 2 inches. Peak solar irradiance during the experiment was 25,084 W/m^2, with the average being 12,553 W/m^2. During pyrolysis of the orange peel, temperatures averaged 290 °C with a peak temperature of 495 °C. Under these conditions the orange peel lost 79.08% of its mass producing mainly a liquid bio-oil. Of the product, 77.64% was liquid, 1.43% was gas and the remainder was char. The results for the solar pyrolysis of the orange peel compared well to literature results of an electric furnace based pyrolysis of orange peel [2]. S. Morales went on to after reviewing his work to note that with an increase in aperture width, the maximum temperature and efficiency of his process would increase [2].

Zeng et al. (2014) investigated the solar pyrolysis of wood in a lab-scale solar reactor, assessing the influence of temperature and gas flow. The pyrolysis temperature ranged from 600 to 2000 °C with a hold time of 12 min, and argon flow rates between 3 Nl/min and 12 Nl/min. The heating rate was constant at 50 °C/s across all experiments conducted [7]. Zeng et al. reported on how the use of arc image furnaces in literature are pronounced for providing a higher liquid yield compared to conventional furnaces. Liquid yield unlike gas and char yield are not significantly dependent on the heat flux density [7]. K. Zeng et al. however were studying solar pyrolysis conditions in order to gain a high gas yield from Beechwood samples. The results showed that for the highest possible gas yield, the temperature needs to be as high as possible. Gas yield constantly increased across the experiments, with a final yield of 51%, but the biggest increase occurred between 600 °C–1000 °C (15–37%). Liquid yield and char yield both decreased with increasing temperature. At 600 °C the liquid yield was 71% dropping to 52% at 1000 °C and further to 41% at 2000 °C. The char remained low throughout the experiment dropping from 14%–8% [7]. Gas flow rate had the opposite effect. At 1200 °C increasing the argon flow rate within the reaction chamber leads to a slight decrease in gas yield and an increase in liquid and char yield. This result would be due to the removal rate of products from the hot zone of the reactor. Zeng et al. has helped show that the main products to be produced form solar pyrolysis are either a liquid or a gas [7]. Char yield never passed 15% within the experiments and so this would need to be taken into consideration if taking this technology forward.

Li et al. conducted experiments similar to K. Zeng et al. as outlined above. Li et al. attempted to produce a pyrolysis gas from pine sawdust, peach pit, grape stalk and grape marc within a temperature range of 800 °C–2000 °C with the use of a solar dish. As expected, gas yield increased with increasing temperature and temperature rate, with liquid yield and char yield reacting oppositely. Li et al. did note that the feedstocks with a higher lignin content provided a higher char yield [8]. In a review conducted by Chintala (2018), the process variables of solar biomass were investigated. Chintala confirmed through citing literature that by increasing the reaction temperature the gas yield increased. Chintala also investigated the effects of biomass particle size and claimed that a larger particle size will increase char yield [9].

Two examples of a high solid yield produced by solar pyrolysis, which are perhaps of a higher relevancy to this study, include work conducted by Ramos et al. and Hans et al. Ramos was able to produce 70 g of biochar from 180 g of wood implying a char yield of 38%. His parabolic solar concentrator had a surface area of 1.37 m^2 and its receiver hit temperatures of above 270 °C. For the conversion to efficiently take place, the process occurred over 5 h during peak day time hours [10]. Hans et al. also managed to produce a solid fuel. Hans et al. took agricultural wastes such as wheat straw and pyrolyzed them in a solar driven reactor for 90 min at 500 °C. The solid fuel produced gained energy density, increasing it from 16.9 MJ/kg to 24–28 MJ/kg. Details of his design are based in the reference Hans et al. [11].

There are limited reports of modelling solar pyrolysis in the literature, however Sanchez et al. (2018) has published a useful system for modelling and evaluation the thermochemical technique [12]. Sanchez et al. breaks the model down into two scenarios (i) heating of the biomass from an ambient to an operating temperature and (ii) the pyrolysis reactions at the operating temperature, details of which can be found in the reference Sanchez et al. [12]. The model aims to predict the length needed for the pyrolysis reactor for a set feed rate. By varying the operating temperature and hold times, the optimum reactor size can be predicted. The equations for the model are outlined in the reference and the simulation software used was MatLAB. Sanchez et al. ran the model based off of data from Seville, Spain. During optimum conditions the model predicted a maximum char yield from a woody biomass feedstock to be 40.8 wt%. However the system would not constantly be able to run at optimum conditions, leaving the average annual yield to be a meek 10.1 wt% [12]. The model and the predictions made by Sanchez et al. are observed to be accurate and could be applied to the future work of this thesis. The design of the system that Sanchez is basing the model off from however does not seem optimum. The use of better materials and a parabolic dish or trough instead of linear Fresnel would hopefully improve the efficiency during non-optimum conditions. Improving the efficiency for optimum conditions would also increase the overall operating temperature and therefore decrease the char yield—which for the production of a solid fuel would be detrimental. Work conducted by Zeng et al., Li et al. and Soria et al. provided a useful solution for this. By creating effectively a shutter system from a carbon composite, they were able to have a larger control of the operating temperature [7].

Developing solar pyrolysis may not the technology of choice for this work, but some of its principles can be carried on. For a high char yield it would seem that a slow pyrolysis under cooler conditions would be optimal. This would be beneficial as it would help with the simplicity of the design. Building a solar collector to reach temperatures between 200–600 °C would be easier and more economically viable compared to one that needed to reach 2000 °C. Little research has been conducted in the field of integrating solar with lower temperature hydrothermal treatments, but will be investigated in this work.

As can be seen through the review of literature the design of the pyrolysis system is key to what operating conditions can be achieved, and thus what pyrolysis products can be produced. Chintala summarized the key research challenges in their review. These challenges are as follow [9]:

- Uniform distribution of the heat flux
- Heat losses from the surface of the reactor/high wind speeds
- High capital costs
- Reactor design for effective thermochemical conversion, including reactor material
- Variation in solar flux (time of day/season)

However, by modeling a variety of options before setting on a design choice the challenges may be overcome. By selecting the correct solar reactor type, either direct or indirect with respect to the product required and building the system from the correct materials, the system should be as efficient as possible. An efficient system will reduce the overall capital costs (though they may still be high) as there will be no wasted materials [13]. For a small scale system, the solar reactor may only be required to run during peak hours and so the variation in solar flux may be a non-issue. In a large scale reactor, molten salt technology and solar thermal storage may play a role to account for the varying solar conditions.

This report provides a method for calculating the energy demand required for a thermochemical process, and relating that to the area of a solar collector required to produce the needed energy input. The energy demand assessment will report on the energy requirements for the thermochemical technologies that are possible for integration with a solar resource. A theoretical thermodynamic approach following the principals of the first law of thermodynamics forms the basis of the calculator. The energy demand calculator is then applied to Uganda. A quantitative and qualitative review of the waste biomass in Uganda is also reported. It is essential that the feedstock selected must be a waste stream and not disrupt current practice. The qualitative biomass review of Uganda will form the basis on to which the quantitative review of the biomass will be formed. The quantitative review will normalize the biomass by reporting the higher heating values (HHVs), showing which biomass sources will make for a feasible feedstock within Uganda.

2. Energy Demands of the Integrated Approach to a Concentrated Solar Powered Thermochemical Process for the Production of a Biofuel

2.1. Methodology of the Energy Demand Calculator

The energy demand calculator created has the ability to estimate either the area of the solar collector required to process a set mass of biomass in a given time, the mass of biomass that can be processed in a set time with set solar collector area or the time required to process a given mass of biomass with a set collector area. The energy demand calculator is based on Equations (1)–(4):

$$A = \frac{100 \times P}{\eta \times G} \quad (1)$$

$$P = \frac{Q}{3.6 \times 10^6} \times \frac{60}{t} \times 1000 \quad (2)$$

$$t = \frac{1.66 \times Q}{\eta \times G \times A} \quad (3)$$

$$m = \frac{\eta \times t \times G \times A}{1.66 \times C \times \Delta T} \quad (4)$$

The symbols and units for the expressions can be shown in Table 1.

Table 1. Symbol, definitions and units for Equations (1)–(4).

Symbol	Definition	Unit
A	Area of mirror	m²
P	Power	W
η	Efficiency	%
G	Solar Irradiance	W/m²
t	Time	min
Q	Heat	J
m	mass	Kg
T	Temperature	K
C	Specific heat capacity	J/kg/K

The model is able to deal with various thermochemical techniques such as steam treatment, HTC and pyrolysis. When steam treatment is required the heat of vaporization for water must be included. Heat of vaporization is not included for HTC or pyrolysis as water is never vaporized. During steam treatment only water is included in the mixture as the solar collector is only required to boil the water to produce steam, vice versa for pyrolysis, only biomass is included in the mixture. HTC requires the heating of biomass plus water at a ratio that is able to be input into the calculator. The calculator requires the specific heat capacity of water, the biomass (if it is steam treatment, specific heat of the biomass is 0) and the operating temperature. The other variables are able to be entered depending on the data acquired and the data required. Heat of HTC and pyrolysis are too be included for their respective calculations. The 'Enthalpy Value' calculator estimates the value to be included within the model. Current enthalpy data have been included within the model, however this can be varied.

2.2. Assumptions of the Calculator

There are a number of assumptions that have been made to allow this calculator to work, and to allow for a comparison of technologies. The system the work is based on is a small-scale system to be used in rural communities in developing countries i.e., rural Uganda. The technology will be simple and relatively low tech. Thermal efficiencies/heat transfer/heat losses and other detailed formulas and data have not been included within the calculator, instead an overall efficiency % is to be input into the calculator. The assumption is that the losses will be similar whether the system be for steam treatment, HTC or pyrolysis. We are assuming for a constant solar irradiance, whereas in reality as proven by monitoring the solar irradiance in Uganda, it does fluctuate. The solar irradiance input should be set as the minimum solar irradiance required for the system to work. The system is only accounting for the energy required for the thermochemical process, it is not taking into account pre or post processing. Whether a dry or wet feedstock is required, drying pre or post processing is assumed to be done in air using the sun's natural light to do so. During the qualitative review of Ugandan biomass, coffee beans were observed to be drying on a black material during the day time. The biomass or biofuel is assumed to be dried under these conditions and so no extra energy input is required. The final assumption would be that the mixture is completely heterogeneous, fully mixed, and heating is uniform and complete. All of the mixture is to be processed.

2.3. Example Scenarios of the Energy Demand Calculator

An example scenario has been put together to show example results for the energy demand model. The example is based off of processing 1 kg of biomass with a specific heat capacity of 2000 J/K/kg (an example specific heat capacity value). Table 2 shows the variables that have been kept the same for the example model. Assuming a biomass to water ratio of 1:10 for steam treatment and HTC, as well as a solar irradiance of 900 W/m^2 for 6 h. These values are all able to be changed within the model. Table 3 shows the variables that have been altered subject to which thermochemical technology is being modelled. The heat of reaction values have been taken from literature and are able to be altered within the model [14,15].

Table 2. Constant variables for the energy demand example simulation.

Variable	Value
Mass of biomass (kg)	1
Biomass to water ratio (1:x)	10
Mass of water (kg)	10
Cp of example biomass (J/kg/K)	2000
Cp of water (J/kg/K)	4186
Temperature Start (K)	293
Time (mins)	360
SI (W/m^2)	900
Efficiency (%)	25
Width (m)	1

Table 3. Dependent variables of the energy demand example simulation.

Variable	Value
Steam Temperature End (K)	373
HTC Temperature End (K)	573
Pyrolysis Temperature End (K)	773
Steam Additional Heat Value (MJ/K/kg)	2.256
HTC Additional Heat Value (MJ/K/kg)	−2.4
Pyrolysis Additional Heat Value (MJ/K/kg)	3.0

Results to this example of how the model operates are shown in Table 4.

Table 4. Results of the example energy demand simulation.

Symbol	Steam	HTC	Pyrolysis
Q (MJ)	25.9	9.88	3.96
E (kWh)	7.197	2.745	1.100
P (kW)	1.199	0.457	0.183
A 100 (m^2)	1.333	0.508	0.204
A eff (m^2)	5.331	2.033	0.815
L (m)	5.331	2.033	0.815

From the results shown in Table 4 it is clear that the energy required to vaporize the 10 kg of water to produce steam is significantly higher than the energy required for HTC and pyrolysis. The higher energy demand of the process leads to a larger collector area required, and with a given width of 1 m for a parabolic trough collector, a longer trough is required. One of the main factors effecting the results are the additional heat values. The heat of vaporization for 10 kg of water is approximately a factor of 10 larger than the given heat of pyrolysis for 1 kg of biomass. Pyrolysis has been shown to require the smallest area of collector. The main factor attributing to this is the mass of water required for HTC and steam treatment of biomass.

Example 2 shows the process in reverse. Finding how much biomass can be produced with a fixed collector size. Constant variables are shown in Table 5 and the dependent variables are kept constant with Table 3.

Table 5. Constant variables for the energy demand in the second example simulation.

Variable	Value
Biomass to water ratio (1:x)	10
Cp of example biomass (J/kg/K)	2000
Cp of water (J/kg/K)	4186
Temperature Start (K)	293
Time (mins)	360
SI (W/m^2)	900
Efficiency (%)	25
Width (m)	1
Length (m)	5.33

Table 6 shows the results produced by the calculator with a set collector size which are applicable for all thermochemical process. A value of 5.33 m^2 has been selected to show that this will result in 1 kg of biomass being able to be treated via a steam treatment. The results show that under the conditions shown in Table 5, 25.9 MJ of energy are able to be produced.

Table 6. Group results from the second example simulation.

Variable	Value
A eff (m^2)	5.33
A 100 (m^2)	1.33
P (kW)	1.20
E (kWh)	7.20
Q (MJ)	25.90

Table 7 shows the amount of biomass than can be processed depending on which thermochemical treatment is used. The specific heat capacity value varies between processes. Steam treatment only requires the heating of water, so the heat capacity of water is used. Equally pyrolysis is the heating of dry biomass, so the specific heat value of biomass is used (same example value). HTC is a mixture of water and biomass as defined by the user, in this case 10:1.

Table 7. Independent results from the second example energy demand simulation.

Variable	Steam	HTC	Pyrolysis
ΔT (K)	80	280	480
Cp mixture (J/kg/K)	4186	3987	2000
ΔH (MJ/kg)	2.256	−2.4	3.0
Mass of biomass total (kg)	1	2.62	6.54

From Table 7, it is clear that under the set conditions of the example, the 25.9 MJ will be able to steam treat 1 kg of biomass, hydrothermally carbonize 2.62 kg of biomass or pyrolyse 6.54 kg of biomass.

3. Overview of the Qualitative Assessment of Biomass in Uganda

The biomass assessment of Uganda reports on what source and use of energy is most needed by those in rural areas, what biomass is readily available, how it is used, and what goes to waste. The aim of the assessment was to discover if there was a readily available biomass waste stream that could be used to produce a useful fuel.

In order to achieve an overall and fair assessment of the country, the maximum distance was covered within the time available. Homesteads, schools, farms, plantations and factories were all visited, as well as general observations made. From interviews and observations with the local people, especially those in schools, a cook fuel/solid fuel for heat would ultimately be the best source of fuel to provide Uganda and other developing countries. Currently the main source of fuel used is untreated forest wood which has a number of disadvantages; deforestation, poor efficiency and high emissions. The potential feedstocks that were identified from the assessment of Uganda include: agricultural and forestry residue (including water hyacinth), sewage sludge and municipal solid waste (MSW). These feedstocks were shown to be largely going to waste, suggesting that if incorporated into the system they would not currently disrupt the lifestyles of the local people.

Literature assessment of the thermochemical technology available and comparing that with concentrated solar technology showed that the possible and practical integration options included a hydrothermal/steam treatment or pyrolysis with either a linear Fresnel or parabolic trough.

4. Quantitative Analysis of the Energy Stored within Waste Biomass Available in Uganda

4.1. Agriculture Residue

Agricultural data have been sourced from the Ugandan Bureau of Statistics and are reported in metric tonnes × 10^6. Table 8 reports the five highest agriculture crops grown. Waste from the various crops has then been calculated by subtracting the edible part of the crops [16].

Table 8. Top five reported crops grown in Uganda with their waste % and resulting waste from crop available for energy use.

Biomass	National Production (Metric Tons × 10^6)	Waste %	Solid Waste (Metric Tons × 10^6)
Plantain	4.297	40	1.719
Cassava	2.894	30	0.868
Maize	2.362	40	0.945
Sweet potato	1.819	7	0.127
Beans	0.929	10	0.093

Table 8 shows plantain to be significantly higher than the other four crops, almost double the second largest waste producer maize. Sweet potato and beans, though largely produced actually produce very little waste. The 7% waste production from sweet potato is attributed to the skin and other fibers produced during the flour production process. It has the potential to be eaten whole and therefore produce even less waste.

4.2. Forestry Residue

It is unfeasible to produce an accurate and precise amount of forestry residue available in Uganda. However, the amount of forest in Uganda is known. In 2005 the UN reported on the amount of forest in Uganda. The total land cover of all forest in Uganda in 2005 was 18% (3.6 million hectares), down from 24% in 1990 [17,18]. The results of the report are currently 14 years out of date and therefore are likely to have decreased by a significant amount due to constant deforestation within the country. From the report, private land owners with large plantations may benefit from a concentrated solar driven thermochemical production of biofuel with forestry residue as a feedstock. This would include private owners such as schools and manufactures whom rely on firewood.

The global forest watch reports that in 2010 Uganda had a tree cover of 5.61 million hectares, equivalent to 23% of land cover. In 2015 the global forest watch reports Uganda to have 4.87 million hectares of forest land. Their data were taken from ESA Climate Change Initiative—Land Cover led by U. C. Louvain (2017) and can be taken as more reliable. The data are collected through the use of google maps and NASA imaging [19]. Figure 1 may however show a better representation of how the forest is split within Uganda.

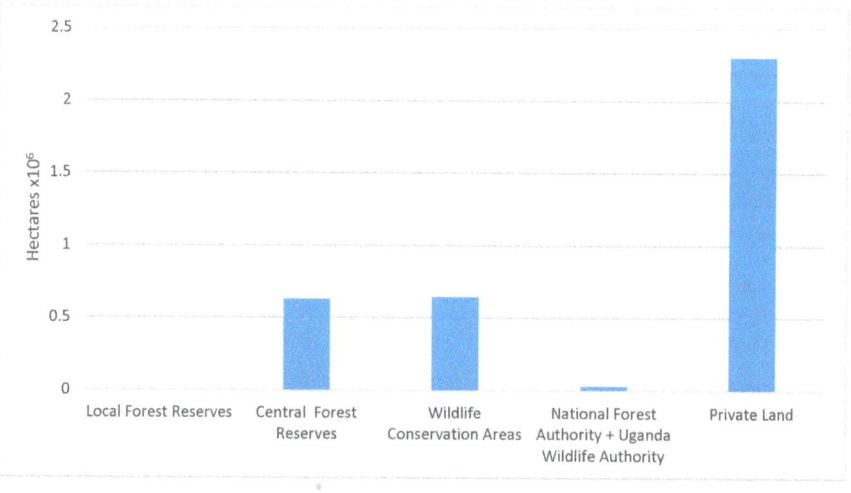

Figure 1. Forest area in Uganda.

4.3. Animal Waste

As with the forestry residue in Uganda, the precise number of cattle and other farmed animals cannot be accurately reported [20]. However, the amount of waste produced per animal can be reported and is so. The data shown in Figure 2 report the average waste per 1000 kg of live animal mass [21]. The waste produced can vary greatly depending on a number of factors such as animal breed, diet, animal age, animal environment and animal productivity [20–22]. The Ankole–Watusi is the most prevalent cattle type in Uganda and is on the larger side of cattle breeds weighing over 600 kg on average fully grown and would produce similar manure to the dairy cattle. Two of the Ankole–Watusi would be able to produce the manure shown in Figure 2. The large black pig is most prevalent in Uganda. Weighing over 300 kg fully grown, four large black pigs should produce a similar amount of manure to two Ankole–Watusi.

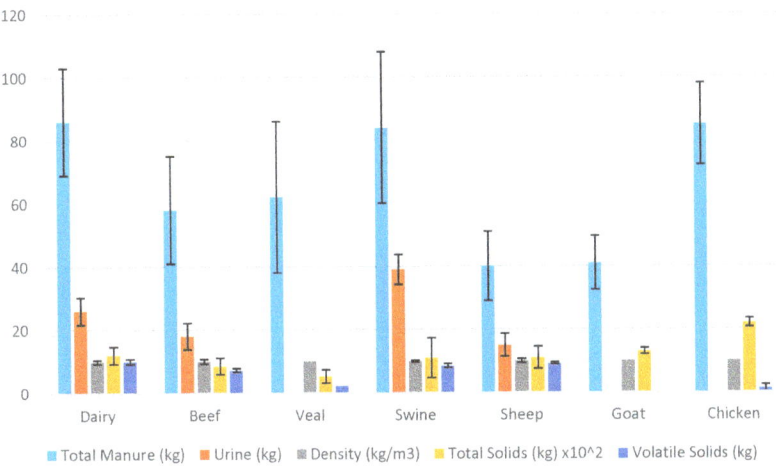

Figure 2. Fresh manure production and characteristics per 1000 kg of live animal mass per day.

4.4. Water Hyacinth

Water hyacinth (*Eichhornia crassipes* (Mart.) Solms) is a free floating herb that resides in fresh water ecosystems. The herb generally grows to 40 cm in length, but it can reach heights of 1 m. The ideal conditions for water hyacinth growth include a pH of 7, temperatures between 15–30 °C, low salinity (fresh water), plenty of sun light availability, high availability of nitrogen/phosphorous/potassium within the water and low disturbance. Under these ideal conditions, water hyacinth is able to double in water coverage in between 6–15 days. Water hyacinth's impressive reproduction rate leads it to be easily farmed. In China farmed fertilized irrigation channels were able to yield up to 750 tonnes/ha/year (extrapolated data). A rate of 200 tonnes/ha/year is much more likely in the tropics i.e., Uganda [23,24].

The quantity of water hyacinth available in Uganda is uncertain. Uganda has 4,152,000 hectares of open water/swamp area, more than enough for water hyacinth farming or for natural growth. Closed ponds however would make the most ideal farms to prevent the spread of water hyacinth to unwanted areas. Uganda has experienced problems with water hyacinth in the past, with some areas of Lake Victoria becoming very over run with the herb. An uncontrolled spread of the herb has many detrimental effects to the ecosystem and the people who rely on it such as; obstruction of water ways for boats and irrigation, prevention of fishing through the formation of a thick mat and through being lethal to fish by preventing light and oxygen reaching them and being a breeding ground for disease by hosting disease carrying insects. Water hyacinth can be controlled using mechanical, chemical and biological methods. Mechanical, though slow and expensive is the preferred method as the herb is then able to be collected and used. Chemical control with the use of herbicides is an effective method but has many detrimental effects for the environment. Biological control with insects and fungi is available, but only in use with other control methods [23,24].

4.5. Energy Content Available within Common Types of Biomass Found within Uganda

Calorific value represents the energy content of biomass and is measured by determining the heat produced during complete combustion. Figure 3 represents a comparison of HHV literature values for common biomass found within Uganda, that are the focus of this report [25,26].

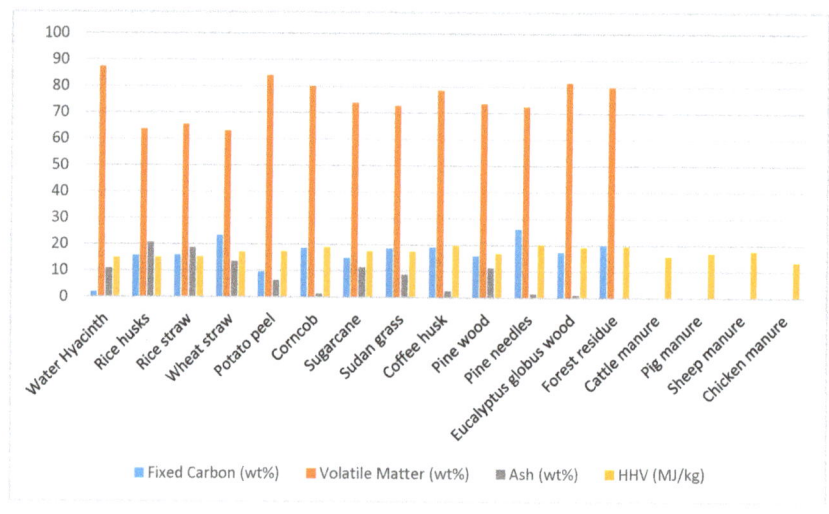

Figure 3. Characteristics of common Ugandan biomass [dry basis].

The HHV values of biomass shown in Figure 3 range from approximately 12–20 MJ/kg. Agricultural waste will average an approximate 15–20 MJ/kg with forestry biomass being towards 20 MJ/kg on a dry basis. HHV values for manure have been reported between 13–20 MJ/kg [25–27].

Upgrading biomass through thermochemical treatments such as HTC and pyrolysis increase the HHV. Uzun et al. (2017) reports a corn cob has a HHV of 18.77 MJ/kg as shown in Figure 3. Machado et al. (2018) reports that after HTC treatment with subcritical water at 250 °C the HHV of corn stover rose to 24.57 MJ/kg [28]. 24.74 MJ/kg was reported by S. Hoekman et al. (2013) [29]. Raveendran et al. (1996) pyrolysed corn cob at 500 °C within a packed bed pyrolyser producing a HHV of 28.6 MJ/kg or 26.4 MJ/kg from de-ashed corn cob [30]. These values compare with the IEA definition of sub-bituminous coal which reports to have a HHV of between 17.4–23.9 MJ/kg [31].

5. Calculator Utilisation

The energy demand calculator is a tool which theoretically determines the amount of energy required or produced from the solar/biomass processes. It does not take into account the practicality of solutions. For example, wanting to produce a solar pyrolysis system requiring high temperatures of 500 °C in a rural location of a developing country would require huge amounts of mirrors. Solar energy is dilute, whilst chemical energy is concentrated, thus you need a vast amount of solar energy to make a significant amount of chemical energy. This being the primary reason as to why high temperature thermochemical treatments such as pyrolysis (or gasification) not being a practical solution. A large central receiver (power tower) solar system would be required to generate the heat needed for the process. This would increase the cost and the complexity of the system drastically. With that stated a low temperature steam treatment (with a low biomass to water ratio) or low temperature HTC (ideally a low water to biomass ratio) are ideal thermochemical treatments. With temperatures between 100–250 °C a simple linear Fresnel concentrating method will be able to meet the temperatures required by a low temperature technique.

Uganda has been selected as the case study for this report, the model can be made applicable to all biomass types. Shown below are five practical examples based on common waste biomass streams from Uganda. Each example aims to hydrothermally carbonise 10/15/20 kg of biomass sample at 200 °C in a location with 750 W/m^2 of direct irradiance. Table 9 shows the constant input variables and Table 10 shows the specific heat capacity of the different biomasses used for comparison. However, it should be noted that heat capacities do vary within biomasses and will be dependent on how much moisture is in the biomass. Reported heat capacities are approximations for an example and lab tested heat capacities should be used for live projects.

Table 9. Constant variables for biomass comparison.

Variable	Value
Mass of biomass (kg)	10/15/20
Biomass to water ratio (1:x)	10
Mass of water (kg)	100
Cp of water (J/kg K)	4186
Temperature start (K)	293
Temperature of HTC (K)	493
Time (mins)	360
SI (W/m^2)	750
Efficiency (%)	25
Mirror collector width (m)	3

Table 10. Specific heat capacities of common biomasses in their representative countries.

Country	Biomass	Specific Heat Capacity (J/kg/K)
Uganda	Rice husk	1377 [32]
Uganda	Sugar cane (bagasse)	1500 [33]
Uganda	Dairy manure	1993 [34]
Uganda	Wood	1700 [35]
Uganda	Water hyacinth	1455 [36]

Figure 4 shows the length of mirror required for the process. As HTC requires a large body of water in the system, a ratio of 1:10, the specific heat capacity and therefore biomass type has a very small effect on the length of mirror required. The mass has a much greater effect on the amount of energy required as shown in Figure 4 Keeping the constant variables, the same, but increasing the mass of biomass to 15 kg, it is clear that mass has a much larger affect than biomass type. As with every extra 1 kg of biomass added to the system, 10 kg of water is required under current HTC practices.

The results shown in Figure 4 show that for a solar driven HTC process, the mass of biomass being processed has a significantly greater effect on the area of mirror collector required, than the type of biomass being processed.

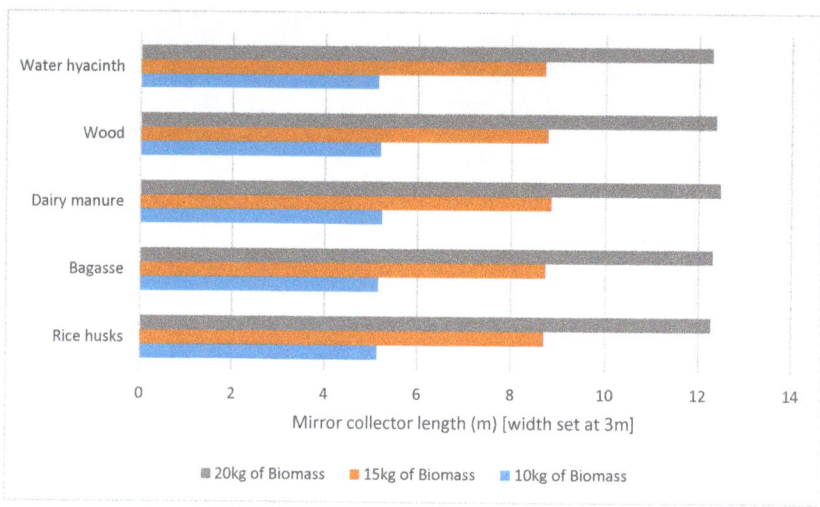

Figure 4. Effect of biomass and biomass mass on the length of mirror required.

6. Conclusions

A calculator has been produced to predict the area of the solar collector required to process a given mass of biomass in a given time, the amount of biomass that can be processed in a set time with set solar collector area or the time required to process a given mass of biomass with a set collector area. The basis of the model is formed from Equations (1)–(3) shown in Section 2.1. A thermodynamic approach with an efficiency factor was deemed as the most practical method for the model. Comparing steam treatment, HTC and pyrolysis in two examples, pyrolysis was shown to be the least energy demanding process. The main factor contributing to this was the amount of water that required heating during steam treatment and HTC. Though pyrolysis is the least energy demanding process, the feedstock selected and practicality should be considered before which thermochemical process is selected for an integrated system.

Uganda is a country rich in biomass waste. From the quantitative review accessing the amount of biomass waste available, it is shown that from the top five most produced crops in Uganda, there is approximately 3.75×10^6 metric tons of waste available. The majority of this waste is left to rot or to go back into the soil if the waste is produced at the farm. Agricultural waste will average an approximate 15–20 MJ/kg as shown in Figure 2. There are 2.3 million hectares of private forest land in Uganda. Forestry residue has been reported to have a HHV of approximately 20 MJ/kg. Within the private forest available, there will be a significant amount of forestry residue available that may be utilized for biofuel. From the qualitative review of biomass, it was stated that the majority of small communities owned their own livestock (individuals, collection of families, schools). Cattle, swine and chicken manure are shown in Table 4 to have the greatest potential as a feedstock. Though this will vary greatly depending on the conditions the animals are kept in. Conditions of the animals will alter the HHV of their waste as literature reports did vary between 13–20 MJ/kg. Water hyacinth has the potential to be a source of feedstock that will not affect the current supply chain. The quantity in Uganda is unknown, but it has the potential to be farmed and to be used as a feedstock for biofuel for those living near areas of contained water. The biomasses reported all have the potential for upgrading into a solid fuel for use within Uganda and other developing countries to replace firewood or other damaging fuels.

A low temperature steam treatment or low temperature HTC are ideal thermochemical treatments for combining with concentrated solar. With temperatures between 100–250 °C a simple linear Fresnel concentrating method will be able to meet the temperatures required by a low temperature biomass upgrading technique. As seen in Figure 4, the amount of biomass needed be processed has a much larger effect on the size of the system than the type of feedstock used.

Author Contributions: Conceptualization, T.G., R.C. and A.R.; methodology, T.G.; validation, T.G.; formal analysis, T.G.; investigation, T.G. and O.I.M.; resources, R.C. and A.R.; data curation, T.G.; writing—original draft preparation, T.G.; writing—review and editing, T.G., R.C. and A.R.; visualization, T.G.; supervision, R.C. and A.R.; project administration, R.C. and A.R.; funding acquisition, R.C. and A.R. All authors have read and agreed to the published version of the manuscript.

Funding: This work was supported by Engineering and Physical Sciences Research Council (EPSRC) Grant number: EP/L014912/1, the Global Challenges Research Fund (GCRF) and the United Bank of Carbon (UBOC).

Acknowledgments: This paper is an outcome of a research project on "A Solar–Driven Thermochemical Process for Creating a Biofuel" funded by the Engineering and Physical Sciences Research Council (EPSRC), the Global Challenges Research Fund (GCRF) and the United Bank of Carbon (UBoC). T.G. is sincerely grateful to O.I.M. and CREEC for providing guidance in Uganda, and to R.C. and A.R. for academic council.

Conflicts of Interest: The authors declare no conflict of interest. The funders had no role in the design of the study; in the collection, analyses, or interpretation of data; in the writing of the manuscript, or in the decision to publish the results.

References

1. Weldekidan, H.; Strezov, V.; Town, G. Review of solar energy for biofuel extraction. *Renew. Sustain. Energy Rev.* **2018**, *88*, 184–192. [CrossRef]
2. Morales, S.; Miranda, R.; Bustos, D.; Cazares, T.; Tran, H. Solar biomass pyrolysis for the production of bio-fuels andchemical commodities. *J. Anal. Appl. Pyrolysis* **2014**, *109*, 65–78. [CrossRef]
3. Adinberg, R.; Epstein, M.; Karni, J. Solar Gasification of Biomass: A Molten Salt Pyrolysis Study. *Sol. Energy Eng.* **2014**, *126*, 850–857. [CrossRef]
4. Alonso, E.; Romero, M. Review of experimental investigation on directly irradiated particles solar reactors. *Renew. Sustain. Energy Rev.* **2015**, *41*, 53–67. [CrossRef]
5. Yao, C.; Epstein, M. Maximizing the output of a solar-driven tubular reactor. *Sol. Energy* **1996**, *57*, 283–290. [CrossRef]
6. Wieckert, C.; Palumbo, R.; Frommherz, U. A two-cavity reactor for solar chemical processes: Heat transfer model and application to carbothermic reduction of ZnO. *Energy* **2004**, *29*, 771–787. [CrossRef]
7. Zeng, K.; Flamant, G.; Gauthier, D.; Guillot, E. Solar pyrolysis of wood in a lab-scale solar reactor: Influence of temperature and sweep gas flow rate on products distribution. *Energy Procedia* **2015**, *69*, 1849–1858. [CrossRef]
8. Li, R.; Soria, J.; Mazza, G.; Gauthier, D.; Rodriguez, R.; Flamant, G. Product distribution from solar pyrolysis of agricultural and forestry biomass residues. *Renew. Energy* **2016**, *89*, 27–35. [CrossRef]
9. Chintala, V. Modelling and evaluating a solar pyrolysis system. *Renew. Sustain. Energy Rev.* **2018**, *90*, 120–130. [CrossRef]
10. Ramos, G.; Perez-Marquez, D. Design of semi-static solar concentrator for charcoal production. *Energy Procedia* **2014**, *54*, 2167–2175. [CrossRef]
11. Hans, G.; Marta, B.; Marco, C.; Marina, C.; Enrico, E.; Elvis, K. Solar biomass pyrolysis with the linear mirror. *Smart Grid Renew. Energy* **2015**, *6*, 179–186.
12. Sanchez, M.; Clifford, B.; Nixon, J.D. Modelling and evaluating a solar pyrolysis system. *Renew. Energy* **2018**, *116*, 630–638. [CrossRef]
13. Zeng, K.; Gauthier, D.; Soria, J.; Mazza, G.; Flamant, G. Solar pyrolysis of carbonaceous feedstocks: A review. *Sol. Energy* **2017**, *156*, 73–92. [CrossRef]
14. Funke, A.; Ziegler, F. Heat of reaction measurements for hydrothermal carbonization of biomass. *Bioresour. Technol.* **2011**, *102*, 7595–7598. [CrossRef] [PubMed]
15. Daugaard, D.E.; Brown, R.C. Enthalpy for Pyrolysis for Several Types of Biomass. *Energy Fuels* **2003**, *17*, 934–939. [CrossRef]
16. Uganda Bureau of Statistics. Uganda Bureau of Statistics. 9 November 2018. Available online: https://www.ubos.org/explore-statistics (accessed on 18 November 2018).
17. Ministry of Water and Environment. *State of Uganda's Forestry 2016*; FAO: Kampala, Uganda, 2016.

18. Castren; Tuukka; Katila, M.; Lindroos, K.; Salmi, J. Private financing for sustainable forest management and forest productions in developing countires—Trends and drivers. *PROFOR*. 2014. Available online: https://www.cbd.int/financial/doc/wb-forestprivatefinance2014.pdf (accessed on 18 November 2018).
19. Global Forest Watch Systems Status. Global Forest Watch. World Resources Institute. 2015. Available online: https://www.globalforestwatch.org/dashboards/country/UGA?category=land-cover (accessed on 18 November 2018).
20. Barker, J.C.; Hodges, S.C.; Walls, F.R. Livestock Manure Production Rates and Nutrient Content. In *2002 North Carolina Agricultural Chemicals Manual*; North Carolina State University: Raleigh, NC, USA, 2002.
21. Engineering Practices Subcommittee of the ASAE Agricultural Sanitation and Waste Management Committee. *ASAE D384.1 FEB03: Manure Production and Characteristics*; American Society of Agricultural Engineers: St. Joseph, MI, USA, 2003.
22. Ogejo, J.A. Manure Production and Characteristics. 26 October 2015. Available online: https://articles.extension.org/pages/15375/manure-production-and-characteristics (accessed on 18 November 2018).
23. Heyze, V.; Tran, G.; Hassoun, P.; Regnier, C.; Bastianelli, D.; Lebas, F. Feedipedia–Animal Feed Resources Information System. INRA, CIRAD, AFZ and FAO. 13 October 2015. Available online: https://www.feedipedia.org/node/160 (accessed on 18 November 2018).
24. Hasan, M.R.; Chakrabarti, R. *Use of Algae and Aquatic Macrophytes as Feed in Small-Scale Aquaculture: A Review*; FAO Fisheries and Aquaculture Technical Paper 531; FAO: Rome, Italy, 2009; pp. 53–65.
25. Uzun, H.; Yildiz, Z.; Goldfarb, J.L.; Ceylan, S. Improved prediction of higher heating value of biomass using an artificial neural network model based on proximate analysis. *Bioresour. Technol.* **2017**, *234*, 122–130. [CrossRef] [PubMed]
26. Fantini, M. Biomass Availability, Potential and Characteristics. In *Biorefineries: Targeting energy, High Value Products and Waste Valorisation*; Springer: Cham, Switzerland, 2017; pp. 21–54.
27. Matheri, A.N.; Ntuli, F.; Ngila, J.C.; Seodigeng, T.; Zvinowanda, C.; Njenga, C.K. Quantitative characterization of carbonaceous and lignocellulosic biomass for anaerobic digestion. *Renew. Sustain. Energy Rev.* **2018**, *92*, 9–16. [CrossRef]
28. Machado, N.T.; Castro, D.A.R.; Santos, M.C.; Araujo, M.E.; Luder, U.; Herklotz, L.; Werner, M.; Mumme, J.; Hoffmann, T. Process analysis of hydrothermal carbonization of corn Stover withsubcritical H_2O. *J. Supercrit. Fluids* **2018**, *136*, 110–122. [CrossRef]
29. Hoekman, S.; Broch, A.; Robbins, C.; Zielinska, B.; Felix, L. Hydrothermal carbonization (HTC) of selected woody and herbaceous biomass feedstocks. *Biomass Convers.* **2013**, *3*, 113–126. [CrossRef]
30. Raveendran, K.; Ganesh, A. Heating value of biomass and biomass pyrolysis products. *Fuel* **1996**, *75*, 1715–1720. [CrossRef]
31. World Nuclear Association. World–Nuclear. August 2018. Available online: http://www.world-nuclear.org/information-library/facts-and-figures/heat-values-of-various-fuels.aspx (accessed on 22 November 2018).
32. Chiriac, R.; Gauthier, G.; Dupont, C. Heat capacity measurements of various biomass types and pyrolysis residues. *Fuels* **2014**, *115*, 644–651.
33. Baxter, L. Research Gate. 2016. Available online: https://www.researchgate.net/post/What_is_the_specific_heat_capacity_of_bagasse_and_sub-bituminous_coal (accessed on 22 July 2019).
34. Nayyeri, M.A.; Kianmehr, M.H.; Arabhosseini, A.; Hassan-Beygi, S.R. Thermal properties of dairy cattle manure. *Int. Agrophys.* **2009**, *23*, 359–366.
35. British Standard Instituation. *Building Materials and Products. Hygrothermal Properties. Tabulated Design Values*: BS EN 12524:2000; BSI: London, UK, 2000.
36. Munjeri, K.; Ziuku, S.; Maganga, H.; Siachingoma, B.; Ndlovu, S. On the potential of water hyacinth as a biomass briquette for heating applications. *Int. J. Energy Environ. Eng.* **2016**, *7*, 37–43. [CrossRef]

© 2020 by the authors. Licensee MDPI, Basel, Switzerland. This article is an open access article distributed under the terms and conditions of the Creative Commons Attribution (CC BY) license (http://creativecommons.org/licenses/by/4.0/).

Article

Forecasting the Structure of Energy Production from Renewable Energy Sources and Biofuels in Poland

Jarosław Brodny [1], Magdalena Tutak [2,*] and Saqib Ahmad Saki [3]

1. Faculty of Organization and Management, Silesian University of Technology, 44-100 Gliwice, Poland; jaroslaw.brodny@polsl.pl
2. Faculty of Mining, Safety Engineering and Industrial Automation, Silesian University of Technology, 44-100 Gliwice, Poland
3. Mining Engineering Department, University of Engineering and Technology, Lahore 54890, Pakistan; saqibahmadsaki@uet.edu.pk
* Correspondence: magdalena.tutak@polsl.pl; Tel.: +48-322372528

Received: 25 April 2020; Accepted: 15 May 2020; Published: 17 May 2020

Abstract: The world's economic development depends on access to cheap energy sources. So far, energy has been obtained mainly from conventional sources like coal, gas and oil. Negative climate changes related to the high emissions of the economy based on the combustion of hydrocarbons and the growing public awareness have made it necessary to look for new ecological energy sources. This condition can be met by renewable energy sources. Both social pressure and international activities force changes in the structure of sources from which energy is produced. This also applies to the European Union countries, including Poland. There are no scientific studies in the area of forecasting energy production from renewable energy sources for Poland. Therefore, it is reasonable to investigate this subject since such a forecast can have a significant impact on investment decisions in the energy sector. At the same time, it must be as reliable as possible. That is why a modern method was used for this purpose, which undoubtedly involves artificial neural networks. The following article presents the results of the analysis of energy production from renewable energy sources in Poland and the forecasts for this production until 2025. Artificial neural networks were used to make the forecast. The analysis covered eight main sources from which this energy is produced in Poland. Based on the production volume since 1990, predicted volumes of renewable energy sources until 2025 were determined. These forecasts were prepared for all studied renewable energy sources. Renewable energy production plans and their share in total energy consumption in Poland were also examined and included in climate plans. The research was carried out using artificial neural networks. The results should be an important source of information on the effects of implementing climate policies in Poland. They should also be utilized to develop action plans to achieve the objectives of the European Green Deal strategy.

Keywords: energy; renewable energy sources; climate policy; forecast; the European Green Deal

1. Introduction

One of the basic factors that has a significant impact on the development of the world economy and the entire civilization is access to large amounts of cheap energy. Energy is one of the basic resources that determine economic, social and political development of individual countries and regions [1–7]. The world's dynamic economic development results in an energy demand that has been growing rapidly in the last dozen or so years. In order to meet these needs, especially by developing countries, energy produced from conventional sources is essential. However, such production generates huge amounts of harmful substances emitted into the environment [8,9]. Pollution caused by energy production from these sources contaminates water, soil and air. Various types of gases are particularly dangerous, including greenhouse gases and dust [10–13]. Combined with emissions from other sectors of the world economy, these emissions are becoming a real threat to life on earth. That is why it is crucial to take measures to reduce emissions of harmful substances. In order to achieve noticeable effects in the surrounding ecosystem, global actions need to be taken into account.

The initiator of such activities has been the United Nations (UN) for many years. At climate summits, it calls for faster and more decisive actions to protect the environment. The European Union (EU) is an increasingly active participant in this process.

At the last UN Conference of the Parties (COP25) climate summit in December 2019, which took place in Madrid, Spain, the European Commission presented a new European climate strategy called the European Green Deal [14]. This strategy assumes that by 2050 the EU economy should become a zero-emission economy, i.e., climate neutral [15,16]. It is associated with, among others, a significant increase in the share of renewable energy sources (RES) in the energy mix of the EU Member States.

These assumptions should be considered immensely ambitious. So far, no region of the world or country has taken such decisive actions in the field of climate and environmental protection.

This strategy is furthest reaching in terms of climate protection since the commitments made under the Kyoto Protocol [17], which should be recognized as the most important factor stimulating the development of renewable energy both in the world and the EU.

One of the most essential areas of economic activity in the EU is meeting the energy needs of its inhabitants [18,19]. The forecast is that in the perspective of the next 25–30 years, energy demand in the EU countries will be systematically increasing [20].

The growing demand for energy in the EU countries and the need to protect the environment, including meeting the requirements of the European Green Deal strategy, means that the EU needs to develop and implement a common climate policy that is acceptable to all countries.

In order to reconcile these seemingly contradictory goals, which focus on the increase in energy production while limiting the negative impact of this process on the environment, energy transition is a must. Conventional energy sources must be replaced and supplemented by RES.

At the same time, the increase in energy production from RES should be large enough to meet the growing demand and additionally allow the reduction of production from conventional sources. Undoubtedly, this task is really demanding and requires many activities in the political, economic and social sphere. Also, such a transition requires large financial outlays, especially in the scope of unavoidable investments. The unit value of energy produced from RES is low, but large investments are needed to obtain this energy [21–23]. Such activities, especially in countries where the energy industry is based on conventional raw materials (hard coal and lignite, oil, gas), require both political will and social acceptance.

However, it seems that the environmental awareness of societies, especially in the EU, is at a level that creates an opportunity to conduct such changes.

For many years, the EU's energy policy has been based on an integrated approach to the issue of energy security of countries and the competitiveness of the economy as well as environmental and climate protection [24,25]. The importance and role of RES in the energy production structure have been reported to be growing in the EU countries. The result is an increasing share of energy obtained from RES in the energy mix of the EU countries.

Also, from a political point of view, more and more countries tend to accept the presented strategy. As in any such project, the essence is in the details. Nevertheless, the EU countries are generally aware that the implementation of the European Green Deal strategy is a must.

This strategy raises a lot of controversy, especially in Poland, in which the economy largely uses energy produced from hard and brown coal. A similar situation is also reported for the Czech Republic and Hungary.

In Poland, more than 91% of gross available energy is obtained from conventional sources (fossil fuels). In 2018, the most important energy resource was hard coal, the share of which in the production of this energy was 47.8%, and of brown coal 29% [26].

According to data from the International Renewable Energy Agency (IRENA), electricity production from RES in Poland in 2018 accounted for only 11.2% of total energy production [27]. Most of this energy was obtained from biomass, wind and biogas. The share of solar energy has currently been found to be small. However, since 2012, it has been characterized by a significantly growing trend (Figure 1) [27]. It should also be noted that the share of RES in total energy consumption increased from 2.5% in 1990 to 11.28% in 2018 (Figure 2) [26]. However, this is still a much weaker result than that achieved by the EU countries.

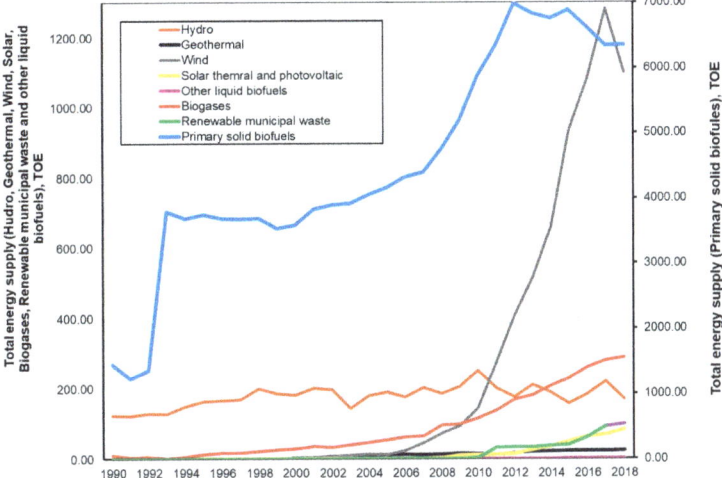

Figure 1. Total energy supply from renewable energy sources (RES) in Poland (own elaboration based on data from [26]).

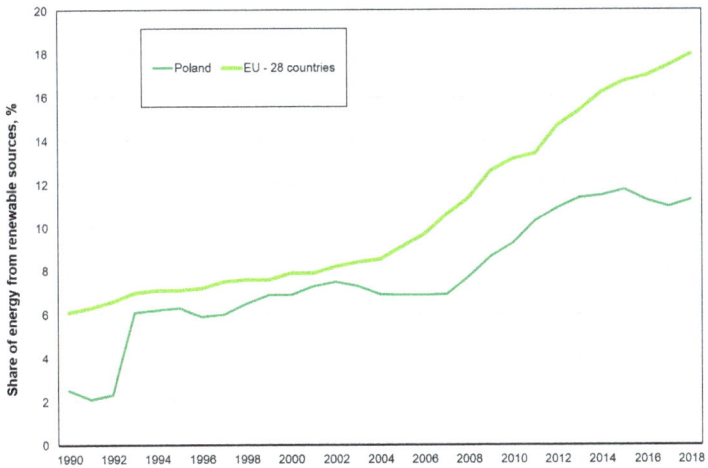

Figure 2. Percentage share of electricity produced from RES in Poland and the EU countries (own elaboration based on data from [26]).

The data presented in Figure 2 is extremely unfavorable for Poland. The economy based on conventional raw materials has a very negative impact on the environment. In the context of negative climate change, which is the result of such an economy, it becomes inevitable to replace conventional energy sources with RES [28]. The impact of these changes on the emission of harmful substances (greenhouse gases) is shown in Figure 3.

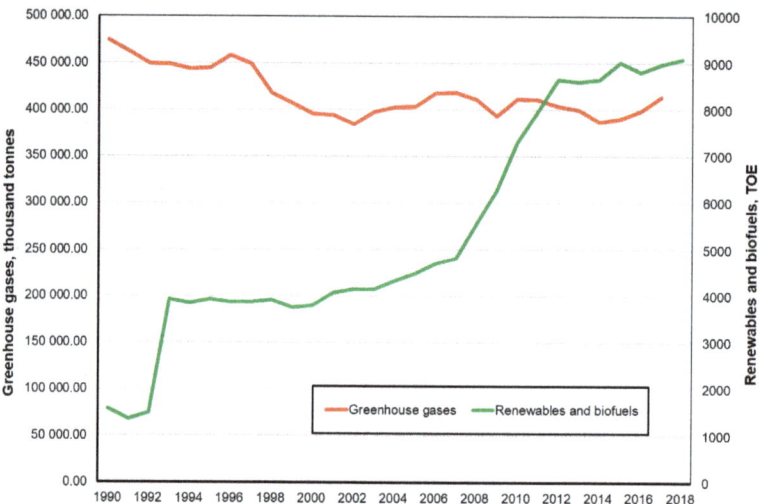

Figure 3. The course of changes in the amount of energy produced from RES and its impact on the emission of harmful gases (own elaboration based on data from [26,29]).

With regard to the foregoing, it is therefore reasonable to state that changing the structure of energy production in Poland is essential. On the one hand, it is required by the EU [30,31] and on the other by public opinion [32,33]. Society is increasingly aware that the huge pollution of the environment has a very negative influence on the climate of the whole Earth and the lives and health of individual citizens.

In general, it can be claimed that the social acceptance of Polish citizens to change the structure of energy production is greater than the political will of the government. The way to achieve the objectives set by the EU in this regard is to increase the share of energy produced from RES in the country's energy mix. It is also important to maintain the reliability of the energy system (energy security) [34] in terms of meeting energy needs, obtaining energy in a cost-effective way and reducing negative impact on the environment [35]. In this context, attention should be paid to the storage of energy produced from RES due to the fact that energy production from RES, as well as demand for it, can be characterized by variability in time [36]. In the context of renewable energy production, it should be borne in mind that both solar and wind energy have virtually zero marginal production costs, but these types of energy are only available when there are favorable weather conditions, i.e., the sun shines or the wind blows. Therefore, it is necessary to store this energy, which is not an easy task [37].

Little research is devoted to the issue of energy production from RES, including forecasts for its production in Poland in the upcoming years.

Onkisz-Popławska et al. presented the prospects for the development of renewable energy in Poland [38]. Igliński et al. showed the state of production of geothermal energy [39] in Poland and biogas-based energy [40]. In turn, in [41], the same authors discussed the state of renewable energy in one of the Polish voivodeships. In [42], the authors present the assessment of the RES penetration and the RES generation ramps (generation variety) within the time horizon until 2025. Bugała et al. in [43] showed the short-term forecast of electric energy generation in photovoltaic systems in Poland.

A number of studies devoted to renewable energy concern the perspectives of their development both in Poland and in the world [44–50].

With regard to forecasting the volume of energy produced from RES, numerous studies concern China [51], Turkey [52], and the United States [53].

Undoubtedly, the results of these papers give a picture of the existing situation in terms of the structure of energy production in the near future.

As already mentioned, there is no such research in the area of predicting energy production from RES in Poland. The only studies on how the production of renewable energy in Poland will look over the next few decades were conducted at the request of the government [54–56] and are very optimistic, despite the signals that the increase in this production in Poland in relation to the increases observed in other EU countries is at a very unsatisfactory level. Thus, according to the authors, such a forecast should be developed by independent researchers. Also, it must be reliable, not based on general plans or approximate estimates, but on current data with the use of modern, advanced methods, which currently include artificial neural networks.

Such forecasts should broaden knowledge in the field of energy production from RES and their perspectives. It is crucial to see when Poland, with the current state of the economy, can achieve the assumed goal of a 15% share of this energy in the total amount of energy produced. According to the original assumptions, this goal should be accomplished in 2020 [30].

Therefore, this article focuses on analyzing the structure and amount of energy produced from RES. Based on the changes taking place in recent years, analyses were carried out to predict energy production from RES until 2025. It was assumed that the results achieved in this period will have a decisive impact on meeting the criteria assumed by the EU. The results should also show the state that can be achieved in 2025 with the current dynamics of change and the policy pursued. In addition, the research also looked at individual sources from which renewable energy is produced in Poland. The analysis of the structure of this production and its prediction should form the basis for developing an energy policy for the coming years.

In order to prepare the forecast of energy production from RES in Poland (until 2025), the method of artificial neural networks was utilized. It belongs to the group of intelligent methods and, according to the authors, its advantages allow it to provide the best results in this type of analysis. Artificial neural networks give the opportunity to build models that can map the complex relationships between input and output data for selected phenomena, the structure and causal relationships, which have not been sufficiently known to build effective mathematical models.

This study characterizes the examined area and discusses the developed research methodology. The forecast takes into account the total production of energy from RES and from selected sources (Hydro, Geothermal, Wind, Solar thermal and photovoltaic, Primary solid biofuels, Other liquid biofuels, Biogases, and Renewable municipal waste). Correlations between energy production from these sources were also shown. In addition, an analysis was also performed, based on which the forecast of the share of renewable energy in total energy consumption by 2025 in Poland was made. For all the presented calculations, error and statistical analyses were conducted, the results of which are presented in this paper.

2. Materials and Methods

2.1. Area of Research

Poland is a country located in Central Europe between the Baltic Sea in the north and the Sudetes and the Carpathian Mountains in the south (Figure 4). From the north, Poland borders with Russia through its Kaliningrad region and Lithuania; from the east with Belarus and Ukraine; from the south with Slovakia and the Czech Republic; from the west with Germany. Most of the northern border of Poland defines the coast of the Baltic Sea. Poland's borders with Ukraine, Belarus and Russia also constitute the external border of the EU and the Schengen area.

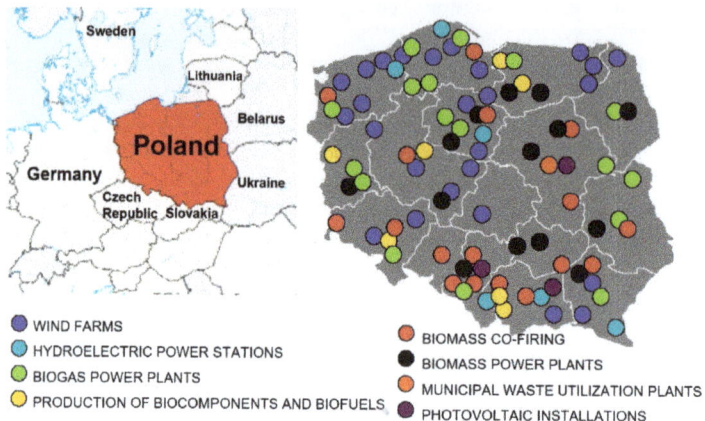

Figure 4. Location of Poland in Europe and RES installations in Poland (own elaboration based on [54]).

As regards the structure of energy production, in 2009, the Polish government adopted the Polish Energy Policy until 2030 [55], which contains the main development directions of the energy sector. In 2019, Poland's draft energy policy until 2040 was adopted. With regard to renewable energy, it presents a plan for the development of RES, expected to reduce the emissivity of the energy sector based mainly on conventional energy sources and the change in the structure of energy production [56].

The Polish energy strategy until 2040 assumes that the use of RES will be significantly affected by technological progress associated with already known methods of generating this energy (e.g., an increase in the use of wind by wind farms or solar radiation by photovoltaic panels), as well as with the development of new production technologies and energy storage.

In order to accomplish the assumed goals, an increase in energy production from RES and its wide use in all sectors of the economy will be inevitable.

This document presents the forecast of renewable energy consumption between 2020–2040. It assumes reaching a level of about 21%–23% share of energy from RES in final energy consumption in 2030 (in power engineering—possible increase in the share to 32%, in heating and cooling—1.1% point y/y increase in the share, in transport—14%), while in 2040, this share is expected to be 28.5%. The assumptions of Poland's energy policy regarding the use of RES until 2040 are shown in Figure 5.

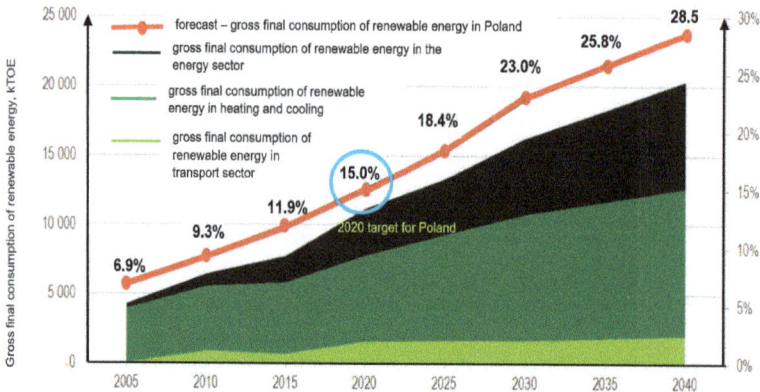

Figure 5. Assumptions for Poland's energy policy regarding the use of energy from RES until 2040 according to [56] (own elaboration based on [56]).

One of the aims of the presented study was to check whether, at the current pace of development of the renewable energy sector, the assumptions presented in Figure 5 are being implemented and to what extent.

2.2. Materials

For the analysis of the current state and prediction of renewable energy production in the perspective until 2025, data from the European Statistical Office [26] and the Data World Bank databases was used [57].

Information on the volume of renewable energy production between 1990–2018 is summarized in Table 1. This data includes energy production from the following RES: total renewables and biofuels, hydro, geothermal, wind, solar thermal and photovoltaic, primary solid biofuels, other liquid biofuels, biogases, and renewable municipal waste.

The variables presented in Table 1 were initially statistically analyzed and their basic statistical parameters were determined (mean, maximum, minimum, standard deviations and coefficient of variation), which is summarized in Table 2.

The analysis of basic statistics on variables that determine the volume of production from RES showed that they are significantly differentiated, which means that they meet the condition of diagnostic features. The values of the coefficient of variation for studied variables are characterized by a considerable range. The highest value of the coefficient of variation was found for the variable other liquid biofuels (202.26%), and the lowest for the variable hydro (16.96%). Skewness for most of the studied variables (except hydro and primary solid biofuels) was shown to have a positively skewed (right-skewed) distribution.

The statistical analysis of the data presented in Table 1 also involved the determination of the Pearson correlation coefficient between studied variables and the correlation matrix between these coefficients. The results are summarized in Table 3.

Table 1. The structure of renewable energy production from selected sources in Poland between 1990–2018 (own elaboration based on data from [26]).

Year	Renewables and Biofuels	Hydro	Geothermal	Wind	Solar Thermal and Photovoltaic	Primary Solid Biofuels	Other Liquid Biofuels	Biogases	Renewable Municipal Waste
					Thousand Tonnes of Oil Equivalent (TOE)				
1990	1581.603	123.783	0	0	0	1448.433	0	9.387	0
1991	1355.546	122.528	0	0	0	1228.480	0	4.538	0
1992	1496.088	129.579	0	0	0	1361.016	0	5.493	0
1993	3925.738	127.945	0	0	0	3796.312	0	1.481	0
1994	3847.349	149.011	0	0	0	3692.247	0	6.091	0
1995	3924.018	162.253	0	0.086	0	3748.519	0	13.160	0
1996	3868.649	166.036	0	0.000	0	3685.225	0	17.388	0
1997	3866.886	168.616	0	0.172	0	3680.997	0	17.101	0
1998	3916.280	198.538	0	0.344	0	3696.355	0	21.042	0
1999	3752.804	185.297	0	0.344	0	3541.846	0	25.174	0.143
2000	3801.557	181.083	2.962	0.430	0	3587.298	0	28.924	0.860
2001	4070.756	199.914	2.866	1.204	0	3831.231	0	35.278	0.263
2002	4142.185	195.959	6.282	5.245	0.024	3902.121	0	32.316	0.239
2003	4148.118	143.680	7.428	10.662	0.024	3918.983	0	38.789	0.334
2004	4320.780	178.997	7.595	12.237	0.096	4061.718	0	46.360	0.310
2005	4486.514	189.242	11.369	11.648	0.143	4166.213	0	53.573	0.717
2006	4694.944	175.588	12.778	22.019	0.263	4323.923	0	62.410	0.717
2007	4823.745	202.247	10.485	44.848	0.358	4394.597	0	64.679	0.836
2008	5559.413	185.052	12.683	71.952	1.29	4750.669	0	96.159	0.215
2009	6265.283	204.225	14.331	92.629	8.001	5190.169	0.757	98.022	0.693
2010	7293.909	251.070	13.447	143.107	10.032	5866.199	0.220	114.574	2.938
2011	7966.178	200.463	12.683	275.542	12.483	6350.626	0.549	136.906	31.958
2012	8644.125	175.129	15.788	408.133	14.906	6987.723	0.065	167.980	32.483
2013	8606.385	209.705	18.582	516.234	24.848	6836.797	0.173	181.356	33.223
2014	8652.992	187.657	20.230	659.985	35.345	6755.398	0.089	207.438	36.878
2015	9019.135	157.540	21.711	933.651	49.892	6883.634	1.520	228.838	39.959
2016	8805.893	183.959	22.213	1082.338	62.934	6620.163	1.793	261.059	61.049
2017	8970.596	220.084	22.584	1281.947	68.695	6340.936	1.888	280.576	92.452
2018	9084.209	169.389	23.671	1100.498	82.762	6347.228	1.937	288.337	98.327

Table 2. Basic statistical parameters of studied variables (own elaboration).

Variable	Mean	Median	Minimum	Maximum	Suma	Standard Deviation	Coefficient of Variation, %	Skewness	Kurtosis
			Thousand Tonnes of Oil Equivalent (TOE)						
Renewables and Biofuels	5341.092	4320.780	1355.546	9084.209	154891.7	2435.631	45.60	0.31	−1.05
Hydro	177.399	181.083	122.528	251.070	5144.6	30.089	16.96	−0.05	0.23
Geothermal	8.955	7.595	0.000	23.671	259.7	8.437	94.22	0.35	−1.30
Wind	230.181	11.648	0.000	1281.947	6675.3	393.830	171.10	1.69	1.54
Solar thermal and photovoltaic	12.831	0.096	0.000	82.762	372.1	23.646	184.29	1.94	2.68
Primary solid biofuels	4517.071	4061.718	1228.480	6987.723	130995.1	1651.150	36.55	−0.16	−0.44
Other liquid biofuels	0.310	0.000	0.000	1.937	9.0	0.627	202.26	1.99	2.48
Biogases	87.739	46.360	1.481	288.337	2544.4	91.368	104.14	1.07	−0.16
Renewable municipal waste	14.986	0.334	0.000	98.327	434.6	27.761	185.25	2.00	3.40

Table 3. Linear correlations of the variables (own elaboration).

Variable	Renewables and Biofuels	Hydro	Geothermal	Wind	Solar Thermal and Photovoltaic	Primary Solid Biofuels	Other Liquid Biofuels	Biogases	Renewable Municipal Waste
Renewables and Biofuels	1.00	0.55	0.92	0.83	0.79	0.98	0.69	0.93	0.78
Hydro	0.55	1.00	0.47	0.23	0.21	0.59	0.19	0.40	0.22
Geothermal	0.92	0.47	1.00	0.83	0.81	0.87	0.73	0.94	0.78
Wind	0.83	0.23	0.83	1.00	0.98	0.73	0.91	0.96	0.96
Solar Thermal and Photovoltaic	0.79	0.21	0.81	0.98	1.00	0.68	0.93	0.93	0.97
Primary Solid Biofuels	0.98	0.59	0.87	0.73	0.68	1.00	0.58	0.86	0.68
Other Liquid Biofuels	0.69	0.19	0.73	0.91	0.93	0.58	1.00	0.84	0.88
Biogases	0.93	0.40	0.94	0.96	0.93	0.86	0.84	1.00	0.92
Renewable Municipal Waste	0.78	0.22	0.78	0.96	0.97	0.68	0.88	0.92	1.00

The analysis of correlation between variables was carried out for the level of statistical significance of $p < 0.05$. The analysis of the results showed that the studied variables are marked by different values of this coefficient, and all correlations are positive. The smallest correlation was reported to occur between hydro and other liquid biofuels and amounted to 0.19. In most cases, correlation coefficients were found to have achieved high values.

Graphic relationships with the determined parameters of the statistical analysis between selected variables are presented in Figure 6.

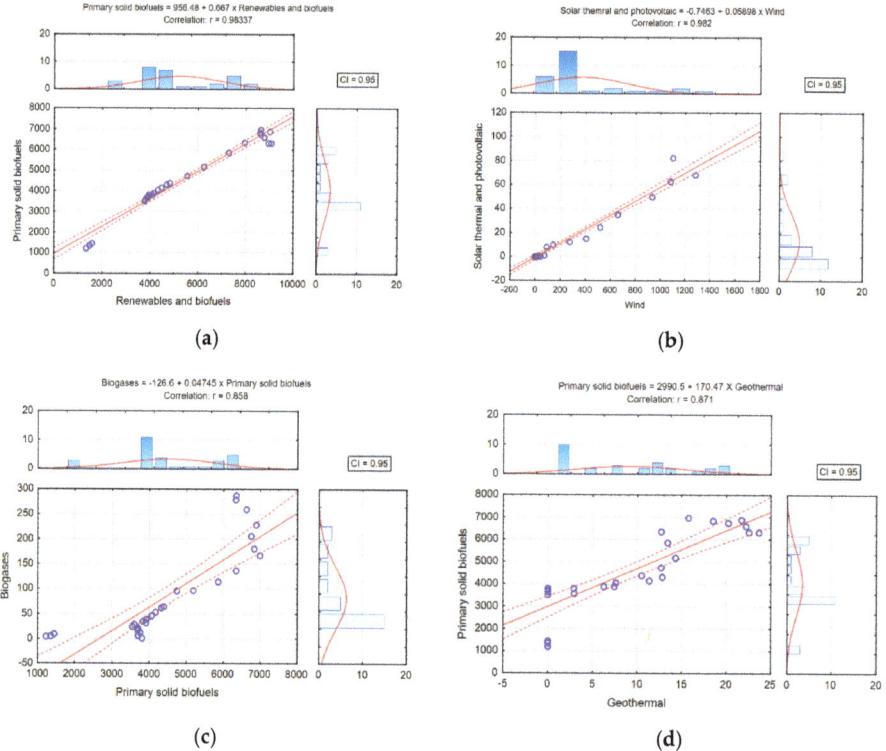

Figure 6. The relationship between selected variables determining the volume of energy production from RES in Poland. (**a**) Correlation between renewables and biofuels and primary solid biofuels; (**b**) correlation between wind and solar energy; (**c**) correlation between primary solid biofuels and biogases; (**d**) correlation between geothermal and primary solid biofuels (own elaboration).

The results constituted the input material for the basic analysis aimed at determining the predicted values of energy production from RES in Poland until 2025 and the values of the share of energy from RES in gross final energy consumption.

2.3. Methods

Artificial neural networks (ANN) intensively developed for several decades are a universal approximation system that allows mapping multidimensional data sets. They have the ability to both learn and adapt to changing environmental conditions and to generalize acquired knowledge, being in this respect an artificial intelligence system [58–64].

The neural network is a simplification of the structure of the human brain and is used in many disciplines of science. The main advantage of these networks is the possibility of obtaining solutions to complex problems that are difficult to solve by other conventional methods. They are frequently used as a forecasting tool, also for short-term forecasts [65–67]. The main elements of artificial neural networks are three layers: input, hidden and output (Figure 7).

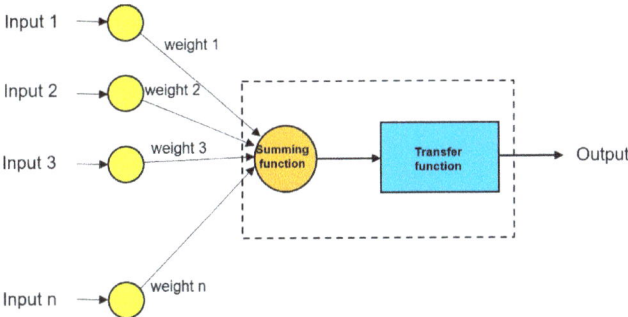

Figure 7. The basic structure of a neural network (own elaboration).

One of the most commonly used neural network model for forecasting is the Multi-layer perceptron (MLP) network, which consists of three layers: the first—Input layer, the last—Output layer and the middle—Hidden layer.

With regard to the neural network operation (Figure 8), signals carrying input data appear at its entry. From the input layer, x_i signals are sent to all hidden neurons of the Y layer. Each hidden neuron has a specific number of entries, and each entry has a weight w_i^X associated with it. Inside the hidden layer neurons, based on information from the input layer neurons xi and weights w_{ij}, the aggregated input value is calculated, which is the sum of the weighted inputs $\Sigma\, x_i w_i^X$. In turn, the neuron activation functions allow for the determination of the output values of hidden layer neurons y_i and the output values of the output layer neurons z. These values are then added together, resulting in the signal s_i [68]:

$$s_i = \sum_{j=1}^{n} w_i^x \cdot x_i \tag{1}$$

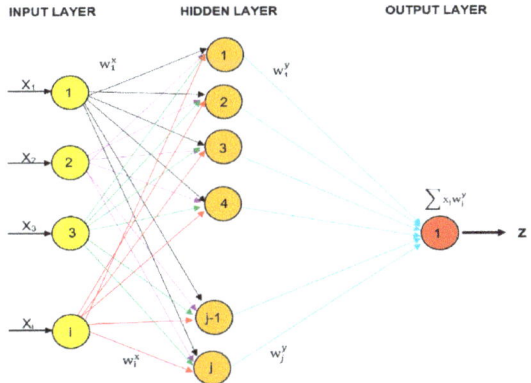

Figure 8. Construction of the Multi-layer perceptron (MLP) network with one hidden layer (own elaboration).

The activation function can be either a linear or non-linear function. The most commonly used activation functions are [69]:

Logistic function:
$$\varphi(x) = \frac{1}{1-e^{-1}} \qquad (2)$$

Hyperbolic tangent function:
$$\varphi(x) = \frac{e^x - e^{-x}}{e^x + e^{-x}} \qquad (3)$$

Exponential function:
$$\varphi(x) = e^{-x} \qquad (4)$$

Linear function:
$$\varphi(x) = x \qquad (5)$$

For the studied process of forecasting energy production from RES using artificial neural networks, the research procedure consisted of the following stages (Figure 9):

1. Database generation.
2. Division of the data set into the training and test sets and determination of the minimum number of neurons in the input and hidden layers.
3. Selection of the best network (criterion of correlation coefficient between actual and predicted values).
4. Prediction.

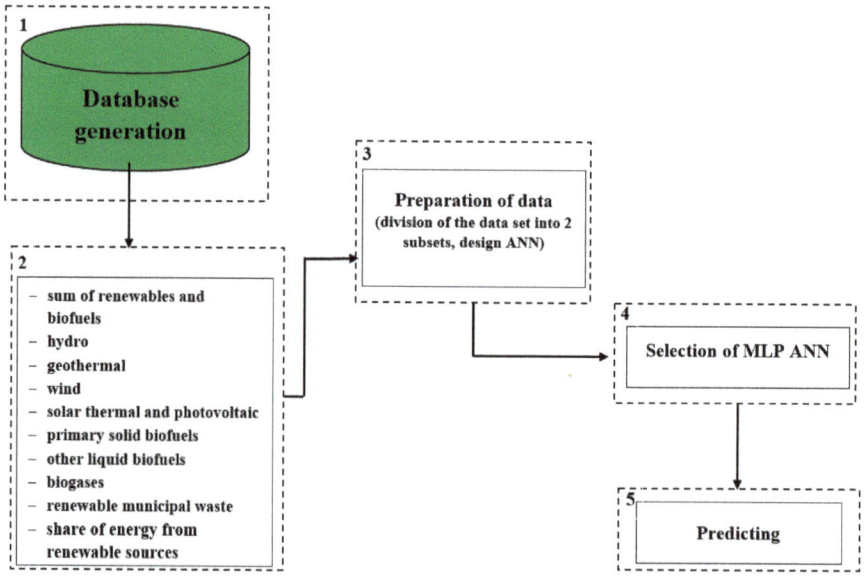

Figure 9. Diagram of the research procedure (own elaboration).

2.4. Error Estimation Methods

For the obtained forecasts of the total renewable energy production, and based on the sources included in the analysis, the mean error, mean percentage error, mean absolute error and the mean absolute percentage error were determined. These errors were determined from the following relationships [70,71]:

Mean error (ME):

$$ME = \frac{1}{n}\sum_{i=1}^{n}(X_{A,i} - X_{P,i}) \tag{6}$$

Mean percentage error (MPE):

$$MPE = \frac{1}{n}\sum_{i=1}^{n}\left(\frac{X_{A,i} - X_{P,i}}{X_{A,i}}\right) \times 100\% \tag{7}$$

Mean absolute error (MAE):

$$MAE = \frac{1}{n}\sum_{i=1}^{n}|X_{A,i} - X_{P,i}| \tag{8}$$

Mean absolute percentage error (MAPE):

$$MAPE = \frac{1}{n}\sum_{i=1}^{n}\left|\frac{X_{A,i} - X_{P,i}}{X_{A,i}}\right| \times 100\% \tag{9}$$

where $X_{A,i}$ and $X_{P,i}$ represent the observed and predicted values.

3. Results

The data set presented in Table 1, characterizing the amount of renewable energy production from various sources, was divided into two subsets (by the limited size of the data set): training data set (80% of cases) and test data set (20% of cases).

In order to forecast energy production from RES by 2025, a specific network structure was adopted, and it consists of an input layer, a hidden layer and an output layer.

Table 4 summarizes the structures of neural networks that obtained the highest values of correlation coefficients between the actual and predicted quantities in the training tests. As can be seen, for each of the predicted variables, the best values of these coefficients were obtained by networks of different structure.

Table 4. List of neural network structures used for research together with the results of training tests (own elaboration).

Forecast Variant	Network Structure	Correlation Coefficient		Matching Error		Activation Function—Neurons	
		Learning	Test	Learning	Test	Hidden	Output
Renewables and biofuels	MLP 9-5-1	0.927	0.934	22,473.10	14,457.60	exponential	exponential
Hydro	MLP 8-7-1	0.630	0.621	55.046	129.812	logistic	linear
Geothermal	MLP 8-12-1	0.916	0.959	0.099	1.041	logistic	exponential
Wind	MLP 8-7-1	0.972	0.987	1712.202	731.323	exponential	exponential
Solar thermal and photovoltaic	MLP 8-5-1	0.977	0.988	5.263	2.887	exponential	exponential
Primary solid biofuels	MLP 7-5-1	0.905	0.900	8695.744	1942.004	hyperbolic tangent	exponential
Other liquid biofuels	MLP 9-14-1	0.937	0.609	0.018	0.012	logistic	linear
Biogases	MLP 8-10-1	0.948	0.973	35.107	55.967	hyperbolic tangent	linear
Renewable municipal waste	MLP 8-8-1	0.965	0.978	13.000	3.815	logistic	exponential

The obtained correlation coefficient values for the training data set are at a satisfactory level, especially when taking into account the small amount of data adopted for prediction (29 values defining the predicted variables). Neural networks have a special property, which means that the more data on the predicted variable, the better the network quality, and the more accurate the forecasts.

As already mentioned, the data used for the forecast come from the Eurostat database, which applies to renewable energy production for the years 1990–2018. This data constitutes time series characterized most frequently by non-stationarity (for variables like renewables and biofuels, geothermal, wind, solar thermal and photovoltaic, primary solid biofuels, other liquid biofuels, biogases, and renewable municipal waste, etc.) (Figure 1). The time series, which is characterized by a very small degree of stationarity, applies to the variable hydro energy.

The structures of neural networks determined based on tests were used to perform basic calculations (Table 4).

Based on the analyses carried out, the predicted values of energy production from RES were determined in the perspective until 2025. The time horizon of the forecast covered the period from 2019 to 2020. The analysis involved the determination of predicted energy production values from individual studied RES and the percentage share of this energy in gross final energy consumption

Figure 10 presents the results of the forecast for the total amount of energy produced from RES together with the actual values (until 2025). In turn, Figures 11–18 show the actual and predicted volume of renewable energy production from selected RES.

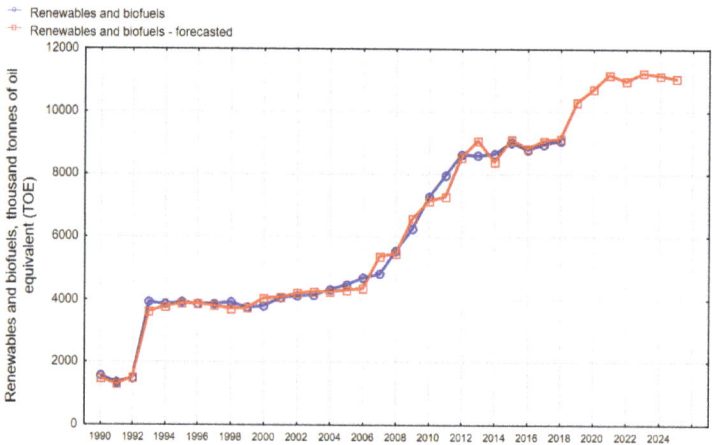

Figure 10. Actual and predicted volume of renewable energy production in Poland (own elaboration).

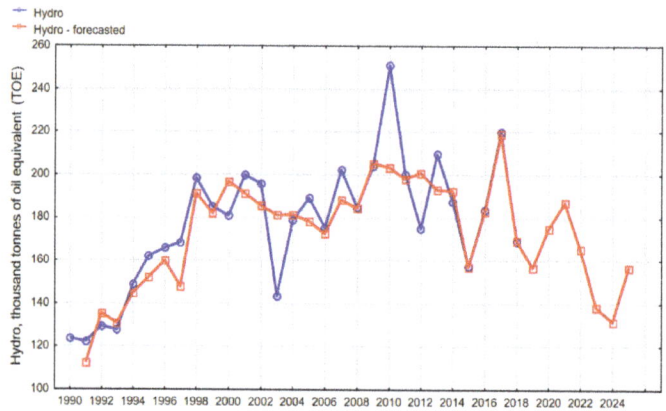

Figure 11. Actual and predicted volume of hydro energy production in Poland (own elaboration).

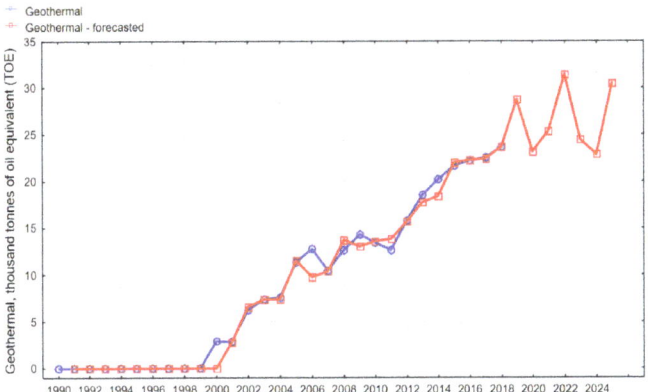

Figure 12. Actual and predicted volume of geothermal energy production in Poland (own elaboration).

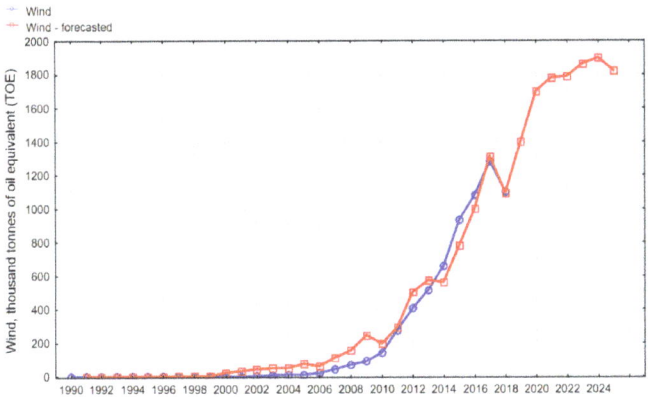

Figure 13. Actual and predicted volume of wind energy production in Poland (own elaboration).

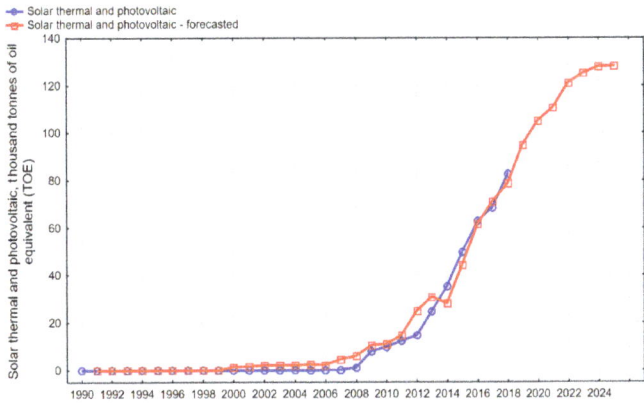

Figure 14. Actual and predicted volume of solar thermal production from photovoltaic energy in Poland (own elaboration).

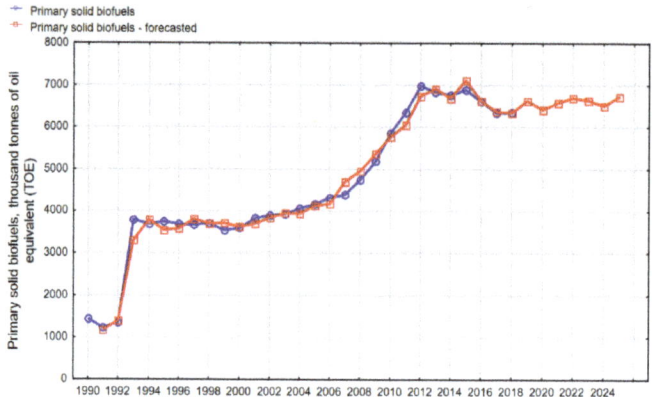

Figure 15. Actual and predicted volume of primary solid biofuels energy production in Poland (own elaboration).

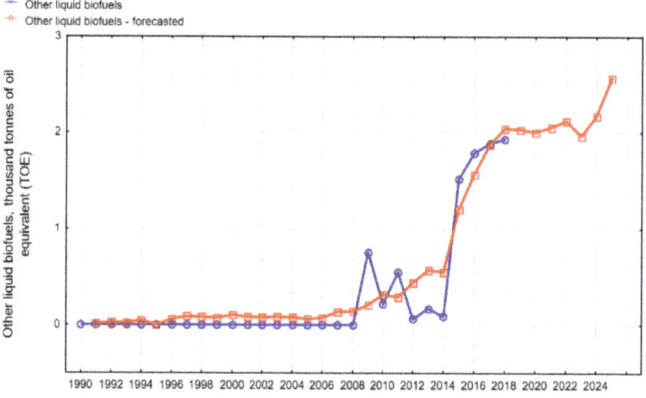

Figure 16. Actual and predicted volume of other liquid biofuel energy production in Poland (own elaboration).

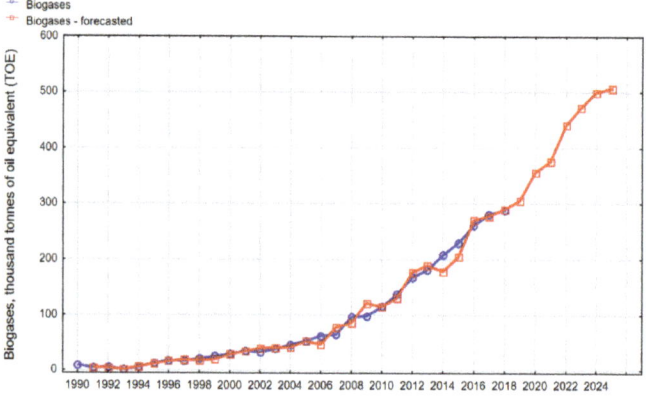

Figure 17. Actual and predicted volume of biogas energy production in Poland (own elaboration).

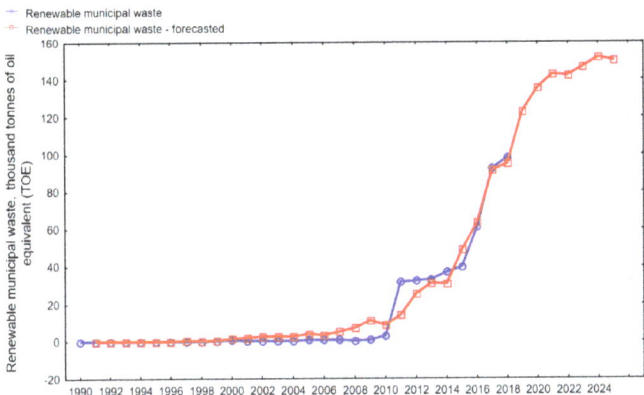

Figure 18. Actual and predicted volume of renewable municipal waste energy production in Poland (own elaboration).

Based on the results, it can be concluded that the most accurate was the mapping of total actual production of renewable energy (Figure 10), wind energy (Figure 13), solar energy (Figure 14), primary solid biofuels energy (Figure 15), biogas energy (Figure 17), and renewable municipal waste energy (Figure 18). The least accurate mapping was achieved for hydro energy production (Figure 11). This is due to the fact that the actual time series fluctuated significantly and there was no clear trend, which, combined with a small amount of data, made it very difficult to provide this forecast. All this can be seen in the results.

The results in the training part indicate that the predicted values are slightly more often overestimated than underestimated in relation to the actual values. This is mainly due to the approximation system of neural network models.

In general, however, it can be stated that the adopted MLP network architecture for the predicted variables allowed for the forecast of renewable energy production from the studied sources with satisfactory accuracy.

Moreover, it can also be stated that in the coming years, an increase in renewable energy production should be expected, practically from all sources. The exception is the production of energy from water. This is due to the fact that Poland is a lowland country with decreasing rainfall. As a consequence, the hydropower potential is relatively low, which results in decreasing financial expenditure on the development of hydropower [72]. According to statistics, the utilization of hydroenergetic potential of power plants in Europe is on average around 47%, and in Poland only 12% [73]. It is also important that Poland's existing hydropower potential is used to a much lesser extent than in the past [74]. Therefore, more decisive actions are needed to increase the use of this potential. One of the barriers limiting the development of hydropower in Poland is also the widespread belief in the harmful effects of river regulation on the natural environment [75]. All these problems make investing in hydropower very risky.

Based on the results, an analysis of the dispersion of the actual and predicted values of studied variables was also performed. The results are shown graphically in Figure 19.

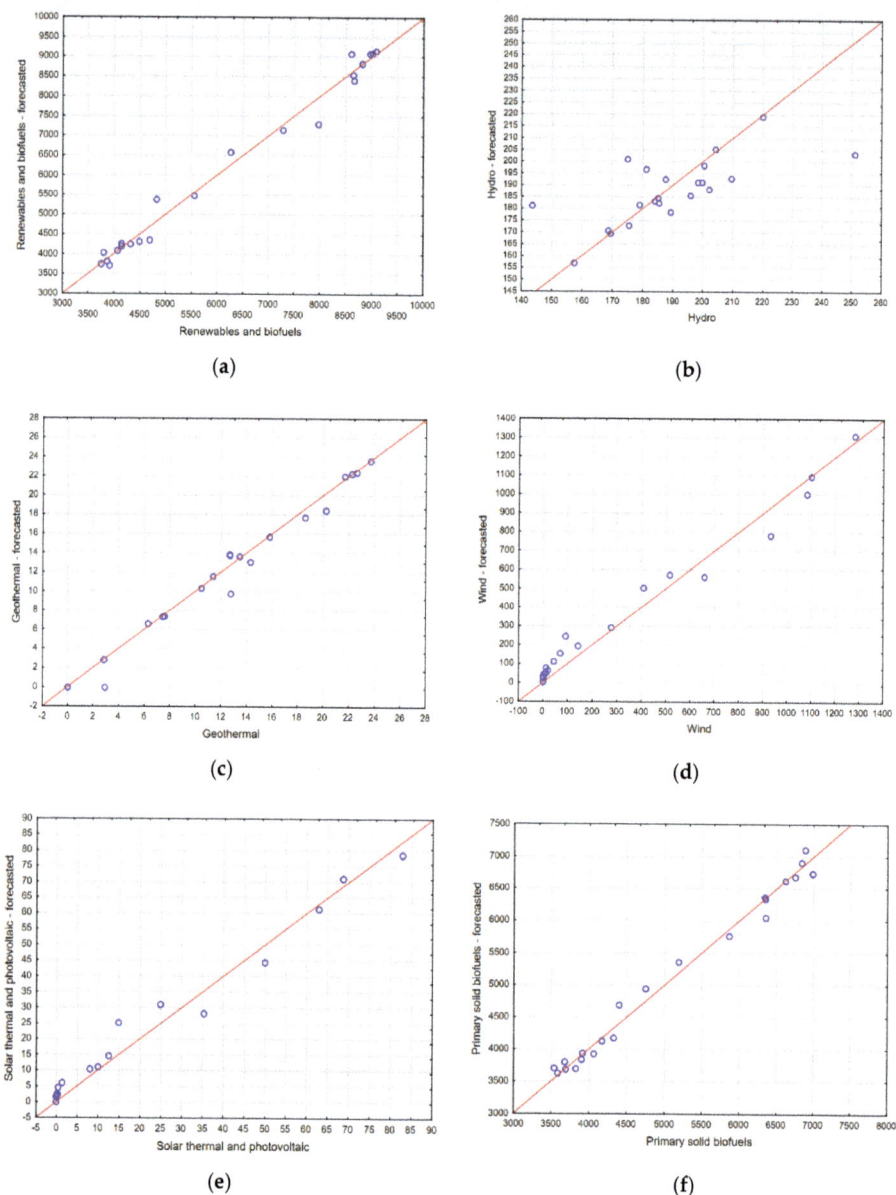

Figure 19. Cont.

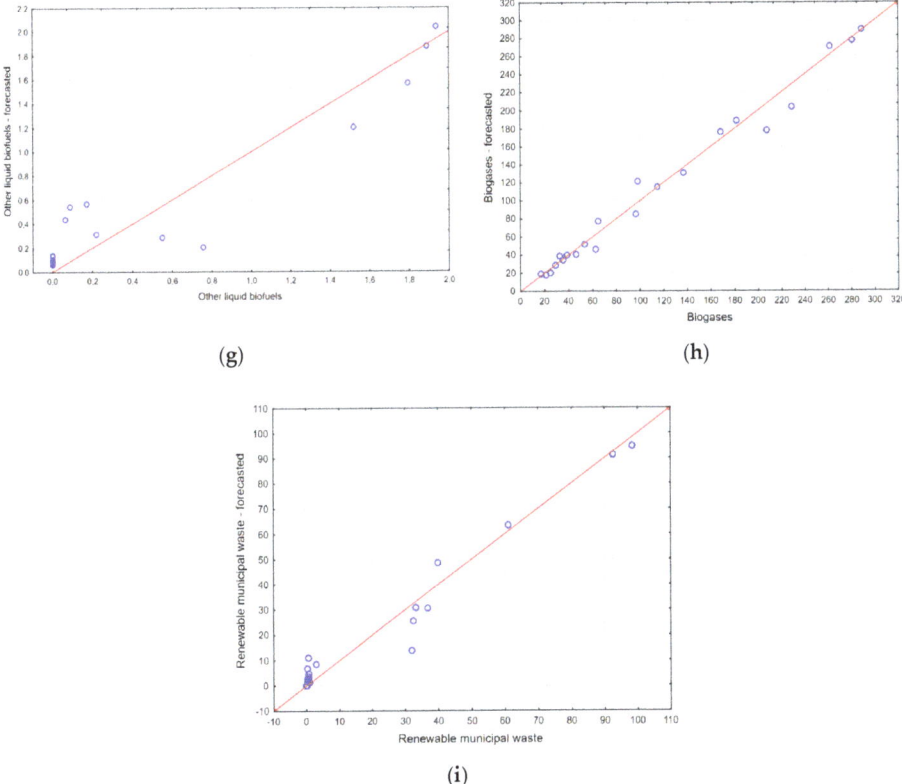

Figure 19. Predicted versus actual values of the volume of renewables and biofuels energy production (**a**), hydro energy production (**b**), geothermal energy production (**c**), wind energy production (**d**), solar thermal and photovoltaic energy production (**e**), primary solid biofuels energy production (**f**), other liquid biofuels energy production (**g**), biogases energy production (**h**) and renewable municipal waste energy production (**i**) (own elaboration).

When analyzing the distribution presented in Figure 19, it can be concluded that the largest dispersion is shown by the results of hydro energy production. For this forecast, the correlation coefficient between the actual and predicted data is the smallest (0.630 for the training sample and 0.621 for the test sample) (Table 4). Despite the attempts to use networks with different configurations (e.g., by increasing the maximum number of neurons in the hidden layer), no better correlation of the forecast with the actual values for this case was obtained.

In order to better visualize the results, Figure 20 presents histograms of the actual and predicted values of studied variables along with the marked density functions.

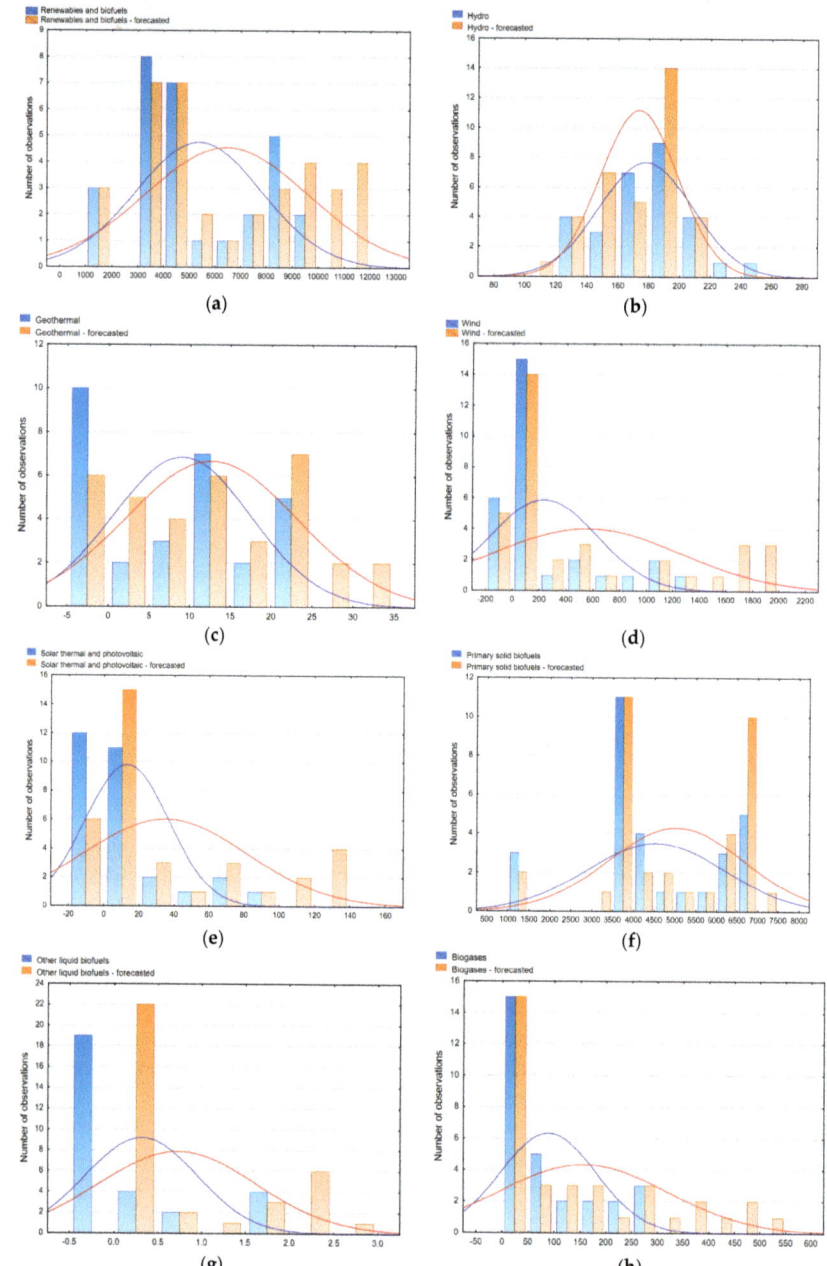

Figure 20. *Cont.*

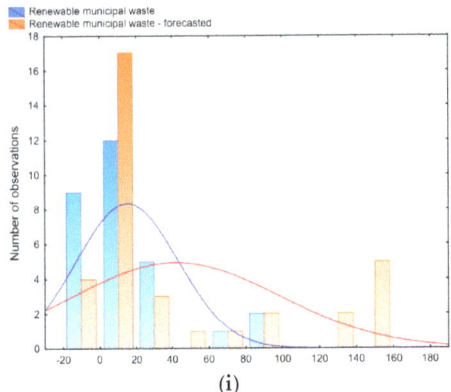

(i)

Figure 20. Histograms of the actual and predicted values of renewables and biofuels energy production (**a**), hydro energy production (**b**), geothermal energy production (**c**), wind energy production (**d**), solar thermal and photovoltaic energy production (**e**), primary solid biofuels energy production (**f**), other liquid biofuels energy production (**g**), biogases energy production (**h**) and renewable municipal waste energy production (**i**) (own elaboration).

Based on these histograms, the courses of both the actual and predicted values of studied parameters were found to be asymmetrical. Also, the determined density functions of the actual and predicted distributions show certain differences, which is confirmed by their recorded dispersion. It is also possible to determine the distribution of the values of studied parameters depending on the number of observations (number of studied years). Moreover, it can be seen that the inclusion of predicted values in these distributions only slightly widens the range of these values. The predicted values most often coincide with the values characterizing the volume of renewable energy production in the examined years.

Moreover, an analysis of errors was made between the actual values of studied parameters and their values obtained from calculations from neural networks (for both training and test data). The values of these errors are summarized in Table 5.

Table 5. Summary of errors (own elaboration).

Forecast Variant	ME	MPE, %	MAE	MAPE, %
Renewables and biofuels	13.90	0.38	163.45	3.07
Hydro	2.80	1.05	9.72	5.43
Geothermal	0.25	4.40	0.49	6.39
Wind	−0.01	−13.77	23.87	16.54
Solar thermal and photovoltaic	−0.05	−3.41	1.73	8.71
Primary solid biofuels	19.46	0.48	126.98	2.93
Other liquid biofuels	0.02	−5.39	0.07	19.19
Biogases	1.11	1.41	6.56	9.75
Renewable municipal waste	0.55	−13.24	2.14	20.74

When analyzing the forecast errors determined in terms of total RES energy production volume and selected sources, it can be stated that they are at an acceptable level.

The highest value of the average MAPE forecast error was 20.74% and concerned the production of energy from renewable municipal waste. The lowest MAPE error value was 2.93% for energy production from primary solid biofuels.

It can therefore be concluded that, despite the limited amount of data, the results obtained are satisfactory and allow the inference process to be carried out in terms of the predicted values of renewable energy production. This, in turn, may become the basis for assessing the effectiveness of operations and determine further directions of work to increase the production of energy from RES.

Pursuant to the Renewable Energy Directive (2009/28/EC) [76], Poland has undertaken that, in 2020, the share of energy from RES in the total amount of energy used will be at least 15%. To determine if and possibly when this goal can be achieved, an additional analysis was conducted to make the forecast of the share of RES in total energy consumption by 2025. In order to prepare this forecast, a new neural network was developed consisting of an input layer, a single hidden layer and an output layer.

Parameters characterizing this network are presented in Table 6, and the designated forecast in Figure 21.

Table 6. Parameters of the network structure to predict the share of energy from RES in gross final energy consumption by 2025 (own elaboration).

Forecast Variant	Network Structure	Correlation Coefficient		Matching Error		Activation Function—Neurons	
		Learning	Test	Learning	Test	Hidden	Output
Share of energy from renewable sources	MLP 10-13-1	0.874	0.868	0.048	0.049	sinusoidal	exponential

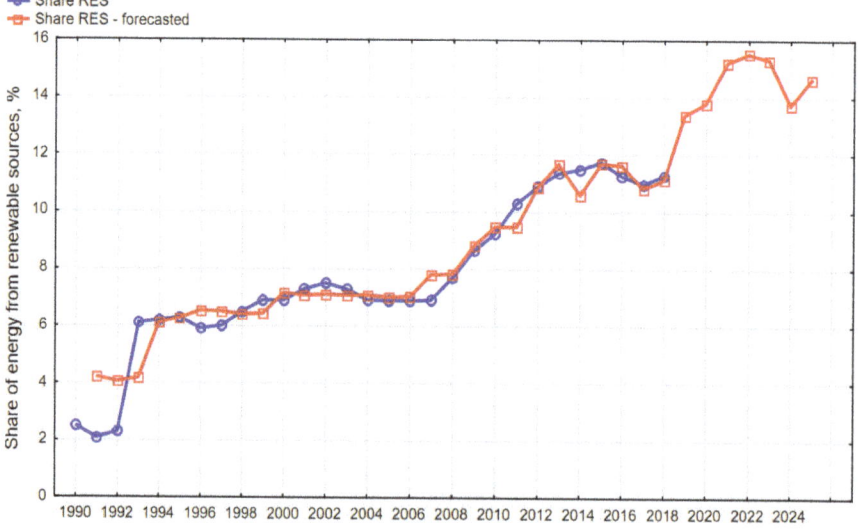

Figure 21. Actual and predicted share of energy from RES in gross final energy consumption (own elaboration).

An analysis of the dispersion of the actual and predicted values of the actual and predicted variables was also made, which is presented graphically in Figure 22, while Figure 23 presents a histogram of these values with the density function marked.

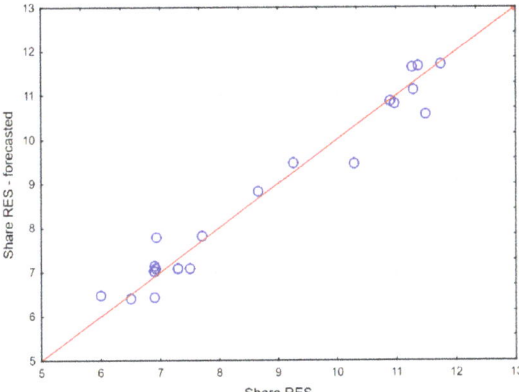

Figure 22. Predicted versus actual values of the share of energy from RES in gross final energy consumption (own elaboration).

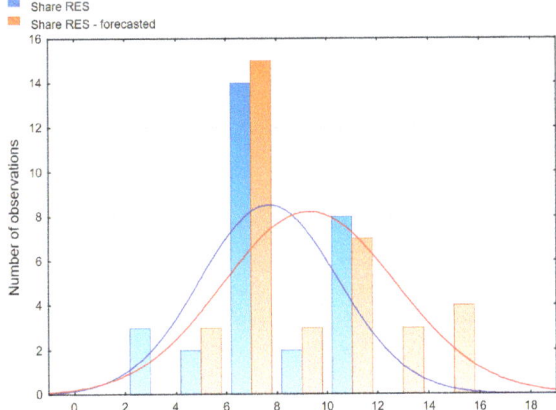

Figure 23. Histogram of the actual and predicted values of the share of energy from RES in gross final energy consumption (own elaboration).

The results of this forecast indicate that Poland will not achieve the assumed 15% target for the share of energy from RES in gross final energy consumption in 2020. This goal can only be met in 2021, which is not a negative result (share of around 15.34%). Obviously, from the point of view of implementing this plan, the forecast shows that in 2020, the share of energy from RES in gross final energy consumption will be at the level of 13.81%. However, the forecast for the coming years is quite worrying. Although in 2022 the share will increase to over 15.53%, in the following year there will be a slight decrease to the level of 13.76% and another increase in 2025 to 14.65%.

Based on the results, an error analysis was also made between the actual values of studied parameters and their values obtained from the calculations (Table 7).

Table 7. Summary of errors (own elaboration).

Forecast Variant	ME	MPE, %	MAE	MAPE, %
Share of energy from renewable sources	−0.08	−5.67	0.47	10.72

When studying the forecast errors in terms of the value of the share of energy from RES in gross final energy consumption, it can be stated that they are at an acceptable level. The MAPE error value is less than 11% and the MAE error value is 0.47.

4. Discussion

Based on the results, it can be stated that within the predicted total amount of energy produced from RES in Poland by 2025, there should be a slight increase in this production (Figure 10). However, this growth seems to be not very dynamic. In principle, between 2021–2025, a constant level of this production can be noted, which, with the overall growing energy consumption, is not a favorable trend.

When analyzing the structure of RES in Poland, it is immensely diverse. This energy is mainly obtained from primary solid biofuels. Unfortunately, a significant stabilization of energy production from this source can be observed (Figure 15). The development of this area depends on many factors, including agriculture, where stagnation has been observed in recent years. Thus, this area needs to be stimulated so the growth can be more dynamic. This would significantly improve the total value of energy produced from RES.

Also, in recent years, the share of energy produced from wind has increased significantly. It is clear that the investments made, especially by the private sector, bring measurable effects (Table 1). The forecast for wind energy production is also exceptionally optimistic (Figure 13). In general, this is the result of large investments that have been made in this respect and favorable climate conditions in Poland.

Vastly dynamic development has also been reported in the field of energy production from solar thermal and photovoltaic energy (Figure 14). In this case, the state policy regarding co-financing of solar and photovoltaic installations and an increase in the number of sunny days in Poland encourage investments in this area. It seems that it is currently one of the most promising areas of energy production from RES.

Energy production from biogases (Figure 17) and renewable municipal waste (Figure 18) also have positive development prospects. In both cases, these sources show high growth dynamics. This is associated with the development and implementation of new technologies in the field of waste utilization and a new policy in the field of waste segregation. The implementation of the circular economy and the increase in public awareness of the sustainable development economy are very conducive to the development of this sector. There are also large reserves in this area, which, with appropriate incentives (also financial), can affect the increase in energy produced.

For these sources, high consistency of the forecast with the actual production values can also be noted (Figures 19 and 20). It can therefore be assumed that the energy production values determined in the forecasting process will be achieved.

However, in the case of energy produced from other liquid biofuels, it can be seen that after a period of stabilization, a large increase in energy produced from this source is expected (Figure 16). In other words, the amount of energy produced from other liquid biofuels is currently relatively small, but it is constantly increasing. There is also a lot of potential in the area of raw materials needed to produce this energy. However, tax exemptions and greater state support are essential in this case.

As regards obtaining energy from geothermal sources, great development prospects can be observed. Despite the disturbances in the production of this energy, an upward trend can be noted (Figure 12). Due to favorable geographical conditions, this energy sector should grow more and more dynamically in the coming years.

For energy produced from water, it is reported to have the weakest development prospects (Figure 11). As already mentioned, the deteriorating hydrological conditions have significantly worsened the investment climate in this area. Rainfall forecasts for Poland in the coming years are also not optimistic. Due to the high costs of hydroelectric plants, no increase in energy produced from water should be expected in the near future.

When analyzing the presented structure of sources from which renewable energy is produced in Poland together with the development perspectives, it can be stated that it has great growth potential. With special emphasis on solar and wind energy, as well as other liquid biofuels, biogas and geothermal sources, the development opportunities are really huge. Both geographical and atmospheric conditions are favorable for this to be achieved. Therefore, it seems reasonable to create favorable legal conditions for investing in RES. In this respect, both private and large state-owned enterprises show great interest in this sector. A comprehensive development strategy for this sector should result in a large increase in energy produced from RES.

It is also worth referring the results to the forecasts presented as part of the EUCALC Explore Sustainable European Futures [77] project, under which a multifunctional calculator was developed. This tool makes it possible to calculate electricity production in Poland (as well as in the EU and its Member States) by 2050. Certain differences were reported between the results of forecasts presented in the article and those obtained from the calculations presented in [77]. This is due to the calculation algorithms and input data used. One of the differences concerns geothermal energy. According to the forecast, in 2025, it should be produced at slightly over 30 thousand tonnes of oil equivalent, while according to the calculator, such energy in Poland will not be produced at all. In this regard, the forecast seems to be more reliable, as today (and also in previous years) in Poland, certain amounts of energy from this source have been produced. The outlook in this area is also rather optimistic.

Some slight differences were also observed for the production of energy from water. According to the forecast presented in the article, this production in 2025 will fall below the level of 160 thousand tonnes of oil equivalent, while according to the results shown in [77], energy from this source will be generated at around 183 thousand tonnes of oil equivalent. As noted by the authors, due to the current conditions, for example, associated with rainfall deficit and watercourse desiccation, as well as the restrictions discussed earlier, the decrease in energy production from this source is more probable. However, these slight differences do not significantly affect the overall positive assessment of this tool.

The analysis of the percentage share of energy from RES in its total consumption in Poland (Figure 21) shows that it will be difficult to increase this share in the near future. Unless more decisive measures are taken to promote energy from RES, it will be difficult to meet the requirements of the European Green Deal strategy. It is also key that investments are intended for those sources that have the greatest development prospects.

In order to effectively direct the necessary changes, it is sensible to conduct research to broaden the knowledge of the current state of Polish energy, in particular, renewable energy. It is undisputable that energy produced from RES is definitely more ecological and constitutes a real alternative to production from conventional sources. The analyses carried out and the results obtained should support this process and broaden the knowledge of the structure of renewable energy production in Poland. In this context, the use of the methodology of artificial neural networks seems fully justified.

5. Conclusions and Further Directions

In the context of climate change observed for many years, caused by greenhouse gas and other harmful substance emissions, the use of conventional energy sources should become an absolute priority of the energy and climate policy of the EU countries, including Poland.

In the case of Poland, it also has symbolic significance, as it is one of the very few countries in Europe in which the energy sector is based to a large extent on conventional energy sources (hard coal and lignite). Although the unit costs of energy production from these sources are low, the environmental impact of this production is immensely negative. The emission of this sector in Poland is especially high. Growing public awareness and increasingly restrictive climate strategies adopted by the EU mean that also in Poland, it is necessary to change the structure of energy production. The use of conventional energy sources should be limited and replaced by RES. The process of energy transition, however, is costly and meets with great resistance on the part of society, especially the one associated with the conventional energy sector.

In order to change this, it is obligatory to develop a coherent, transparent and practicable policy on decarbonizing the economy and alternative solutions. In this respect, solutions implemented by the EU can be very helpful, including those reducing bureaucracy.

When analyzing the development perspectives of the energy sector, it is impossible not to mention its high dependence on various political, economic and demographic factors, among other aspects. In addition, these factors may differ significantly at different times. In this respect, very dynamic technological development (Industry 4.0), the construction of smart cities and factories, and many other changes that await us are of great importance. These changes can undoubtedly affect the forecasts. With their high dynamics, it will be difficult to determine the real demand for both electricity and heat.

The best example is the currently prevailing pandemic associated with the SARS-CoV-2 coronavirus, which has significantly disrupted all forecasts to date, and not only in the energy sector. It seems that in the context of this pandemic, energy from conventional sources has returned. In the long run, however, renewable energy should start to dominate the market. This energy also seems to be more resistant to economic factors and crises as opposed to conventional energy. For Poland, it can be a very important advantage, as after the current crisis, there will be even more pressure on the development of this sector of the economy.

While today the basic stimulus for the economy are still ongoing production and investment processes, after the eradication of the coronavirus, the key to success will be access to cheap energy supplied to plants restarting their business after a few weeks of stagnation or reduction of production. Undoubtedly, this will provide a huge opportunity for the development of RES from sources with low fixed costs, including wind and solar farms. The new social approach to ecology will probably also be important for this development. It is clear that this pandemic will increase ecological awareness, and this is undeniably positive news for RES.

Presented in December 2019, the European Green Deal concept assumes achieving climate neutrality in the perspective of the next 30 years (by 2050). This concept, with very ambitious assumptions and goals, is a great opportunity for the inhabitants of Europe and individual countries both in terms of improving the quality of the environment and achieving a high technological level. Currently, it is difficult to assess whether all the assumed goals of this strategy can be accomplished, but the very adoption of such an ambitious plan is both a great challenge and opportunity for the world to reduce environmental devastation. The introduction of changes related to the implementation of this strategy requires the approval of all EU countries, which will not be easy to achieve.

The European Green Deal requires political commitment to climate protection and is consistently paving the way for a more sustainable future. However, the basis of its success will be the financial activities that need to be implemented. The Just Transition Fund will have to support countries and initiatives that will increase the use of RES at the expense of conventional sources.

These solutions are a great chance for a civilization leap for the Polish economy. They create an opportunity to finance some of the activities related to energy transition and decarbonization of the economy. In the case of Poland, this transition from conventional to alternative energy is associated with large investments. In this respect, the EU assistance will be needed. The basis of the energy transition is the development of RES, which will allow the Polish economy to become more competitive. Nevertheless, this process requires both strong political actions and financial resources.

The research results presented in the article clearly show that there is a lot to be done in Poland regarding the development of RES. The forecasts are not very optimistic, and without more decisive actions, the share of energy produced from RES will be hard to reach, as assumed in the plans at the level of 18.4% in 2025. In 2030, it should be 23%. Compared to other European countries, Poland's achievements are currently not satisfactory, and thus achieving and maintaining the goals set in the plans is unrealistic for the time being.

For the further development of this sector in Poland, analyses of the structure of sources from which renewable energy is produced and forecasts for its production up to 2025 are significant. Focusing on several sources that have the best development perspectives seems obvious. The results clearly indicate these sources and should be achieved in the coming years. Obviously, these forecasts did not take into account the revolutionary changes that may occur in the near future and change these forecasts.

Undoubtedly, the coronavirus pandemic, which is currently spreading around the world, will be of great significance in this respect. Time will tell what effects this phenomenon will have on the global economy. However, a large global economic slowdown can already be observed. This is also associated with a decrease in energy demand, which significantly distorts any forecasts regarding the share of individual sources in its production. Perhaps, this is another signal that may encourage a more decisive approach to climate protection and to increase the pace of producing modern zero-emission energy and building the entire global economy.

Thus, it is reasonable to state that in order to increase public awareness and broaden the knowledge in the field of RES, research is inevitable to obtain new information.

Undoubtedly, this study has generated the knowledge that should be used to create and later implement climate policies both in Poland and Europe.

Author Contributions: Conceptualization, M.T. and J.B.; methodology, J.B and M.T.; software, M.T.; and J.B.; formal analysis, J.B. and M.T.; investigation, J.B. and M.T.; resources, J.B.; M.T. and S.A.S.; data curation, J.B. and M.T.; writing of the original draft preparation, J.B. and M.T. and S.A.S.; writing of review and editing, J.B.; M.T.; visualization, M.T.; supervision, M.T. and J.B.; project administration, M.T. and J.B.; funding acquisition, J.B. All authors have read and agreed to the published version of the manuscript.

Funding: This publication was funded by The Silesian University of Technology, grant number 13/030/RGJ19/0050 (Rector's Grants in Research and Development) and by the statutory research performed at Silesian University of Technology, Department of Production Engineering, Faculty of Organization and Management (13/030/BK-20/0059).

Conflicts of Interest: The authors declare no conflicts of interest.

Abbreviations

The following abbreviations are used in this manuscript:

ANN	Artificial neural network
COP	Conference of the Parties
EU	European Union
MAE	Mean absolute error
MAPE	Mean absolute percentage error
ME	Mean error
MPE	Mean percentage error
MLP	Multi-layer perceptron
RES	Renewable energy sources
TOE	Thousand tonnes of oil equivalent
UN	United Nations

References

1. Chiou-Wei, S.Z.; Chen, C.-F.; Zhu, Z. Economic growth and energy consumption revisited—Evidence from linear and nonlinear Granger causality. *Energy Econ.* **2008**, *30*, 3063–3076. [CrossRef]
2. Kanagawa, M.; Nakata, T. Assessment of access to electricity and the socio-economic impacts in rural areas of developing countries. *Energy Policy* **2008**, *36*, 2016–2029. [CrossRef]
3. Brodny, J.; Tutak, M. Analyzing Similarities between the European Union Countries in Terms of the Structure and Volume of Energy Production from Renewable Energy Sources. *Energies* **2020**, *13*, 913. [CrossRef]
4. Karekezi, S.; Kimani, J. Have power sector reforms increased access to electricity among the poor in East Africa? *Energy Sustain. Dev.* **2004**, *8*, 10–25. [CrossRef]
5. Nouni, M.; Mullick, S.; Kandpal, T. Providing electricity access to remote areas in India: Niche areas for decentralized electricity supply. *Renew. Sustain. Energy Rev.* **2008**, *34*, 430–434. [CrossRef]
6. Oseni, M. Household's access to electricity and energy consumption pattern in Nigeria. *Renew. Sustain. Energy Rev.* **2012**, *16*, 990–995. [CrossRef]

7. Slough, T.; Urpelainen, J.; Yang, J. Light for all? Evaluating Brazil's rural electrification progress. *Energy Policy* **2015**, *86*, 315–332. [CrossRef]
8. Dogan, E.; Seker, F. Determinants of CO_2 emissions in the European Union: The role of renewable and non-renewable energy. *Renew. Energy* **2016**, *94*, 429–439. [CrossRef]
9. Sharma, S.S. Determinants of carbon dioxide emissions: Empirical evidence from 69 countries. *Appl. Energy* **2011**, *88*, 376–382. [CrossRef]
10. Bose, B.K. Global Warming: Energy, Environmental Pollution, and the Impact of Power Electronics. *IEEE Ind. Electron. Mag.* **2010**, *4*, 6–17. [CrossRef]
11. Saboori, B.; Sulaiman, J. Environmental degradation, economic growth and energy consumption: Evidence of the environmental Kuznets curve in Malaysia. *Energy Policy* **2013**, *60*, 892–905. [CrossRef]
12. Shafiei, S.; Salim, R.A. Non-renewable and renewable energy consumption and CO_2 emissions in OECD countries: A comparative analysis. *Energy Policy* **2014**, *66*, 547–556. [CrossRef]
13. Ezzati, M.; Kammen, D.K. The Health Impacts of Exposure to Indoor Air Pollution from Solid Fuels in Developing Countries: Knowledge, Gaps, and Data Needs. *Environ. Health Perspect.* **2002**, *110*, 1057–1068. [CrossRef]
14. COP25 Summary Report. Available online: https://www.ieta.org/resources/Documents/IETA-COP25-Report_2019.pdf (accessed on 10 April 2020).
15. European Commission. Available online: https://ec.europa.eu/info/strategy/priorities-2019-2024/european-green-deal_en (accessed on 10 April 2020).
16. European Commission. Available online: https://ec.europa.eu/regional_policy/en/newsroom/news/2020/01/14-01-2020-financing-the-green-transition-the-european-green-deal-investment-plan-and-just-transition-mechanism (accessed on 10 April 2020).
17. UNFCCC. Kyoto Protocol Reference Manual on Accounting of Emissions and Assigned Amount United Nations Framework Convention on Climate Change. 2008. Available online: http://unfccc.int/kyoto_protocol/items/3145.php (accessed on 10 April 2020).
18. Haas, R.; Panzer, C.; Resch, G.; Ragwitz, M.; Reece, M.; Held, A. A historical review of promotion strategies for electricity from renewable energy sources in EU countries. *Renew. Sustain. Energy Rev.* **2011**, *15*, 1003–1034. [CrossRef]
19. Streimikiene, D.; Ciegis, R.; Grundey, D. Energy indicators for sustainable development in Baltic States. *Renew. Sustain. Energy Rev.* **2007**, *11*, 877–893. [CrossRef]
20. Tomaszewski, K. Energy policy of the European Union in the context of economic security issues. *Przegląd Politol.* **2018**, *1*, 133–145. [CrossRef]
21. Browne, O.; Poletti, S.; Young, D. How does market power affect the impact of large scale wind investment in 'energy only' wholesale electricity markets? *Energy Policy* **2015**, *87*, 17–27. [CrossRef]
22. 2018 Cost of Wind Energy Review. Available online: https://www.osti.gov/servlets/purl/1581952 (accessed on 10 April 2020).
23. Reuter, W.F.; Szolgayová, J.; Fuss, S.; Obersteiner, M. Renewable energy investment: Policy and market impacts. *Appl. Energy* **2012**, *97*, 249–254. [CrossRef]
24. Kilinc-Ata, N. The evaluation of renewable energy policies across EU countries and US states: An econometric approach. *Energy Sustain. Dev.* **2016**, *31*, 83–90. [CrossRef]
25. Haas, R.; Resch, G.; Panzer, C.; Busch, S.; Ragwitz, M.; Held, A. Efficiency and effectiveness of promotion systems for electricity generation from renewable energy sources—Lessons from EU countries. *Energy* **2011**, *36*, 2186–2193. [CrossRef]
26. Eurostat Database. Available online: https://ec.europa.eu/eurostat/web/energy/data/database (accessed on 10 April 2020).
27. International Renewable Energy. Available online: https://www.irena.org/ (accessed on 10 April 2020).
28. Zsiborács, H.; Baranyai, N.H.; Vincze, A.; Zentkó, L.; Birkner, Z.; Máté, K.; Pintér, G. Intermittent Renewable Energy Sources: The Role of Energy Storage in the European Power System of 2040. *Electronics* **2019**, *8*, 729. [CrossRef]

29. Eurostat Database. Available online: https://ec.europa.eu/eurostat/statistics-explained/index.php/Environment (accessed on 10 April 2020).
30. Directive (EU) 2018/2001 of the European Parliament and of the Council of 11 December 2018 on the Promotion of the Use of Energy from Renewable Sources. Available online: https://eur-lex.europa.eu/legal-content/EN/TXT/?uri=uriserv:OJ.L_.2018.328.01.0082.01.ENG&toc=OJ:L:2018:328:TOC (accessed on 10 April 2020).
31. In Focus: Renewable Energy in Europe. Available online: https://ec.europa.eu/info/news/focus-renewable-energy-europe-2020-mar-18_en (accessed on 10 April 2020).
32. Directions of Energy Development in Poland. Opinions about Energy Sources and Their Use. Available online: https://www.cbos.pl/SPISKOM.POL/2015/K_017_15.PDF (accessed on 10 April 2020).
33. Mroczek, B.; Kurpas, D. Social Attitudes Towards Wind Farms and other Renewable Energy Sources in Poland. *Med. Środowiskowa-Environ. Med.* **2014**, *17*, 19–28.
34. Nyga-Łukaszewska, H.; Aruga, K.; Stala-Szlugaj, K. Energy Security of Poland and Coal Supply: Price Analysis. *Sustainability* **2020**, *12*, 2541. [CrossRef]
35. Brown, T.; Schäfer, M.; Greiner, M. Sectoral Interactions as Carbon Dioxide Emissions Approach Zero in a Highly-Renewable European Energy System. *Energies* **2019**, *12*, 1032. [CrossRef]
36. Gyalai-Korpos, M.; Zentkó, L.; Hegyfalvi, C.; Detzky, G.; Tildy, P.; Hegedűsné Baranyai, N.; Pintér, G.; Zsiborács, H. The Role of Electricity Balancing and Storage: Developing Input Parameters for the European Calculator for Concept Modeling. *Sustainability* **2020**, *12*, 811. [CrossRef]
37. Kaldellis, J.K.; Zafirakis, D. Optimum energy storage techniques for the improvement of renewable energy sources-based electricity generation economic efficiency. *Energy* **2007**, *32*, 2295–2305. [CrossRef]
38. Oniszk-Popławska, A.; Rogulska, M.; Wiśniewska, G. Renewable-energy developments in Poland to 2020. *Appl. Energy* **2003**, *76*, 101–110. [CrossRef]
39. Igliński, B.; Buczkowski, R.; Kujawski, R.; Cichosz, M.; Piechota, G. Geoenergy in Poland. *Renew. Sustain. Energy Rev.* **2012**, *16*, 2545–2557. [CrossRef]
40. Igliński, B.; Buczkowski, R.; Cichosz, M. Biogas production in Poland—Current state, potential and perspectives. *Renew. Sustain. Energy Rev.* **2015**, *50*, 686–695. [CrossRef]
41. Igliński, B.; Kujawski, R.; Buczkowski, R.; Cichosz, M. Renewable energy in the Kujawsko-Pomorskie Voivodeship (Poland). *Renew. Sustain. Energy Rev.* **2010**, *14*, 1336–1341. [CrossRef]
42. Andrychowicz, M.; Olek, B.; Przybylski, J. Review of the methods for evaluation of renewable energy sources penetration and ramping used in the Scenario Outlook and Adequacy Forecast 2015. Case study for Poland. *Renew. Sustain. Energy Rev.* **2017**, *74*, 703–714. [CrossRef]
43. Bugała, A.; Zaborowicz, M.; Boniecki, P.; Janczak, D.; Koszela, K.; Czekała, W.; Lewicki, A. Short-term forecast of generation of electric energy in photovoltaic systems. *Renew. Sustain. Energy Rev.* **2018**, *81*, 306–312. [CrossRef]
44. Kupczyk, A.; Mączyńska, J.; Redlarski, G.; Tucki, K.; Bączyk, A.; Rutkowski, D. Selected Aspects of Biofuels Market and the Electromobility Development in Poland: Current Trends and Forecasting Changes. *Appl. Sci.* **2019**, *9*, 254. [CrossRef]
45. Piwowar, A.; Dzikuć, M. Development of Renewable Energy Sources in the Context of Threats Resulting from Low-Altitude Emissions in Rural Areas in Poland: A Review. *Energies* **2019**, *12*, 3558. [CrossRef]
46. Mardani, A.; Jusoh, A.; Zavadskas, E.K.; Cavallaro, F.; Khalifah, Z. Sustainable and Renewable Energy: An Overview of the Application of Multiple Criteria Decision Making Techniques and Approaches. *Sustainability* **2015**, *7*, 13947–13984. [CrossRef]
47. Momete, D.C. Measuring Renewable Energy Development in the Eastern Bloc of the European Union. *Energies* **2017**, *10*, 2120. [CrossRef]
48. Muradin, M.; Foltynowicz, Z. Potential for Producing Biogas from Agricultural Waste in Rural Plants in Poland. *Sustainability* **2014**, *6*, 5065–5074. [CrossRef]
49. Nasirov, S.; Silva, C.; Agostini, C.A. Investors' Perspectives on Barriers to the Deployment of Renewable Energy Sources in Chile. *Energies* **2015**, *8*, 3794–3814. [CrossRef]

50. Xu, X.; Niu, D.; Qiu, J.; Wu, M.; Wang, P.; Qian, W.; Jin, X. Comprehensive Evaluation of Coordination Development for Regional Power Grid and Renewable Energy Power Supply Based on Improved Matter Element Extension and TOPSIS Method for Sustainability. *Sustainability* **2016**, *8*, 143. [CrossRef]
51. Liu, T.; Xu, G.; Cai, P.; Tian, L.; Huang, Q. Development forecast of renewable energy power generation in China and its influence on the GHG control strategy of the country. *Renew. Energy* **2011**, *36*, 1284–1292. [CrossRef]
52. Celiktas, M.S.; Kocar, G. From potential forecast to foresight of Turkey's renewable energy with Delphi approach. *Energy* **2010**, *35*, 1973–1980. [CrossRef]
53. Daim, T.; Harell, G.; Hogaboam, L. Forecasting Renewable Energy Production in the US. *Foresight* **2012**, *14*, 225–241. [CrossRef]
54. The Polish Investment and Trade Agency. Available online: https://www.paih.gov.pl/sektory/odnawialne_zrodla_energii (accessed on 10 April 2020).
55. Energy Policy of Poland Until 2030. Available online: http://www.lse.ac.uk/GranthamInstitute/wp-content/uploads/laws/1564%20English.pdf (accessed on 10 April 2020).
56. Ministry of State Assets. Available online: https://www.gov.pl/web/aktywa-panstwowe/zaktualizowany-projekt-polityki-energetycznej-polski-do-2040-r (accessed on 10 April 2020).
57. The Data World Bank. Available online: https://databank.worldbank.org/reports.aspx?source=2&series=EG.FEC.RNEW.ZS&country=POL (accessed on 10 April 2020).
58. Drew, P.J.; Monson, R.T. Artificial neural networks. *Surgery* **2000**, *127*, 3–11. [CrossRef] [PubMed]
59. Chen, F.D.; Li, H.; Xu, Z.H.; Hou, S.X.; Yang, D.Z. User-friendly optimization approach of fed-batch fermentation conditions for the production of iturin A using artificial neural networks and support vector machine. *Electron. J. Biotechnol.* **2015**, *18*, 273–280. [CrossRef]
60. Li, H.; Chen, F.D.; Cheng, K.W.; Zhao, Z.Z.; Yang, D.Z. Prediction of Zeta Potential of Decomposed Peat via Machine Learning: Comparative Study of Support Vector Machine and Artificial Neural Networks. *Int. J. Electrochem. Sci.* **2015**, *10*, 6044–6056.
61. Zhao, M.; Li, Z.; He, W. Classifying Four Carbon Fiber Fabrics via Machine Learning: A Comparative Study Using ANNs and SVM. *Appl. Sci.* **2016**, *6*, 209. [CrossRef]
62. Hopfield, J.J. Artificial neural networks. *IEEE Circuits Devices Mag.* **1988**, *4*, 3–10. [CrossRef]
63. Dayhoff, J.E.; DeLeo, J.M. Artificial neural networks. *Cancer* **2001**, *91*, 1615–1635. [CrossRef]
64. Lula, P.; Tadeusiewicz, R. *Wprowadzenie do Sieci Neuronowych*; StatSoft: Kraków, Poland, 2001.
65. Tealab, A.; Hefny, H.; Badr, A. Forecasting of nonlinear time series using ANN. *Future Comput. Inform. J.* **2017**, *2*, 39–47. [CrossRef]
66. Tutak, M.; Brodny, J. Forecasting Methane Emissions from Hard Coal Mines Including the Methane Drainage Process. *Energies* **2019**, *12*, 3840. [CrossRef]
67. Krzemień, A. Dynamic fire risk prevention strategy in underground coal gasification processes by means of artificial neural networks. *Arch. Min. Sci.* **2019**, *64*, 3–19.
68. Jasiński, T.; Marszal, A.; Bochenek, A. *Selected Applications Artificial Neural Networks on the Currency Market, Forward Market and in Spatial Economy*; Politechnika Łódzka: Łódź, Poland, 2016.
69. Bhardwaj, S.; Chandrasekhar, E.; Padiyar, P.; Gadre, V.M. A comparative study of wavelet-based ANN and classical techniques for geophysical time-series forecasting. *Comput. Geosci.* **2020**, *138*, 104461. [CrossRef]
70. Yu, C.; Li, Y.; Zhang, M. Comparative study on three new hybrid models using Elman Neural Network and Empirical Mode Decomposition based technologies improved by Singular Spectrum Analysis for hour-ahead wind speed forecasting. *Energy Convers. Manag.* **2017**, *147*, 75–85. [CrossRef]
71. Jahangir, H.; Golkar, M.A.; Alhameli, F.; Mazouz, A.; Ahmadian, A.; Elkamel, A. Short-term wind speed forecasting framework based on stacked denoising auto-encoders with rough ANN. *Sustain. Energy Technol. Assess.* **2020**, *38*, 100601. [CrossRef]
72. Bajkowski, S.; Górnikowska, B. Hydroenergetics against other sources of renewable energy. *Przegląd Nauk. Inżynieria I Kształtowanie Środowiska* **2013**, *59*, 77–87.
73. Kułagowski, W. Hydroenergetics in Poland—current state, development prospects. *Gospod. Wodna* **2001**, *3*, 119–123.

74. Kowalczyk, K.; Cieśliński, R. Analysis of the hydroelectric potential and the possibilities of its use in the Pomeranian voivodeship. *Water-Environ. Rural Areas* **2018**, *18*, 69–86.
75. Kalda, G. Analysis of the hydropower industry in Poland. *J. Civ. Eng. Environ. Archit.* **2014**, *61*, 81–92. [CrossRef]
76. Directive 2009/28/EC of the European Parliament and of the Council of 23 April 2009 on the Promotion of the Use of Energy from Renewable Sources and Amending and Subsequently Repealing Directives 2001/77/EC and 2003/30/EC. Available online: https://eur-lex.europa.eu/legal-content/EN/ALL/?uri=CELEX:32009L0028 (accessed on 10 April 2020).
77. EUCALC. Available online: http://www.european-calculator.eu/ (accessed on 10 May 2020).

© 2020 by the authors. Licensee MDPI, Basel, Switzerland. This article is an open access article distributed under the terms and conditions of the Creative Commons Attribution (CC BY) license (http://creativecommons.org/licenses/by/4.0/).

MDPI
St. Alban-Anlage 66
4052 Basel
Switzerland
Tel. +41 61 683 77 34
Fax +41 61 302 89 18
www.mdpi.com

Energies Editorial Office
E-mail: energies@mdpi.com
www.mdpi.com/journal/energies